HEARING:

ANATOMY, PHYSIOLOGY, AND DISORDERS OF THE AUDITORY SYSTEM

Second Edition

HEARING:
ANATOMY, PHYSIOLOGY, AND DISORDERS OF THE AUDITORY SYSTEM
Second Edition

A. R. Møller
School of Behavioral and Brain Sciences
University of Texas at Dallas
Texas

ELSEVIER

AMSTERDAM • BOSTON • HEIDELBERG • LONDON
NEW YORK • OXFORD • PARIS • SAN DIEGO
SAN FRANCISCO • SINGAPORE • SYDNEY • TOKYO

Academic Press is an imprint of Elsevier

Academic Press is an imprint of Elsevier
30 Corporate Drive, Suite 400, Burlington, MA 01803, USA
525 B Street, Suite 1900, San Diego, California 92101-4495, USA
84 Theobald's Road, London WC1X 8RR, UK

This book is printed on acid-free paper. ⊗

Front cover design concept by Milda Dorsett.

Library of Congress Cataloging-in-Publication Data

Møller, Aage R.
  Hearing : anatomy, physiology, and disorders of the auditory system/A.R.
Moller, -- 2nd ed.
    p. cm.
  Includes bibliographical references and index.
  ISBN-13: 978-0-12-372519-6 (casebound : alk. paper)
  ISBN-10: 0-12-372519-4 (casebound : alk. paper)
  1. Hearing--Physiological aspects. 2. Hearing disorders--Pathophysiology.
3. Ear--Anatomy. I. Title.
  RF290.M58 2006
  617.8--dc22   2006014244

British Library Cataloguing-in-Publication Data
A catalogue record for this book is available from the British Library.

ISBN 13: 978-0-12-372519-6
ISBN 10: 0-12-372519-4

For information on all Academic Press publications
visit our Web site at www.books.elsevier.com

Printed in the United States of America
06 07 08 09 10 11 9 8 7 6 5 4 3 2 1

Working together to grow
libraries in developing countries

www.elsevier.com | www.bookaid.org | www.sabre.org

ELSEVIER   BOOK AID
            International   Sabre Foundation

Contents

Preface

This book is intended for otologists, audiologists, neurologists and researchers in the field of hearing. The book will also be of interest to psychologists and psychiatrists who treat patients with tinnitus and other hyperactive auditory disorders. The book provides the basis for a broad understanding of the anatomy and function of the ear and the auditory nervous system, and it discusses the cause and treatment of hearing disorders. Most books on hearing focus either on the anatomy and function of the ear, the auditory nervous system or on peripheral or central hearing disorders. This book covers both anatomy and physiology of the ear and the nervous system. The book also provides a comprehensive coverage of disorders of the auditory system emphasizing the interaction between pathologies of the middle ear and the cochlea and the function of the nervous system and vice versa. Hyperactive disorders of the auditory nervous system and the role of expression of neural plasticity in causing auditory symptoms are also topics of the book. An extensive list of references makes it possible for the reader to find original work on the different subjects.

Understanding of the anatomy and the function of the auditory system together with knowledge about the pathophysiology of the auditory system are essential for all clinicians who are involved in diagnosis and treatment of disorders of the auditory system. The book prepares the clinician and the clinical researcher for the challenges of the modern clinical auditory discipline. The book also provides basic information about the auditory system in a form that is suitable for the scientist who does basic research on the auditory system. The book thus aims at cross-fertilization between clinicians, clinical researchers and basic scientists. It is my hope that such knowledge can guide basic auditory research into clinically relevant questions.

The book is the third edition of books on the auditory system, the first, *Auditory Physiology*, published in 1983

by Academic Press, and the second, *Hearing: Its Physiology and Pathophysiology*, published in 2000, also by Academic Press.

The book has 11 chapters that are organized in three sections. Chapters from earlier editions have been re-organized and most parts have been re-written and new information has been added. A separate chapter is devoted to an extended coverage of hyperactive disorders, most importantly tinnitus, the cause and treatment of which is discussed in detail. A new chapter describes cochlear and brainstem implants and hearing conservation programs are discussed in an appendix.

The four chapters of Section I cover anatomy and physiology of the middle ear and the cochlea, including a chapter on the electrical potentials that are generated by the cochlea. Section II has two chapters that cover anatomy and physiology of the nervous system. Both the classical and the less known non-classical (extralemniscal) auditory pathways are covered extensively. The latter is involved in some forms of tinnitus and may be activated in other disorders also. A third chapter is devoted to evoked potentials from the nervous system. The neural generators of the ABR are discussed in detail. The anatomy and physiology of the acoustic middle-ear reflex is covered in a fourth chapter in this section.

The final section (Section III) discusses disorders of the auditory system. Two chapters regard hearing impairment and hyperactive disorders, focusing on tinnitus, its etiology, and treatment. These two chapters stress the role of expression of neural plasticity. A third chapter in this section concerns cochlear implants and auditory brainstem implants. The basic design and function of the processors in these modern auditory prostheses are described and the physiologic basis for the function of these prostheses is discussed. An appendix discusses hearing conservation programs.

Acknowledgements

I want to thank Hilda Dorsett for help with the new artwork and for revising some of the illustrations from the first edition of the book and Karen Riddle for transcribing many of the revisions of the manuscript. I also want to thank Johannes Menzel, Senior Publishing Editor, Elsevier Science, and Heather Furrow and John Donahue, Project Managers, Elsevier, Burlington, MA for their excellent work on the book.

I would not have been able to write this book without the support from the School of Behavioral and Brain Sciences at the University of Texas at Dallas.

Last but not least I want to thank my dear wife, Margareta B. Møller, MD, DMedSci., for her support during writing of this book and for her valuable comments on earlier versions of the manuscripts for this book.

Dallas, November, 2005
Aage R. Møller

Introduction

It is now recognized that disorders of one part of the auditory system often affect the function of other parts of the auditory system. This is especially apparent with regard to hyperactive disorders such as tinnitus and hyperacusis, but even noise induced hearing loss and presbycusis are not isolated cochlear phenomena, for the auditory nervous system is involved in these disorders. Expression of neural plasticity and a complex series of events seem to be necessary in order that such pathologies become manifest. This means that it is no longer valid to divide disorders of the auditory system in to peripheral and central disorders. This book therefore takes an integrated approach to disorders of the auditory system.

While most disorders of the auditory system have detectable morphologic abnormalities, hyperactive disorders lack such detectable morphologic changes, and even other objective signs are often absent. Symptoms such as tinnitus, hyperacusis, and phonophobia even involve physiological abnormalities in other parts of the central nervous system than the classical auditory pathways. A part of the auditory nervous system, known as the non-classical, or extralemniscal, auditory pathways, seems to be involved in some of these hyperactive disorders, and that may also cause abnormal activation of structures of the limbic system, which can explain why patients with tinnitus often present with symptoms of affective disorders such as fear and depression. The role of the non-classical auditory nervous system may have much wider importance than previously known. This book provides a thorough description of the anatomy and physiology of this part of the auditory nervous system and it discusses how their function can change and cause different symptoms. The book also covers less common disorders such as bilirubinemia and cortical lesions and it discusses vestibular Schwannoma and their diagnosis.

Because of the complexity of many disorders of the auditory system the clinician must have a thorough understanding of the basic functions of the entire auditory system and the interactions between the peripheral and the central portions of the auditory system that may occur in various hearing disorders.

Cochlear implants now provide an effective way to treat severe hearing loss. The implementation of cochlear and brainstem implants requires a thorough knowledge not only about the function of such devices but also an understanding of the way sounds are normally coded and processed in the nervous system is a prerequisite for understanding how such prostheses can provide useful hearing. The more recent addition to auditory prostheses, namely auditory brainstem (cochlear nucleus) implants, present an even greater challenge for the clinician and there are ample possibilities to do important research in this area. A separate chapter in the book deals with cochlear and brainstem (cochlear nucleus) implants and the physiological basis for their success is discussed.

Cochlear implants and auditory brainstem implants do not provide the same coding of sounds in the nervous system as provided by the normal ear and expression of neural plasticity is essential for the success of such prostheses. Thus, optimal implementation of such prostheses requires understanding of basic auditory physiology.

The advent of these new aspects in treatment of disorders of the auditory system should not detract attentions from classical problems such as hearing loss from middle ear and cochlear pathologies. Also in these areas of hearing new knowledge has contributed to better understanding of pathologies of the auditory system. The surprising research results that show that exposure to sound can reduce presbycusis and that noise induced hearing loss is affected by pre-exposure to sound are examples of signs of a greater complexity of disorders of the auditory system than previously assumed. The results indicate that the auditory nervous system is involved in disorders that earlier were assumed to be caused solely by morphological changes in the cochlea. This means that altered function of the nervous system caused by altered input contributes to the symptoms and signs of such disorders. This book provides insight into the physiologic basis for the involvement of the auditory nervous system in disorders that earlier were assumed to only involve the ear. The role of expression of neural plasticity in creating the symptoms and signs of these disorders is discussed.

Understanding how electrical potentials are generated in the auditory nervous system is a prerequisite for correct interpretation of clinical tests that make use of such recordings. The book describes the various electrical potentials that are generated in the auditory nervous system, what anatomical structures generate the different components of such far-field potentials as the ABR and the MLR, and how these potentials are affected by different types of pathologies.

Prevention of hearing loss is important and audiologists and otolaryngologists play important roles in reducing the risk of noise induced hearing loss. The basis for that is discussed in several chapters in the book. The practical and legal aspects of hearing conservation are covered in an appendix.

THE EAR

The ear as a sensory organ is far more complex than other sensory organs. The sensory cells are located in the cochlea but the cochlea not only serves to convert sound into a code of neural impulses in the auditory nerve but it also performs the first analysis of sounds that prepare sounds for further analysis in the auditory nervous system. This analysis consists primarily of separating sounds into bands of frequencies before they are coded in the discharge pattern of individual auditory nerve fibers. The separation of sounds is accomplished by the properties of the basilar membrane and the sensory cells that are located along its length. The cochlea is more frequency selective for weak sounds than louder sounds, which facilitates detection of weak sounds. The cochlea also compresses the amplitudes of sounds, which makes it possible to code sounds within the very large range of sound intensities that is covered by normal hearing. Without such amplitude compression the ear could not detect and analyze sounds in the intensity range of normal hearing.

The cochlea is fluid filled and that means that sounds must be converted into vibrations of fluid in order to activate the sensory cells. Direct transfer of sound to a fluid is ineffective. The middle ear facilitates the transfer of sound to the cochlea by acting as a transformer that matches the impedance of the air to that of the cochlea. The middle ear is the only part of the entire auditory system where medical or surgical interventions can remedy hearing loss from disease processes or trauma.

During the past decade or so, our understanding of the function of the cochlea has changed in a fundamental way and its function now appears far more complex than perceived earlier. Earlier it was believed that the basilar membrane was a linear system where properties determined at one sound intensity were directly applicable to all sound intensities. More recently, it has become evident that the frequency selectivity of the basilar membrane depends on the sound intensity. Earlier it was believed that the function of hair cells was limited to transducing the vibration of the basilar membrane into a neural code. The discovery that hair cells also can change their length in response to sound and thus interact with the vibration of the basilar membrane in addition to being transducers radically changed our perception of the function of the cochlea. The best description of the function of outer hair cells is that they act as "motors" that counteract the frictional losses of energy in the cochlea. This particular function of outer hair cells increases the sensitivity of the ear by approximately 50dB. The discovery of the active role of outer hair cells explains how the loss of outer hair cells causes hearing loss. The interaction between the hair cells and the basilar membrane vibration makes the cochlea more complex than other sensory organs. Extensive research during many years has resulted in more knowledge being accumulated about the function of the cochlea than of any other sensory organ.

1

Anatomy of the Ear

1. ABSTRACT

1. The ear consists of the outer ear, the middle ear and the inner ear.
2. The outer ear consists of the pinna and the ear canal.
3. The skin of the ear canal is innervated by four cranial nerves: the trigeminal; the facial; the glossopharyngeal; and the vagus nerves.
4. The middle ear consists of the tympanic membrane and three ossicles: malleus; incus; and stapes.
5. Two muscles are attached to the ossicles: the tensor tympani to the manubrium of malleus; and the stapedius to the stapes. The tensor tympani muscle is innervated by the trigeminal nerve and the stapedius muscle is innervated by the facial nerve.
6. The cochlea in humans has a little more than $2\frac{1}{2}$ turns.
7. The cochlea has three fluid-filled compartments: the scala tympani; scala media; and the scala vestibuli. The basilar membrane separates the scala media from the scala tympani and Reissner's membrane separates the scala vestibuli from the scala media.
8. The ionic composition of the fluid in scala tympani and scala vestibuli (perilymph) is similar to that of extracellular fluid (high contents of sodium, low contents of potassium), while the fluid in scala media (endolymph) is similar to intracellular fluid (high contents of potassium, low contents of sodium).
9. The fluid space of scala tympani and scala vestibuli communicates with the cerebrospinal fluid space through the cochlear aqueduct. The fluid space in scala media communicates with the endolymphatic sac through the endolymphatic canal.
10. Hair cells are organized along the basilar membrane in one row of inner hair cells and 3–5 rows of outer hair cells.
11. The hair cells of the cochlea differ from vestibular hair cells in that they lack a kinocilium.
12. Each inner hair cell is innervated by many (type I) auditory nerve fibers, while each (type II) nerve fiber innervates many outer hair cells.
13. Efferent nerve fibers terminate directly onto outer hair cells while other efferent fibers terminate on the dendrites of the type I fibers that innervate the inner hair cells.

2. INTRODUCTION

The ear (Fig. 1.1) consists of three parts: the outer ear; the middle ear; and the inner ear. The inner ear consists of two parts: the vestibular apparatus for balance; and the cochlea for hearing. The outer ear and the middle ear conduct sound to the cochlea, which separates sounds with regard to frequency before they are transduced by the hair cells into a neural code in the fibers of the auditory nerve.

3. OUTER EAR

The different parts of the external ear, "the auricle," have specific names (Fig. 1.2). The groove called the

(A)

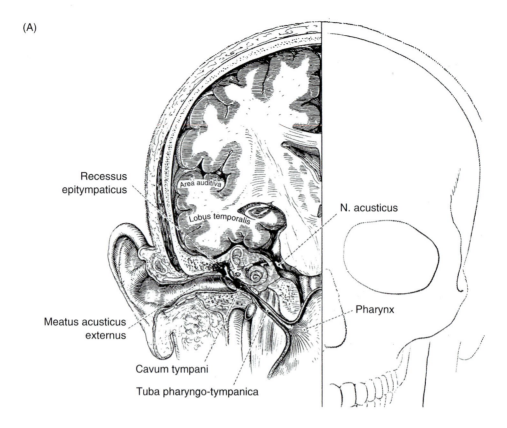

Recessus
epitympaticus

Area auditiva

Lobus temporalis

N. acusticus

Meatus acusticus
externus

Pharynx

Cavum tympani

Tuba pharyngo-tympanica

(B)

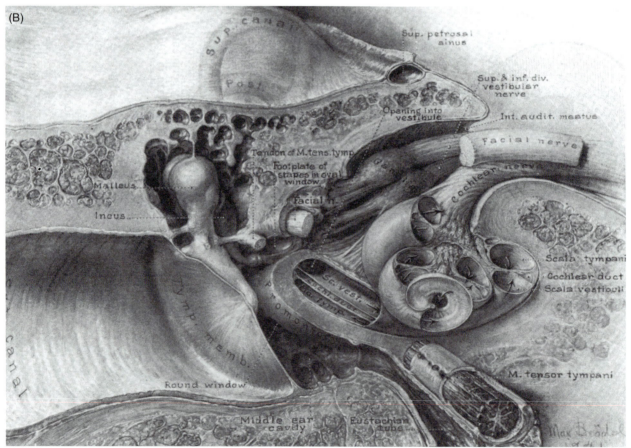

FIGURE 1.1   (A) Localization of the ear in the head (after Melloni, 1957). (B) Cross-section of the human ear (reprinted from Brodel, 1946).

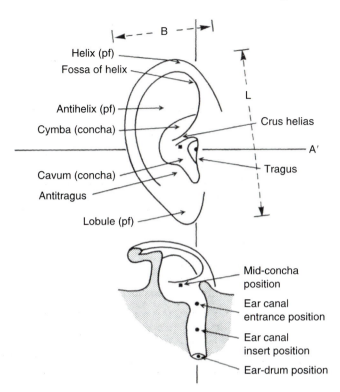

FIGURE 1.2   Schematic drawing of the human external ear showing components that are of importance for sound conduction: pinna flange (helix, antihelix, and lobule), concha (cymba and cavum), and ear canal (reprinted from Shaw, E. A. C. 1974. The external ear. In: Keidel, W. D. and Neff, W. D. (eds) *Handbook of sensory physiology* V(1). New York: Springer-Verlag, pp. 455–490, with permission from Springer).

concha is acoustically the most important. The outer ear enlarges in older individuals, especially in men.

3.1. Ear Canal

The ear canal has a length of approximately 2.5 cm and a diameter of approximately 0.6 cm. It has the shape of a lazy S. The most medial part is a nearly circular opening in the skull bone, and the outer part is cartilage. The outer cartilaginous portion of the ear canal is also nearly circular in young individuals but with age the cartilaginous part often changes shape and attains an oval shape. In addition to changing its shape with age, the lumen of the ear canal often becomes smaller with age, and in avid swimmers, it may become very narrow.

The ear canal is covered by skin that secrets cerumen (wax) and it has hairs on its surface. There are no sweat glands in the ear canal. Since the skin is not rubbed naturally, as exposed skin on other parts of the body, it must self clean dead cells and cerumen. Two types of cells contribute to secretion of cerumen, namely sebaceous cells located close to the hair follicles and ceruminous glands. The sebaceous glands cannot secrete actively but form their secretion by passive breakdown of cells. Two kinds of cerumen exist, dry and wet.

BOX 1.1

CERUMEN

The dry type is found mostly in people in the Orient and in Mongolians while the wet cerumen is found mostly in Caucasians, Africans and Hispanic people [60, 143]. The kind of cerumen is genetically related and chromosome 16 has been identified as carrying the cerumen locus.

Accumulation of cerumen in the ear canal to an extent that it becomes occluded is a common cause of hearing impairment. Cerumen may also cover the tympanic membrane, which causes hearing loss. Individuals who attempt to clean their ear canals by cotton swaps often push cerumen deeper into the ear canal. The cerumen is supposed to become dry and leave the ear canal. The secreted cerumen has a slight anti-bacterial and anti-fungal property and it may act as an insect repellant.

The outer layer of the skin (epidermis) in the ear canal, together with that of the tympanic membrane migrates outwards. The migration helps heal small injuries and move scars outwards as well as transporting cerumen out of the ear canal. It has been suggested that failure in this migration of the epidermis may cause several kinds of pathology such as development of cholesteatoma and it may play a role in causing inflammation of the ear canal.

The skin of the ear canal has an unusual nerve supply. Its sensory receptors (including bare axons) are innervated by four different cranial nerves (CN), namely the sensory portion of the mandibular division of the trigeminal nerve (CN V), the facial nerve (CN VII) the glossopharyngeal nerve (CN IX) and the auricular branch of the vagal nerve (CN X), which supplies the posterior wall of the ear canal and the tympanic membrane. This nerve branch is a part of Arnold's nerve, which also receives contributions from the glossopharyngeal nerve. The innervation of the ear canal by the glossopharyngeal nerve explains why many people cough when the skin of the inner part of the ear canal is touched. The innervation by the glossopharyngeal and the vagal nerve explain why mechanical stimulation of the ear canal can affect the heart and blood circulation and cause sensitive individuals to faint when the ear canal is cleaned for wax.

4. MIDDLE EAR

The middle ear consists of the tympanic membrane that terminates the ear canal (Fig. 1.3) and the three small bones (ossicles), the malleus, the incus and the stapes (Fig. 1.3 and Fig. 1.4). Two small muscles, the tensor tympani muscle and the stapedius muscle, are also located in the middle ear. The manubrium of malleus is imbedded in the tympanic membrane and the head of the malleus is connected to the incus that in turn connects to the stapes, the footplate of which is located in the oval window of the cochlea. The *chorda tympani* is a branch of the facial nerve (the nervous intermedius) that travels across the middle ear cavity (Fig 1.4). It carries taste fibers and probably also pain fibers. The Eustachian tube connects the middle ear cavity to the pharynx.

4.1. Tympanic Membrane

The tympanic membrane (Fig. 1.5) is a slightly oval, thin membrane that terminates the ear canal. It is cone-shaped, with an altitude of 2 mm with the apex pointed inward. Seen from the ear canal, the membrane is slightly concave and is suspended by a bony ring. Normally it is under some degree of tension. Its surface area is approximately 85 mm². The main part of the tympanic membrane, the pars tensa with an area of approximately 55 mm² (Fig. 1.5), is composed of radial and circular fibers overlaying each other. These fibers are comprised of collagen and they provide a lightweight stiff membrane that is ideal for converting sound into vibration of the malleus. A smaller part of the tympanic membrane, the pars flaccida, located above the manubrium of malleus, is thicker than the pars tensa and its fibers are not arranged as orderly as

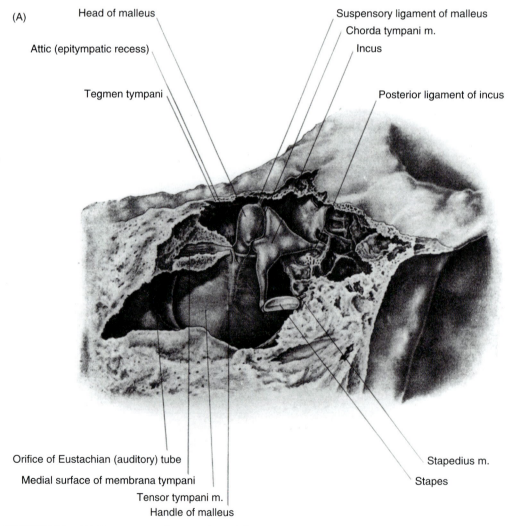

(A)
Head of malleus
Attic (epitympatic recess)
Tegmen tympani
Suspensory ligament of malleus
Chorda tympani m.
Incus
Posterior ligament of incus
Orifice of Eustachian (auditory) tube
Medial surface of membrana tympani
Tensor tympani m.
Handle of malleus
Stapedius m.
Stapes

FIGURE 1.3   (A) Cross-section showing the middle ear (reprinted from Brodel, 1946, with permission from Elsevier).

(B)

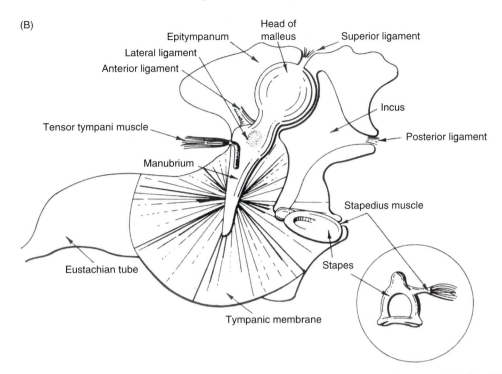

FIGURE 1.3 *(Continued)*   (B) Schematic drawing of the human middle ear seen from inside the head (from Møller, 1972, with permission from Elsevier).

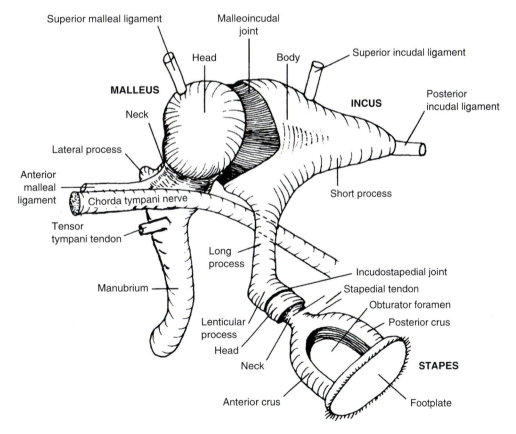

FIGURE 1.4   The ossicular chain as it is normally placed within the middle-ear cavity (adapted from Tos, 1995, with permission from Thieme Medical Publishers).

the collagen fibers of the pars tensa. The tympanic membrane is covered by a layer of epidermal cells, continuous with the skin in the ear canal. This outer layer of the tympanic membrane migrates from its center outwards and this moves small injuries and scars and transports small foreign bodies out into the ear canal. Small holes in the tympanic membrane usually heal spontaneously.

4.2. Ossicles

The middle-ear bones are suspended by several ligaments (Figs 1.3 and 1.4). The manubrium of the malleus is embedded in the tympanic membrane with the tip of the manubrium located at the apex of the tympanic membrane (Fig. 1.5). The head of the malleus is suspended in the epitympanum. The short process of the incus rests in the fossa incudo of the malleus, and it is held in place by the posterior incudal ligament. The long process, also called the lenticular process, of the incus forms one side of the incudo-stapedial joint. The head of the malleus and the incus are fused together in a double saddle joint and the joint between these two bones is regarded to be rigid. The joint between the incus and the stapes is rigid for movement of the stapes towards the cochlea (piston like movements), but the joint is flexible for movements of the stapes that are induced by contraction of the stapedius muscle. The stapes is suspended in the oval window of the cochlea by two ligaments and one ligament is stiffer than the other.

4.3. Middle-ear Muscles

Two small muscles are located in the middle ear. One, the tensor tympani muscle, is attached to the manubrium of the malleus and the other, the stapedius muscle, is attached to the stapes (Figs 1.3 and 1.4). The tensor tympani muscle extends between the malleus and the wall of the middle-ear cavity near the entrance to the Eustachian tube. When contracting, it pulls the manubrium of the malleus inward, displacing the tympanic membrane inwards and stretching the membrane. The stapedius muscle is the smallest striate muscle of the body. It is attached to the head of the stapes and most of the muscle is located in a bony canal. It pulls the stapes in a direction that is perpendicular to its piston-like motion, tilting the stapes so that it rotates around its posterior ligament. The tensor tympani muscle is innervated by the trigeminal nerve (CN V) and the stapedius muscle by the facial nerve (CN VII).

4.4. Eustachian Tube

The Eustachian tube consists of a bony part (the protympanum) that is located close to the middle ear cavity, and a cartilaginous part that forms a closed slit where it terminates in the nasopharynx (Fig. 1.6).

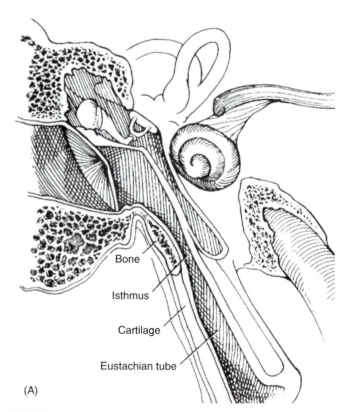

(A)

FIGURE 1.6   (A) Cross-section of the human middle ear to show the Eustachian tube.

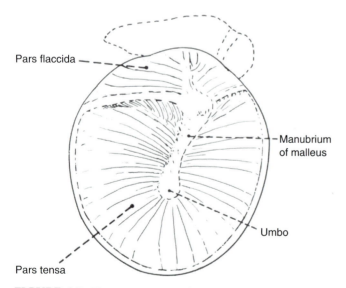

FIGURE 1.5   The tympanic membrane and the position of the malleus and incus (reprinted from Anson and Donaldson, 1973, with permission from Elsevier).

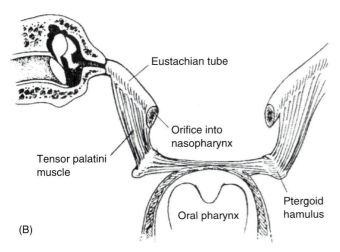

FIGURE 1.6 *(Continued)*   (B) Orientation of the Eustachian tube in the adult. The tensor veli palatini is shown (both reprinted from Hughes, 1985, with permission from Thieme Medical Publishers).

The optimal function of the middle ear depends on keeping the air pressure in the middle-ear cavity close to the ambient pressure. That is accomplished by briefly opening the Eustachian tube. In the adult, the Eustachian tube is 3.5–3.9 cm long and it follows an inferior (caudal) – medially – anterior (ventral) direction in the head, tilting downwards (caudally) by approximately 45 degrees to the horizontal plane (Fig. 1.6B). The Eustachian tube is shorter in young children and it is directed nearly horizontally.

The cartilaginous part of the Eustachian tube forms a valve that closes the middle ear off from pressure fluctuations in the pharynx such as occurs during breathing and it decreases transmission of a person's voice to the middle-ear cavity. The mucosa inside the Eustachian tube (which really is not a tube except for the bony part) is rich in cells that produce mucus and it has cilia that propel mucus from the middle ear to the nasopharynx. The slit shaped cartilaginous part of the Eustachian tube allows transport of material from the middle-ear cavity to the nasopharynx but not the other way.

The most common way the Eustachian tube opens is by contraction of a muscle, the tensor veli palatini muscle. The tensor veli palatini muscle is located in the pharynx and innervated by the motor portion of the fifth cranial nerve. This muscle contracts naturally when swallowing and yawning, and some individuals have learned to contract their tensor veli palatine muscle voluntarily. The Eustachian tube can also be opened by positive air pressure in the middle ear cavity but not by negative pressure, which in fact may close it harder.

4.5. Middle-ear Cavities

The middle-ear cavities consist of the tympanum (the main cavity) that lies between the tympanic membrane and the wall of the inner ear (the promontorium), a smaller part (the epitympanum) that is located above the tympanum, and a system of mastoid air cells. The head of the malleus is located in the epitympanum (Fig. 1.3). The middle-ear cavity and the Eustachian tube are covered with mucosa. The total volume of the middle-ear cavities is often given to be approximately 2 cm^3, but the size of the middle-ear cavities varies

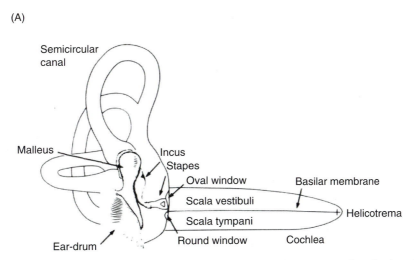

FIGURE 1.7   (A) Schematic drawing of the ear showing the cochlea as a straight tube (reprinted from Møller, 1983, with permission from Elsevier).

(Continued)

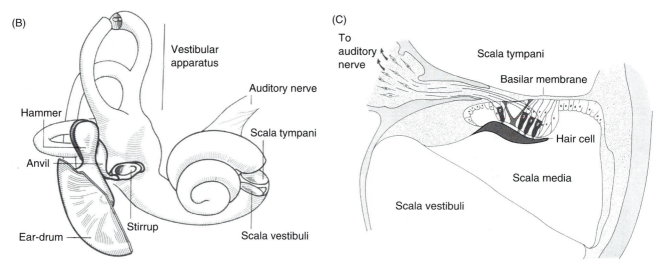

FIGURE 1.7 *(Continued)* (B) Schematic drawing of the human ear. (C) Cross-section of the cochlea ((B) and (C) reprinted from Møller, A.R. 1975. Noise as a health hazard. *Ambio* 4: 6–13, with permission from The Royal Swedish Academy of Sciences).

considerably from person to person and if the volume of the mastoid air cells is included, the total volume can be as large as 10 cm³.

5. COCHLEA

The cochlea is a snail-shaped bony structure that contains the sensory organ of hearing. The cochlea in humans has a little more than 2 1/2 turns. Uncoiled the cochlea has a length of 3.1–3.3 cm. The height of the cochlea is approximately 0.5 cm in humans and similar in small animals such as the chinchilla. The cochlea, together with the vestibular organ, is totally enclosed in the temporal bone, which is one of the hardest bones in the entire body. Together the cochlea and the vestibular organs are often referred to as the labyrinth. The bony structures are known as the bony labyrinth and the content is the membranous labyrinth. The cochlea has three fluid-filled canals: the scala vestibuli; the scala tympani; and the scala media (Fig. 1.8). The scala media, located in the middle of the cochlea, is separated from the scala vestibuli by Reissner's membrane and from the scala tympani by the basilar membrane. The ionic composition of the fluid in the scala media is similar to that of intracellular fluid, thus rich in potassium and low in sodium, while the fluid in the scala vestibuli and scala tympani is similar to that of extracellular fluid such as the cerebrospinal fluid, thus rich in sodium and poor in potassium.

The scala media narrows towards the apex of the cochlea ending just short of the apical termination of the bony labyrinth. An opening near the apical termination of the bony labyrinth, called the helicotrema, allows communication between the scala vestibuli and scala tympani. In humans, the area of this aperture is approximately 0.05 mm². The basilar membrane separates sounds according to their frequency (spectrum) and the organ of Corti, located along the basilar membrane, contains the sensory cells (hair cells) that transform the vibration of the basilar membrane into a neural code.

While the gross anatomy of the cochlea has been known for many years, recent studies of its morphology and function have produced what seems to be an endless series of surprising and intriguing results. In fact, the sensory transduction in the cochlea has attracted more research effort than any other part of the auditory system and the function of the auditory receptor organ is better known than that of any other sensory system.

5.1. Organ of Corti

The organ of Corti contains many different kinds of cells. The sensory cells, the hair cells, so called because of the hair-like bundles that are located on their top, are arranged in rows along the basilar membrane (Fig. 1.9). The hair cells have bundles of stereocilia on their top but the hair cells in the mammalian cochlea have no kinocilia (Fig. 1.10). The hair cells are of two

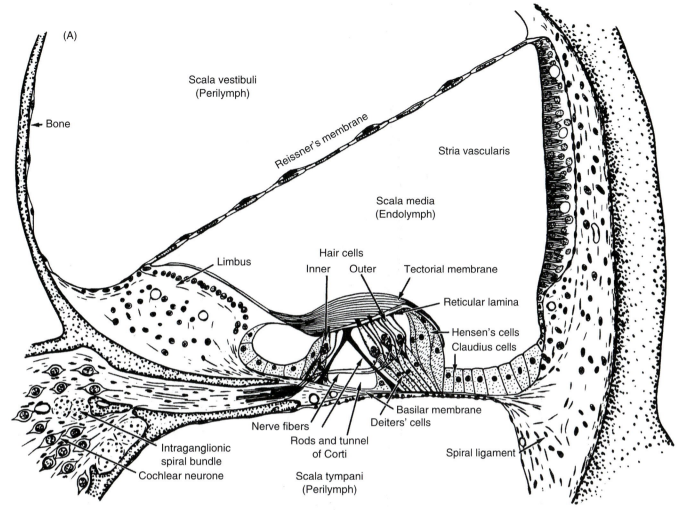

(A)

Scala vestibuli
(Perilymph)

Bone

Reissner's membrane

Stria vascularis

Scala media
(Endolymph)

Limbus

Hair cells
Inner   Outer   Tectorial membrane

Reticular lamina

Hensen's cells
Claudius cells

Basilar membrane
Deiters' cells

Nerve fibers

Rods and tunnel
of Corti

Intraganglionic
spiral bundle
Cochlear neurone

Spiral ligament

Scala tympani
(Perilymph)

FIGURE 1.8   (A) Cross-section through one turn of the cochlea (the second turn of the guinea pig's cochlea) (reprinted from Davis et al., 1953, with permission from the American Institute of Physics).

(Continued)

main types: outer hair cells and inner hair cells. The human cochlea has approximately 12,000 outer hair cells arranged in 3–5 rows along the basilar membrane, and approximately 3,500 inner hair cells arranged in a single row. On each outer hair cell, 50–150 stereocilia are arranged in 3–4 rows that assume a W or V shape (Fig. 1.9) whereas the inner hair cells stereocilia are arranged in flattened U-shaped formations. Between the row of inner hair cells and the rows of outer hair cells is the tunnel of Corti, bordered by inner and outer pillar cells (Fig. 1.8A).

The outer hair cells are different from the inner hair cells in several ways. Outer hair cells are cylindrical in shape (Fig. 1.10A) while the inner hair cells are flask-shaped or pear-shaped (Fig. 1.10B). The stereocilia are linked to each other with specific structures

(cross-links) [112]. The tallest tips of the outer hair cell stereocilia are embedded in the overlying tectorial membrane, whereas the tips of the inner hair cell stereocilia are not. The outer hair cells in the apical region of the cochlea are longer than in the more basal regions, approximately 8 μm[1] long in the apical region and less than 2 μm in the base. The diameter of the longest outer hair cell is thus approximately one tenth of the diameter of a human hair.

Inner hair cells have similar dimension in the entire cochlea and all have approximately the same number

[1]1 μm = 1 micrometer = 1/1,000,000 of 1 m, or 1/1,000 of 1 mm. A human hair is approximately 100 μm; a red blood cell is approximately 7 μm.

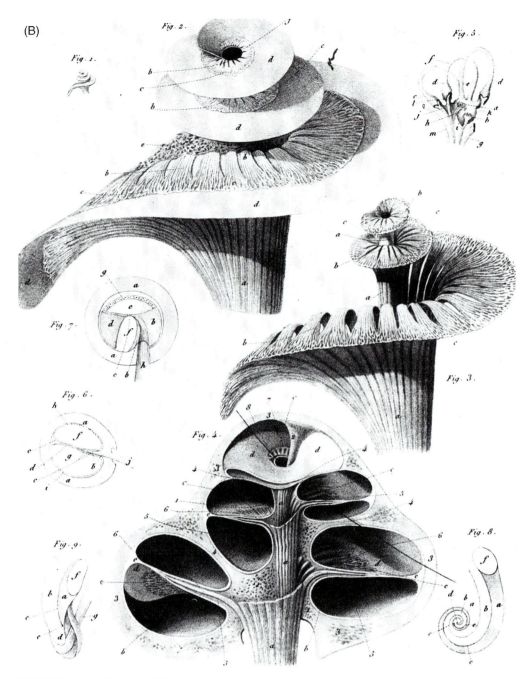

FIGURE 1.8 *(Continued)*   (B) Breschet's drawings of the cochlea, the spiral lamina, and the cochlear nerve. The smaller sketches show differences between the mammalian cochlea and the avian organs of hearing (reprinted from Hawkins, 1988, with permission from Raven Press; after Breschet, 1836).

of stereocilia (approximately 60). The stereocilia on inner hair cells that are located at the base of the cochlea are shorter than stereocilia of hair cells that are located in the apical region of the cochlea.

In addition to hair cells, other types of cells are found in the cochlea. Supporting cells of the organ of Corti are the Deiter's cells and Henson's cells, inner border and inner phalangeal cells. The rest of the cell types will not be mentioned here as they are not thought of as contributing to sound transduction.

The stria vascularis is an important structure located between the perilymphatic and the endolymphatic

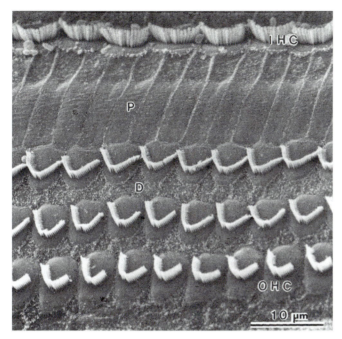

FIGURE 1.9 Scanning electron micrograph of a section of the organ of Corti in a monkey with the tectorial membrane removed to show the organization of the hair cells. One row of inner hair cells (IHC) is visible at the top of the figure and three rows of outer hair cells (OHC) in typical W-shaped formation of stereocilia on the top of the cells are seen. P = pillar cells; D = Deiters' cells (reprinted from Harrison, R. V. and Hunter-Duvar, I. M., 1988. An anatomical tour of the cochlea. In: Jahn, A. F. and Santos-Succhi, J. (eds) *Physiology of the ear*. New York: Raven Press, with permission from Raven Press).

space along the cochlear wall. The stria vascularis has a rich blood supply and its cells are rich in mitochondria, indicating that it is involved in metabolic activity. Many of its intermediate cells have a high content of melanin. The spiral ligament, to which the basilar membrane is attached, supports the stria vascularis.

5.2. Basilar Membrane

The basilar membrane consists of connective tissue and it forms the floor of the scala media. It has a width of approximately 150 μm in the base of the cochlea and it is approximately 450 μm wide at the apex. It is also stiffer in the basal end than at the apex. Due to this gradual change in stiffness, sounds that reach the ear create a wave on the basilar membrane that travels from the base towards the apex of the cochlea. This traveling wave motion is the basis for the frequency separation that the basilar membrane provides before sounds activate the sensory cells that are located along the basilar membrane. As we shall see in Chapter 3 the frequency analysis in

the cochlea is complex, involving interactions between the basilar membrane, the surrounding fluid, and the sensory cells. The outer hair cells interact actively with the motion of the basilar membrane (see p. 46).

5.3. Innervation of Hair Cells

Three types of nerve fibers innervate the cochlea: afferent auditory nerve fibers, efferent auditory fibers (olivocochlear bundle) and autonomic (adrenergic) nerve fibers. The afferent auditory nerve fibers are bipolar cells, the cell bodies of which are located in the spiral ganglion that is located in a bony canal, the Rosenthal's canal (Fig. 1.11). The auditory nerve fibers pass through the habenula perforata before they continue as radial fibers to the inner hair cells. In humans, the auditory nerve has approximately 30,000 afferent nerve fibers. Two types of afferent fibers have been identified. Type I are myelinated and have large cell bodies and comprise 95% of the auditory nerve fibers. Type II (approximately 5% of the auditory nerve) are unmyelinated and have small cell bodies. Details about the anatomy of the auditory nerve will be given in Chapter 4.

The auditory nerve fibers connect to the hair cells via synapses (Fig. 1.12). These connections are different for inner and outer hair cells. Many type I auditory nerve fibers terminate on each inner hair cell while a single type II auditory nerve fiber connects to many outer hair cells (Fig. 1.13). It has been estimated that each inner hair cell receives approximately 20 nerve fibers. The nerve fibers (type II) that make synaptic contact with the outer hair cells cross over the cochlear tunnel to reach the rows of outer hair cells, where each nerve fiber, called an outer spiral fiber, innervates many hair cells and extends apically as much as 0.6 mm (Fig. 1.13) along the outer hair cell region. The inner radial fibers (type I) are thus different from the spiral fibers (type II).

The hair cells also receive different connections from the descending auditory nervous system, the olivocochlear bundle (Rasmussen's bundle). Outer hair cells receive the largest number of such nerve fibers. The efferent fibers (approximately 500–600 in humans [137]) have their cell bodies in the nuclei of the superior olivary complex (SOC) of the brain stem. These fibers are of two kinds: One kind is the medial olivocochlear fibers that are large myelinated fibers that originate in the medial superior olivary (MSO) complex and which terminate on outer hair cells. These fibers mostly originate from cells on the opposite side and thus cross the mid-line (see Chapter 5). Each outer hair cell receives many efferent fibers and each efferent fiber connects to many outer hair cells.

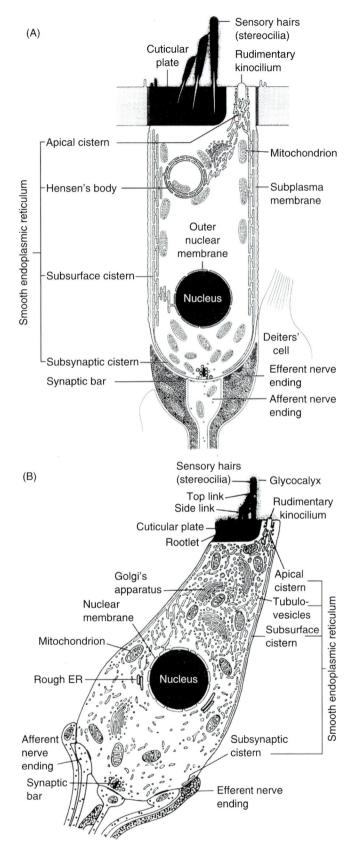

(A)

Sensory hairs
(stereocilia)

Cuticular
plate

Rudimentary
kinocilium

Apical cistern

Mitochondrion

Hensen's body

Subplasma
membrane

Outer
nuclear
membrane

Subsurface cistern

Nucleus

Subsynaptic cistern

Deiters'
cell

Synaptic bar

Efferent nerve
ending

Afferent nerve
ending

Smooth endoplasmic reticulum

(B)

Sensory hairs
(stereocilia)

Glycocalyx

Top link
Side link

Rudimentary
kinocilium

Cuticular plate

Rootlet

Golgi's
apparatus

Apical
cistern

Nuclear
membrane

Tubulo-
vesicles

Mitochondrion

Subsurface
cistern

Rough ER

Nucleus

Afferent
nerve
ending

Subsynaptic
cistern

Synaptic
bar

Efferent nerve
ending

Smooth endoplasmic reticulum

FIGURE 1.10 (A) Schematic drawing of the cross-section of an outer hair cell. (B) Schematic drawing of the cross-section of an inner hair cell (reprinted from Lim, 1986, with permission from Elsevier).

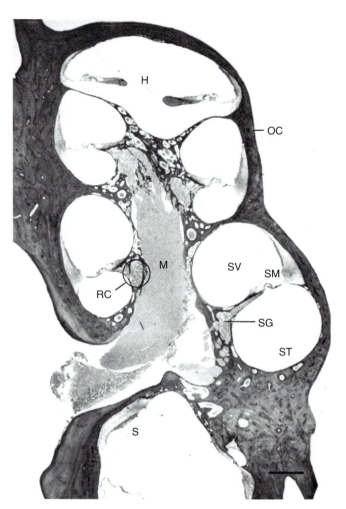

FIGURE 1.11 Spiral ganglion of the auditory nerve shown in a mid-modiolar section of the cochlea in a chinchilla showing the otic capsule (OC), modiolus (M), helicotrema (H), the three cochlear scalae (scala vestibulli [SV], scale media [SM], and scala tympani [ST]), and the saccule (S) portion of the vestibular labyrinth. Rosenthal's canal (RC) (circled) contains the spiral ganglion (SG). Bar = 500 µm (reprinted from Santi, 1988, with permission from Raven Press and courtesy AJ Duvall).

The other kind, the lateral olivocochlear efferent fibers, are small unmyelinated fibers that originate in the lateral nucleus of the superior olivary complex (LSO), mostly on the same side as the ear where they terminate on type I afferent connections that leaves the inner hair cells. The efferent fibers that reach outer hair cells mainly make presynaptic connections while those reaching the inner hair cells make postsynaptic connections (Fig. 1.12). Efferent fibers connect more sparsely to inner hair cells. When the efferent fibers exit the brainstem they first travel with the vestibular nerve and then shift to the cochlear nerve at Ort's anastomosis.

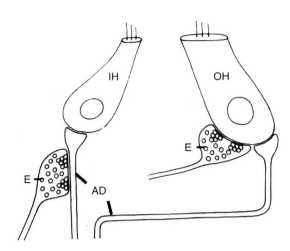

FIGURE 1.12 Schematic drawings of innervation of hair cells by nerve fibers of the auditory nerve. OH = outer hair cells; IH = inner hair cells; AD = afferent dendrite; E = efferent synapse (reprinted from Spoendlin, 1970).

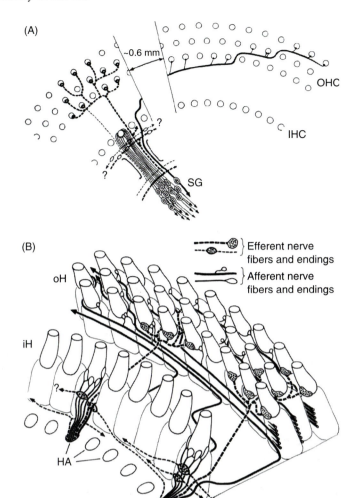

FIGURE 1.13 (A) and (B) Schematic drawings of innervation of hair cells by nerve fibers of the auditory nerve. OH/OHC = outer hair cells; IH/IHC = inner hair cells; SG = spiral ganglion; HA = habenulae openings (reprinted from Spoendlin, 1970).

It is important to notice that the efferent fibers act directly on outer hair cells while efferent fibers only affect the output of inner hair cells by controlling the neural excitation in the nerve fibers that leave inner hair cells. The importance of that will become evident in discussions of the function of outer hair cells compared with that of inner hair cells (Chapter 3).

The inner ear also has an autonomic nerve supply. The autonomic fibers, mostly adrenergic sympathetic nerve fibers, mainly innervate blood vessels but they also contact hair cells [27].

5.4. Fluid Systems of the Cochlea

The fluid system of the cochlea is complex. It is shared with the vestibular organ and consists of two distinctly different systems: the perilymphatic system, in which the ionic composition of the fluid resembles that of the cerebrospinal fluid; and the endolymphatic system, in which the fluid resembles that of intracellular fluid – thus rich in potassium. In the cochlea, the endolymphatic space is separated from the perilymphatic space by Reissner's membrane and the basilar membrane (see Fig. 1.7C). The ionic composition of the perilymph is important for the function of the hair cells.

The fluid space of the perilymphatic system of the inner ear communicates with the cerebrospinal fluid in the skull cavity via the cochlear aqueduct, which connects the perilymphatic space with the cranial fluid space (Figs. 1.14 and 1.15). The duct has a very small diameter, 0.05–0.5 mm, and there is evidence that it

may not be totally open in many adults but it is known to be open in animals. It has been shown in experiments in cats that intracranial pressure (ICP) variations are communicated to the perilymphatic space with a short time constant [16]. If the cochlear aqueduct is closed artificially in such animals, changes in the ICP affect the pressure in the perilymphatic space to a much smaller extent and there is a time lag between changes in the ICP and changes in pressure in the perilymphatic space.

The endolymphatic space communicates with the endolymphatic sac through the endolymphatic duct (Fig. 1.15). The endolymphatic sac is the space between two layers of the dura mater. It is located close to the skull wall, near the porus acousticus (the opening of the internal auditory meatus). The pressure

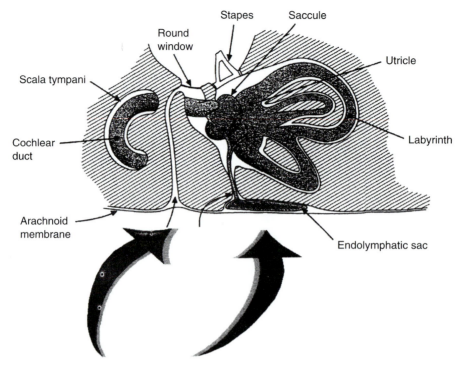

FIGURE 1.14 Schematic drawing of the fluid system of the cochlea (reprinted from Marchbanks, 1996, with permission from Springer).

or rather the volume in the different compartments is kept in balance by mechanisms that are not entirely known but are thought to involve the function of the endolymphatic sac (see p. 231). Reissner's membrane, which separates the endolymphatic space from the perilymphatic space in the cochlea, has a high degree of compliance. Therefore, very small changes in pressure can cause large changes in the volume of the endolymphatic space. An imbalance between the pressures in those two systems can

cause hearing impairment and disturbances of balance (see Chapter 9).

5.5. Blood Supply to the Cochlea

The arterial supply to the cochlea is the labyrinthine artery. It originates in the anterior inferior cerebellar artery (AICA) and follows the eighth cranial nerve in the internal auditory meatus, where it gives off the anterior vestibular artery to the vestibular apparatus (Fig. 1.16). Further into the internal auditory meatus the labyrinthine artery branches to form the vestibular-cochlear artery that supplies parts of the cochlea. The other branch is the spiral modiular artery that serves as a collateral blood supply to the cochlea [4]. The labyrinthine artery is an end-artery with little or no collateral blood supply to the cochlea. In humans, the labyrinthine artery is much longer than in the animals commonly used in experiments related to hearing. This is because the distance between the brainstem and the cochlea is much longer in humans, partly because the subarachnoidal space is much larger. (The auditory nerve, which the artery follows for most of its course, is approximately 2.5 cm long in humans [72, 73] compared to 0.5–0.8 cm in animals [41].)

It is important to note that the labyrinthine artery that runs in the internal auditory meatus is not a single artery but several smaller arterioles, almost like an

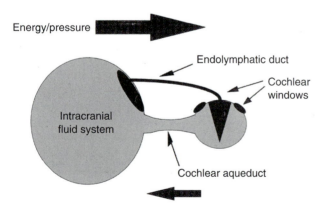

FIGURE 1.15 Schematic drawing of the cochlear fluid systems and their connections with the cerebrospinal fluid in the brain through the cochlear aqueduct (reprinted from Marchbanks, 1996, with permission from Springer).

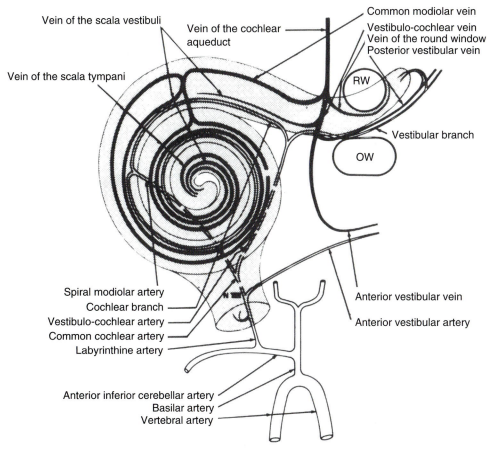

FIGURE 1.16   Arterial blood supply to the cochlea (reprinted from Axelsson and Ryan, 1988, with permission from Raven Press).

arterial plexus. Such a series of parallel small caliber arteries attenuate rapid changes in blood flow (pulsation) and thus contribute to providing a smooth (constant) blood supply to the cochlea and the vestibular system. The small diameter arteries in connection with a distal reservoir function as a low-pass filter that attenuates fast changes in blood flow. This may be of importance for avoiding stimulation of the auditory sensory cells from pulsation of the blood supply to the cochlea.

2

Sound Conduction to the Cochlea

1. ABSTRACT

1. Sound normally reaches the cochlea via the ear canal and the middle ear, but it may also reach the cochlea through bone conduction. Sound that enters the middle-ear cavities can also set the tympanic membrane in motion and thereby reach the cochlea.
2. The sound pressure at the tympanic membrane depends on the acoustic properties of the pinna, ear canal, and the head.
3. The ear canal acts as a resonator, which causes the sound pressure at the tympanic membrane to be higher than it is at the entrance of the ear canal. The gain is largest near 3 kHz (the resonance frequency) where it is approximately 10 dB.
4. In a free sound field, the head causes the sound pressure at the entrance of the ear canal to be different (mostly higher) than it is when measured at the place of the head without the person being present.
5. The effect of the head on the sound pressure at the entrance of the ear canal depends on the frequency of the sound and on the angle of incidence of the sound (direction to the sound source).
6. The difference in time of arrival of a sound at the two ears is the physical basis for directional hearing in the horizontal plane, together with the difference in intensity of the sound at the two ears.
7. The middle ear acts as an impedance transformer that matches the high impedance of the cochlea to the low impedance of air.
8. The gain of the middle ear is frequency dependent and the increase in sound transmission to the cochlear fluid due to improvement in impedance matching is approximately 30 dB in the mid-frequency range.
9. It is the difference between the force that acts on the two windows of the cochlea that sets the cochlear fluid into motion. Normally the force on the oval window is much larger than that acting on the round window because of the gain of the middle ear.
10. The ear's acoustic impedance is a measure of the tympanic membrane's resistance against being set into motion by a sound.
11. Measurements of the ear's acoustic impedance have been used in studies of the function of the middle ear and for recordings of contraction of the middle ear muscles.

2. INTRODUCTION

In the normal ear, sound can be conducted to the cochlea mainly through two different routes, namely: (1) through the middle ear (tympanic membrane and the ossicular chain); and (2) through bone conduction. Bone conduction of airborne sound has little importance for normal hearing but it is important in audiometry where sound applied to one ear by an earphone may reach the other ear by bone conduction (cross transmission).

3. HEAD, OUTER EAR AND EAR CANAL

The ear canal, the pinna and the head influence the sound that reaches the tympanic membrane. The influence of these structures is different for different frequencies and the effect of the head depends on the direction of the head to the sound source.

3.1. Ear Canal

The ear canal acts as a resonator and the transfer function[1] from sound pressure at the entrance of the ear canal to sound pressure at the tympanic membrane has a peak at approximately 3 kHz (average 2.8 kHz [113]) at which frequency the sound pressure at the tympanic membrane is approximately 10 dB higher than it is at the entrance of the ear canal (Fig. 2.1). This regards sounds coming from a source that is located at a distance from the observer (free sound field). The effect of the ear canal is different when sound is applied through headphones or through insert earphones (Fig 2.1).

3.2. Head

In a free sound field the head acts as an obstacle to the propagation of sound waves. Together the outer ear and the head transform a sound field so that the sound pressure becomes different at the entrance of the ear canal compared with the sound pressure that is measured in the place of the head. The effect of the head on the sound at the entrance of the ear canal is related to the size of the head and, the wavelength[2] of sound. This means that the "amplification" is frequency (or spectrum) dependent and, therefore, the spectrum of the sound that acts on the tympanic membrane becomes different from that which can be measured in the sound field in which the individual is located. The sound that reaches the entrance of the ear canal also depends on the head's orientation relative to the direction to the sound source. Depending on its orientation relative to the sound source, the head can function as a baffle for the ear that points towards the sound source or it can act as a shadow for sounds reaching the ear that is located away from the sound source.

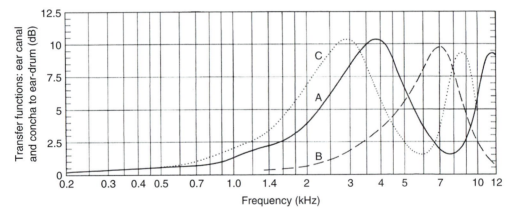

FIGURE 2.1   Effect of the ear canal on the sound pressure at the tympanic membrane: (A) average difference between the sound pressure at the tympanic membrane and that measured at the entrance of the ear canal; (B) difference between the sound pressure at the tympanic membrane and a location in the ear canal that is 1.25 cm from the tympanic membrane (similar to that of an insert earphone); and (C) theoretical estimate of the difference between the sound pressure at the tympanic membrane and that at a point that is the geometric center of the concha (reprinted from Shaw, 1974, with permission from Springer).

[1]The transfer function (or frequency transfer function) of a transmission system is a plot of the ratio between the output and the input, plotted as a function of the frequency of a sinusoidal input signal, known as a Bode plot. Such a plot is not a complete description of the transmission properties of a system unless the phase angle between the output signal and the input signal as a function of the frequency is included. Nevertheless, often only the amplitude function is shown, often expressed in logarithmic measures (such as decibels).

[2]The wavelength of sound is the propagation velocity divided by the frequency. The propagation velocity of sound in air is approximately 340 m/s slightly depending on the temperature and the air pressure. Assuming a propagation velocity of 340 m/s the wavelength of a 1,000 Hz tone is 340/1,000 = 0.34 m = 34 cm.

The results from studies of the effect of the head on the sound pressure at the entrance of the ear canal always refer to a situation where the head is in a free sound field with no obstacles other than the individual on which the measurements are performed. Such a situation occurs in nature with the sound source placed at a long distance and where there is no reflection from obstacles. This is a different situation from an ordinary room where sound reflections from the walls modify the sound field by their reflection of sound. A free sound field can be artificially created in a room with walls that absorb all sound (or at least most of it) and thus avoid reflection. Such a room is known as an anechoic chamber. Anechoic chambers are used for research such as that of the transformation of sound by the head and the ear canal.

3.3. Physical Basis for Directional Hearing

The physical basis for directional hearing in the horizontal plane is differences in the arrival time of sounds that reach the two ears and differences in the intensity at the entrance of the ear canal. The intensity difference is not only a factor of the direction to a sound source in the horizontal plane (azimuth) but it also depends on the frequency (spectrum) of the sound while the difference in arrival time is independent of the frequency of the sound. The differences in the sound that reaches the two ears are processed and discriminated in the central nervous system (see p. 143). The basis for discriminating direction in the vertical plane (elevation) is poorly understood but may have to do with the outer ear's acoustic properties with regard to high frequency sounds. Sound arrives at the two ears with a time difference except when sounds come from a location directly in front of or directly behind the observer. The reason is that the sound travels a different distance to reach the two ears. The difference in arrival time is related to the travel time from a sound source and it has a simple linear relation to the azimuth. The maximal difference in arrival time of the two "ears" in the standard model of the head shown in Fig. 2.2 is approximately 0.6 ms (Fig. 2.3). Values calculated from measurements taken from a hard spherical model of the head (solid line) agree closely with actual measurements made on a live subject.

Information about the difference in arrival time and the difference in sound pressure at the two ears is used by the central auditory nervous system to determine the direction to a sound source in the horizontal plane (azimuth). It is believed that the intra-aural time difference is most important for transient sounds and sounds with most of their energy in the frequency range below 1.5 kHz while it is the difference in the

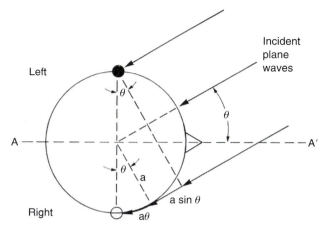

FIGURE 2.2 Schematic drawing showing how a spherical model of the head can be used to study the effect of azimuth of an incident plane sound wave (reprinted from Shaw, 1974, with permission from the American Institute of Physics).

intensity that is most important for high frequency sounds (see p. 142).

A solid sphere the size of a head (Fig. 2.2) has been used as a model of the head in studies of the transformation of sound from a free sound field to that found at the tympanic membrane and how that transformation changes when the head is turned at different angles relative to the direction to the sound source [128]. Such studies have shown that the sound pressure at the tympanic membrane is approximately 15 dB higher than it is in a free sound field in the frequency range 2–4 kHz when a sound source is located directly in front of an observer (Fig. 2.4). A dip occurs

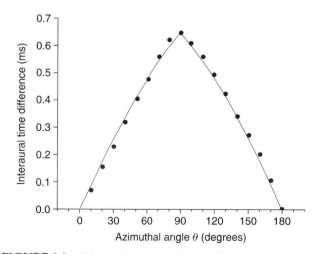

FIGURE 2.3 Calculated intra-aural time difference as a function of azimuths for a spherical model of the head (Fig. 2.2) with a radius of 8.75 cm (solid line), and measured values in a human subject (open circles) (reprinted from Shaw, 1974, with permission from the American Institute of Physics; after Feddersen et al., 1957).

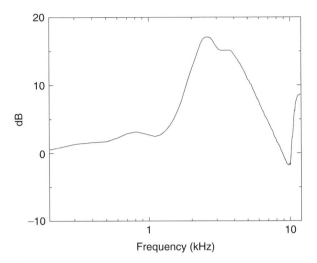

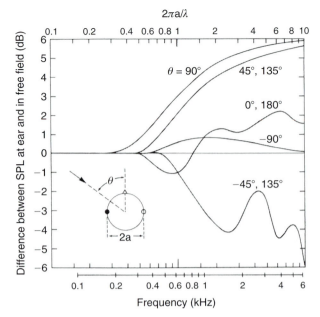

FIGURE 2.4 The combined effect of the head and the resonance in the ear canal and the outer ear, obtained in a model of the human head. The difference in sound pressure measured close to the tympanic membrane and a sound pressure in a free sound field with the sound coming from a source located directly in front of the head (based on Shaw, 1974).

FIGURE 2.5 Calculated differences between the sound pressure (in decibels) in a free field to a point corresponding to the entrance of the ear canal on a model of the head consisting of a hard sphere (Fig. 2.2). The difference is shown as a function of frequency at different azimuths (reprinted from Shaw, 1974, with permission from the American Institute of Physics).

in the transfer function of sound to the tympanic membrane at approximately 10 kHz.

The difference in the intensity of sounds that reach the two ears is a result of the head being an obstacle that interferes with the sound field. The head acts as a shield to the ear that is turned away from the sound source, which decreases the sound that reaches that ear and it acts as a baffle for the ear turned toward the sound source and that increases the sound intensity at that ear. This means that the effect of the head on the transfer of sound to the entrance of the ear canal depend on both the angle (azimuth) to the sound source and the frequency of the sounds (Fig. 2.5).

The difference between the sound pressure in a free field and that which is present at the entrance to the ear canal is small at low frequencies because the effect of the head is small for sound of wavelengths that are long in comparison to the size of the head (Fig. 2.2). In the frequency range between 2.5 and 4 kHz the amplification of sounds by the head and the pinna varies from 8 to 21 dB depending on the angle to the sound source in the horizontal plane (azimuth). The shadow and baffle effects of the head and the outer ear contribute to the difference in the sound intensity experienced at the two ears for sounds that do not come from a source located directly in front (0° azimuth) or directly behind (180°). In a broad frequency range above 1 kHz the intensity of sounds that come from a direction (azimuth) of 45–90° relative to straight ahead is approximately 5 dB higher at the entrance of the ear canal than at the free sound field occupied by the individual (Fig. 2.5).

The transformation of sound from a free sound field to the sound that reaches the tympanic membrane varies between individuals because of differences in the size and shape of the head making the results such as those shown in Fig. 2.5 represent the average person only.

4. MIDDLE EAR

Two problems are associated with transfer of sound to the cochlear fluid. One is related to sounds being ineffective in setting a fluid into motion because of the large difference in the acoustic properties (impedance) of the two media, air and fluid. The other problem is related to the fact that it is the difference between the force that acts at the two windows that causes the cochlear fluid to vibrate. The difference in the impedance of the two media would cause 99.9% of the sound energy to be reflected at the interface between air and fluid and only 0.1% of the energy will be converted into vibrations of the cochlear fluid if sound was led directly to one of the cochlear windows. Both these problems are elegantly solved by the middle ear. The middle ear acts as an impedance transformer that matches the high impedance of the cochlear fluid to the low impedance of air, thereby improving sound transfer to the cochlear fluid. By increasing the sound transmission selectively to the oval window of the

BOX 2.1

STUDIES OF PHYSICAL FACTORS THAT ARE IMPORTANT FOR DIRECTIONAL HEARING

The difference between the sound pressure at the tympanic membranes of the two ears has also been studied using a manikin equipped with microphones in place of the tympanic membrane [106] (Fig. 2.6). The results of such studies are in good agreement with those using a spherical model of the head. This model includes the pinna and the results show that the pinna mostly affects transmission of high frequency sounds. While the studies using a manikin more accurately mimic the normal situation, the results do not include the effect of the absorption of sound on the surface of the normal head.

A change in the direction to a sound source in the vertical plane (elevation) does not cause any change in the inter-aural time difference and determination of the elevation must therefore rely on other factors such as the differences in the spectrum of broad band sounds that reaches the two ears for different elevations [8]. This occurs because the transformation of a sound from the free field to the tympanic membrane depends on the elevation to the sound source. The pinna plays an important role in this dependence of the sound transformation on the elevation of the sound source.

The effect of elevation (angle to the sound source in the vertical plane) on the sound that reaches the two ears is greatest above 4 kHz (Fig. 2.7) [128]. The sound pressure at the tympanic membrane for 0° azimuth and an elevation of 0° falls off above 4 kHz (solid line in Fig. 2.7). With increasing elevation this upper cut off frequency shifts toward higher frequencies (dashed lines in Fig. 2.7). At an elevation of 60° the cut off is above 7 kHz and at that frequency, the sound pressure is more than 10 dB above the value it has at an elevation of 0° [128].

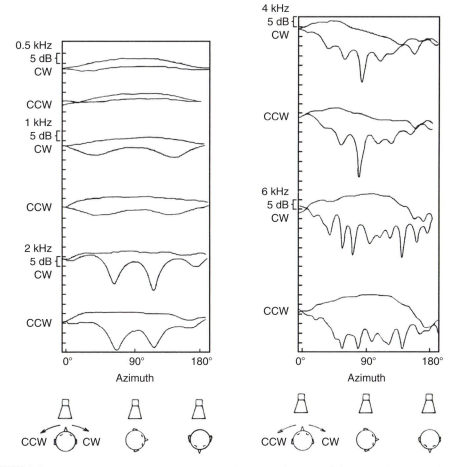

FIGURE 2.6 Sound intensity at the "tympanic membrane" as function of the azimuth measured in a more detailed model of the head (manikin) than the one shown in Fig. 2.2. The difference between the sound intensity at the two ears is the area between the two curves (based on Nordlund, 1962, with permission from Taylor & Francis).

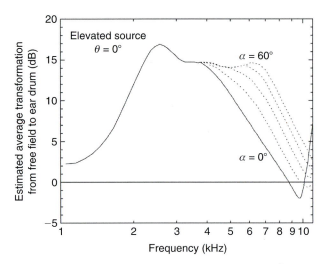

FIGURE 2.7   Effect of elevation on the sound pressure at the tympanic membrane (reprinted from Shaw, 1974, with permission from Springer).

cochlea, the middle ear creates a difference in the force that acts on the two windows of the cochlea and it thus provides an effective transfer of sound to vibration of the cochlear fluid.

4.1. Middle Ear as an Impedance Transformer

Theoretical considerations show that the transmission of sound to the oval window would be improved by 36 dB if the middle ear acted as an ideal impedance transformer with the correct transformer ratio. However, the transformer ratio of the human middle ear is slightly different from being optimal and that causes some of the sound to be reflected at the tympanic membrane and thus lost from transmission to the cochlea.

The impedance transformer action of the middle ear is mainly accomplished by the ratio between the effective area of the tympanic membrane and the area of the stapes footplate, but the lever ratio of the middle ear bones also contributes. The ratio of areas of the

---

BOX 2.2

SOUND DELIVERED BY EARPHONES

The sound delivered to the ear by earphones is not affected by the acoustic properties of the head. This means that spectral filter action of the head, pinna and ear canal is not effective when earphones are used. This is one of the reasons that music and speech sounds differently when listening through ordinary earphones compared to listening in a free sound field. This was recognized as a problem for music delivery when earphones came into frequent use. The problem was solved by modifying the sound spectrum that drives the earphones in a way that imitates the effect of the head [8]. This principle was first applied to the Sony® Walkman type of tape players but later used in modern digital devices that deliver music. The modification of the sound spectrum made music and speech played through earphones sounds similar to what it does in a (natural) free field. Such a correction of the spectrum of the input to earphones is the reason sound produced by earphones can sound natural, giving an impression of "sound space." The effect of turning the head when listening in a free field, however, is absent when listening through earphones.

The earphones that are commonly used for audiometric purposes are either supra-aural headphones and now, more commonly, insert earphones. There are two concerns regarding the use of earphones for hearing testing; one is calibration and the other is that an earphone applied to one ear also conducts sound to the other ear, by bone conduction. This "cross-talk" is different for different earphone types, being much greater for supra-aural headphones than for insert earphone (Fig. 2.8A). This cross transmission is the reason that it may be necessary to mask the better hearing ear when testing the hearing in individuals with large differences between hearing thresholds in the two ears. For frequencies below 1 kHz the attenuation of the cross-transmitted sound is greater than 80 dB for insert earphones. Insert earphones have roughly the same frequency characteristics as supra-aural earphones but concerns about the accuracy of the calibration remain.

Normally, hearing tests are performed in sound insulated rooms but occasionally it is necessary to test the hearing in environments with high ambient noise. In such situations, it is important that the earphone that is used attenuates sounds from the environment. Insert earphones also provide much higher attenuation of external noise than supra-aural headphones (Fig. 2.8B).

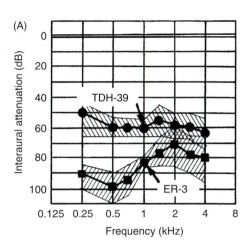

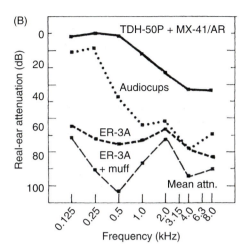

FIGURE 2.8 (A) Average and range of intramural attenuation obtained in six subjects with two types of earphones (TDH 39 and an insert earphone, ER-3) (reprinted from Killion et al., 1985. (B) External noise attenuation of four different earphones often used in audiometry (reprinted from Berger and Killion, 1989, with permission from the American Institute of Physics).

BOX 2.3

MIDDLE EAR'S EFFECTIVENESS IN TRANSFERING SOUND TO THE COCHLEA

The specific impedance of air is 42 cgs units and that of water 1.54×105 cgs units (41.5 dynes/cm^3 and 144,000 dynes/cm^3), thus a ratio of approximately 1:4,000. Transmission of sound to the oval window will therefore be optimal if the middle ear has a transformer ratio that is equal to the square root of 4,000 (equals 63). This assumes that the input impedance to the cochlea is equal to that of water; in fact it is less. Studies in the cat show that the input impedance of the cochlea is lower at low frequencies than at high frequencies. In the middle frequency range the impedance of the cochlea is approximately the same as that of seawater. Rosowski [122] calculated the overall effectiveness of transferring sound from a free field to the cochlear fluid for the cat (Fig. 2.9). Merchant et al. [85] arrived at gain values of approximately 20 dB between 250 Hz and 500 Hz with a maximum of 25 dB at 1 kHz above which the gain decreases at a rate of 6 dB/octave. The results obtained by different investigators differ and show a gain of the middle ear in the range 25–30 dB.

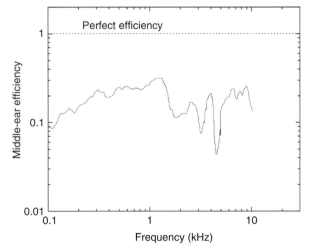

FIGURE 2.9 The efficiency of the cat's middle ear, showing the fraction of sound power entering the middle ear that is delivered to the cochlea (after Rosowski, 1991, with permission from the American Institute of Physics).

BOX 2.4

THE GAIN OF THE MIDDLE EAR

One of the first animal studies that qualitatively measured the gain of the cat's middle ear in transferring sound to the cochlea, was published by Wever, Lawrence and Smith (Fig. 2.10A) [153]. Early studies of the transfer function of the middle ear used pure tones of different frequencies measuring the sound pressure at the tympanic membrane that is required to produce cochlear microphonic (CM^2) potentials of a certain amplitude [153]. Usually the sound pressure that evokes a 10 µV CM response is determined in the frequency range of interest (for instance, from 100 to 10 kHz). Measurements are first done while the middle ear is intact and then repeated after the middle ear is removed surgically and the sound led directly to the oval window (dashes in Fig. 2.10A), or

to the round window (dots in Fig. 2.10A) using a speculum that was attached to the bone of the cochlea. This arrangement ensured that sound only reached one of the two cochlear windows at a time. When the sound is conducted directly to either the round or the oval window a much higher sound level is needed to obtain a 10 µV CM potential than when conducted via the normal route with the middle ear being intact. The difference between the solid curve in Fig. 2.10A and the dotted or the dashed curves (Fig. 2.10B) is a measure of the gain in sound conduction to the cochlea provided by the cat's middle ear. It is seen that the gain of the cat's middle ear is frequency dependent and it is largest in the frequency range between 0.5 and 10 kHz where it is between 35 and 38 dB.

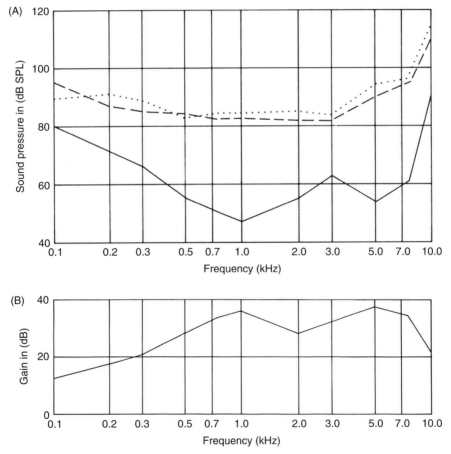

FIGURE 2.10 (A) Illustration of the gain of the middle ear of a cat. Sound pressure needed to produce a CM of an amplitude of 10 mV is shown with the middle ear intact and the sound conducted to the tympanic membrane (solid lines), and after removal of the middle ear and the sound conducted to the oval window (dashes) and round window (dots) using a closed sound delivery system (based on Wever, E.G., Lawrence, M., Smith, K.R. 1948. The middle ear in sound conduction. *Arch of Otolaryng.* 48, 12-35, with permission from Archives of Otolaryngology Head and Neck Surgery. Copyright © (1948) American Medical Association. All rights reserved). (B) Difference between the dotted-dashed curves and the solid curve in (A) (from Møller, 1983; based on Wever, E.G., Lawrence, M., Smith, K.R. 1948. The middle ear in sound conduction. *Arch of Otolaryng.* 48, 12-35, with permission from Archives of Otolaryngology Head and Neck Surgery. Copyright © (1948) American Medical Association. All rights reserved).

tympanic membrane and that of the stapes is frequency dependent because it is the effective area of the tympanic membrane[3] and not its geometrical (anatomical) area that makes up the transformer ratio.

The middle ear has mass and stiffness that make its transmission properties become frequency dependent. Its efficiency as an impedance transformer thus becomes a function of frequency. Stiffness impedes the motion at low frequencies and mass impedes motion at high frequencies. The friction in the middle ear causes loss of energy that is independent of frequency. The lever ratio may be frequency dependent because the mode of vibration of the ossicular chain is different at different frequencies. The effective area of the tympanic membrane depends on the sound frequency and that contributes to the frequency dependence of middle-ear transmission. Because sound transmission through the middle ear is frequency dependent, it is an oversimplification to express the transformer action as a single number and the transformer ratio of the middle ear must be described by a function of frequency, namely, its transfer function.

Estimates of the gain of the middle ear by different investigators vary and there are systematic differences between results obtained in humans and in animals. The total efficiency of the human middle ear is approximately 10 dB less than ideal for frequencies up to approximately 0.2 kHz and its highest efficiency is attained around the frequency 1 kHz where it is approximately 3 dB below that of an ideal impedance transformer. This means that the middle ear transmits approximately one-third of the sound energy to the cochlea in this frequency range and less above and below this range [122]. Above 1.5 kHz the efficiency (in percentage of energy transferred to the cochlea) varies between 20% at 4 kHz and 20% (Fig. 2.9), corresponding to losses between 5 and 25 times (7 and 14 dB), respectively.

In the experiments described above sound was led to only one of the two windows of the cochlea at a time. If sound is led to the middle-ear cavity, a different situation arises because sound then will reach both the oval window and the round window with about the

same intensity. (Hearing loss without the middle ear is discussed in Chapter 9.)

Direct measurements of the sound transmission through the middle ear as the function of the frequency have also been performed both in anesthetized animals and in human cadaver ears. The transfer function of the middle ear has been studied in anesthetized cats by measuring the vibration amplitude of the stapes using microscopic techniques with stroboscopic illumination [44] or by using a capacitive probe to measure the vibration of the round window (Fig. 2.11) [104].

4.2. Transfer Function of the Human Middle Ear

The middle ear in humans is different from those of animals, which are usually used in auditory experiments, and that makes it important to distinguish between results obtained in humans and animals. How to "translate" the results of experiments in animals

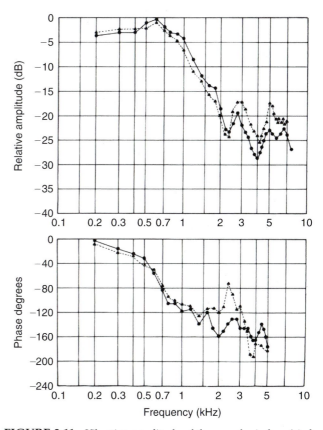

FIGURE 2.11   Vibration amplitude of the round window (circles and solid lines) and the incus (triangles and dashed lines) of the ear of a cat, for constant sound pressure at the tympanic membrane. The vibration amplitude was measured using a capacitive probe (from Møller, 1983; based on Møller, 1963, with permission from the American Institute of Physics).

---

[2]The CM is generated in the cochlea and its amplitude is closely related to the volume velocity of the cochlear fluid. The CM in response to pure tones is a sinusoidal waveform the amplitude of which increases with the increase in the sound pressure of the sound that elicits the CM. Recording of the CM is often used to determine changes in sound transmission of the middle ear. The generation of the cochlear microphonic potential (CM) is discussed in detail in Chapter 4.

[3]The effective area of a membrane like the tympanic membrane is the area of a rigid, weightless piston that transfers sound in the same way as the membrane.

into estimates of sound transmission in humans will be discussed below.

Some of the earliest studies of the frequency transfer function of the middle ear were done in human cadaver ears by von Békésy in 1941 [6].[4] Measurements of the transfer function of the human middle ear are limited to studies in cadavers. The ratio between the vibration amplitude of the ossicles (the umbo and the stapes) in human cadaver ears and the sound pressure

close to the tympanic membrane (Fig. 2.12) reveals transfer functions that are similar to those obtained in animals [46, 71]. The vibration amplitude of the ossicles is nearly constant for low frequencies up to the resonance frequency of the middle ear (approximately 900 Hz). These results are similar to those obtained by von Békésy [6] almost 50 years earlier. The similarity between these results and those obtained using modern techniques is remarkable in the light of the

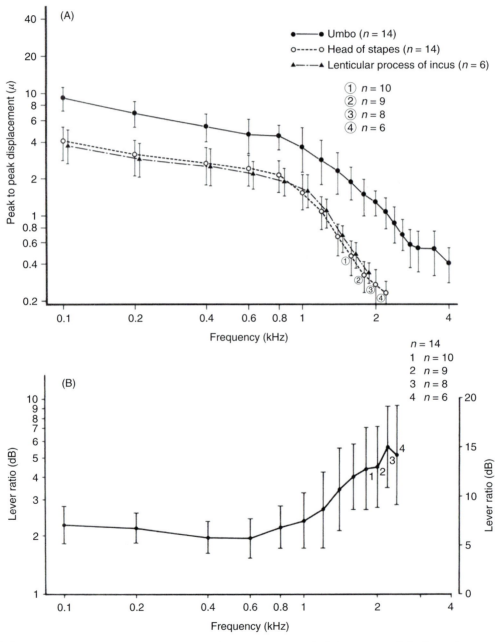

FIGURE 2.12 (A) Average displacements of the umbo, the head of the stapes and the lenticular process of the incus. (B) The lever ratio at 124 dB SPL at the tympanic membrane in 14 temporal bones. Vertical bars indicate one standard deviation (reprinted from Gyo, et al., 1987, with permission from Taylor & Francis).

BOX 2.5

MEASUREMENT OF THE IMPULSE RESPONSE OF THE MIDDLE EAR

Direct measurements of the impulse response of the umbo in awake human volunteers were obtained by applying an acoustic impulse (click sound) to the ear and using laser Doppler shift (laser Doppler vibrometer, LDV) to measure the displacement of the umbo (Fig. 2.13) [139]. Goode et al. [43] used a similar method using commercially available LDV equipment to measure the vibration amplitude of the umbo in human volunteers. Although such measurements do not reflect the transmission properties of the middle ear but rather reflect the ability of the tympanic membrane to transform sound into vibration of the manubrium of the malleus, this method might become a useful clinical method for testing the function of the middle ear.

technical difficulties associated with such measurement at the time that von Békésy did these studies.

The transfer functions of the middle ear shown by Kurokawa and Goode [71] showed a considerable individual variation, attributed mainly to individual variations in the function of the tympanic membrane. The irregularities in the transfer function of the middle ear seen in Fig. 2.12 suggest that the function of the middle ear is more complex than that of a combination of a few elements of mass and stiffness. Several models of the middle ear were developed during the past three or four decades to account for such complexity [97, 121, 164].

4.3. Impulse Response of the Human Middle Ear

Estimation of the impulse response[5] of the cat's middle ear has been obtained by computing the inverse Fourier transform of the frequency transfer functions such as those seen in Fig. 2.11. Such calculations show the displacement of the cochlear fluid in a cat's ear, as it would be in response to a brief sound impulse.

---

[4]All results reported by von Békésy reported in this book were taken from the book *Experiments in Hearing*, G. von Békésy, 1960, McGraw Hill, New York [6]. This book contains translations of original articles by von Békésy, published in the German language. The date (year) of the original publication will be used along with the reference to the 1960 yearbook to give proper credit to the work of von Békésy by emphasizing when the work was first published.

[5]The impulse response of a transmission system such as the middle ear is by definition the response to an infinitely short impulse. In practice the impulse response is obtained by applying a short impulse to the system that is tested. There is a mathematical relationship between the impulse response and the frequency transfer function, and a mathematical operation known as the Fourier transform can convert an impulse response into a transfer function. The inverse Fourier transform convert a transfer function into an impulse response.

4.4. Linearity of the Middle Ear

The assumption that the middle ear functions as linear system was supported by the experimental work by Guinan and Peake [44] who found that the stapes (in the cat) moves in proportion to the sound pressure at the tympanic membrane up to 130 dB SPL for frequencies below 2 kHz and even higher (140–150 dB SPL) for frequencies above 2 kHz.

4.5. Acoustic Impedance of the Ear

The ear's acoustic impedance is a measure of the resistance of the tympanic membrane to be set in motion by sound. Studies of the ear's acoustic impedance can provide important insight into how the middle ear functions, including the role of the different parts of the middle ear in transferring sound into vibration of the cochlear fluid. Studies of the ear's acoustic impedance are also important for studies of middle ear pathology. Measurements of the acoustic impedance of the ear have not only played an important role in scientific examination of the function of the middle ear but are now used routinely in clinical diagnosis of disorders of the middle ear. Tympanometry that is used clinically to assess the function of the middle ear and to determine

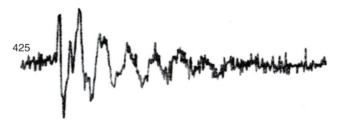

FIGURE 2.13   Impulse response of the umbo obtained in a human individual (reprinted from Svane-Knudsen and Michelsen, 1985, with permission from Springer).

BOX 2.6
=====

CRITERIA FOR LINEAR SYSTEMS

A transmission system must fulfill several criteria in order to be regarded to function as a linear system. The output must increase in the same proportion as the input is increased and if two different input signals (such as two tones with different frequencies) are applied to the input of a system, the output must be the sum of the output of the two signals when applied independently. This is known as the superposition criteria of a linear system. The output of a linear system to which two sinusoidal signals (for instance, tones) are applied only contains energy at the same two frequencies as the input. The transmission properties of a linear system can equally well be determined by using different kinds of input signals in connection with mathematical operations on the results. The properties of a non-linear system cannot be described in a universal way.

---

the air pressure in the middle-ear cavity is a form of measurement of the ear's acoustic impedance. Measurements of changes in the ear's acoustic impedance are used to record the contractions of the middle-ear muscles in studies of the acoustic middle-ear reflex for oto-neurologic diagnosis.

Electrical circuits and mechanical systems are analogous in many ways. Thus in an electrical circuit, electrical current corresponds to vibration velocity and electrical voltage corresponds to mechanical force. The mechanical impedance, Z, is therefore the ratio between force, F, and velocity, V. Mechanical friction corresponds to an electrical resistance, mass (or inertia) corresponds to inductance and a spring (elasticity) to capacitance.

In an acoustic system, volume velocity corresponds to electrical current, sound pressure corresponds to voltage and friction corresponds to electrical resistance (Fig. 2.14C & D). The acoustic impedance of a volume of air corresponds to a capacitor in an electrical circuit and the acoustic impedance of a narrow passage such as that of a narrow tube corresponds to an inductance in an electrical circuit. The acoustic impedance is thus the ratio between sound pressure and volume velocity. In studies of the ear, it is the mechanical impedance of the ear transformed to acoustic impedance by the tympanic membrane that is of interest. A mechanical system such as the middle ear is converted into an acoustic system by a piston or a membrane, such as the tympanic membrane, that converts sound into mechanical force (Fig. 2.14C). If the tympanic membrane acted as an ideal piston the mechanical impedance would be the acoustic impedance divided by the surface area of the piston assuming that appropriate units of measure were used to describe the acoustic and mechanical impedance. How the acoustic impedance of the ear reflects the mechanical properties of the middle ear may be understood by considering a simplified mechanical

model of the middle-ear system equipped with a piston (Fig. 2.14C).

The admittance, Y, is the inverse of the impedance, $1/Z$. It is also known as the compliance, because it is a measure of how easily a current is induced in an electrical system or how easily a mechanical system is set into vibration by an external force. In an electrical circuit, the admittance is the current divided by the voltage. In a mechanical system, the impedance is the velocity divided by the force and in an acoustic system, the admittance is the volume velocity divided by the sound pressure. The admittance may be a complex quantity with a real component, G, and an imaginary component, jB. Like impedance, admittance can also be expressed as an absolute value and phase angle.

The ear's acoustic impedance has been measured in both animals and humans for studies of the function of the middle ear and for pathological studies of the middle ear, but measurements of the absolute value of the ear's acoustic impedance never became a useful clinical diagnostic tool. Instead, measurements of changes in the ear's acoustic impedance came into general use in the clinic for determining the air pressure in the middle-ear cavity (tympanometry) and for recording the response of the acoustic middle-ear reflex.

The acoustic impedance of the human ear has been expressed either as its absolute value and phase angle, or as a real and an imaginary component as a function of the frequency. The resistive (real) component varies very little as a function of the frequency while the imaginary (reactive) component is high at low frequencies and decreases with increasing frequency up to approximately 1 kHz indicating that it is dominated by stiffness below 1 kHz. Both the real and the imaginary components have considerable individual variations (Fig. 2.15) [97] even when obtained in young individuals with normal hearing and no history of

BOX 2.7

BASIC CONCEPTS OF IMPEDANCE

Mechanical and acoustic systems are often described by their electrical analogue circuits because many people are more familiar with electrical circuits than with acoustic and mechanical systems (Fig. 2.14). Per definition the impedance, Z, of an electrical system is the resistance against which an applied voltage induces an electrical current in an electrical circuit. In the simplest of all systems consisting of a single resistor, the impedance is the voltage, E, that is needed to set up a unit current, I, thus using

Ohm's law and knowing the voltage and the current makes it possible to determine the resistance, R: $R = E / I$. When a circuit contains other elements such as capacitors and inductances the impedance must be measured using alternating test signals such as sinusoidal voltage and currents and the impedance becomes dependent on the frequency of the test signals. The impedance of such a circuit can no longer be described by a single number because its impedance becomes a complex quantity that requires two numbers to be described. A complex quantity, such as an impedance, Z, can be described by its real and its imaginary component ($Z = R + jX$, in Fig. 2.14B, where j denote an imaginary quantity). A complex quantity can also be described by its absolute value (length of a vector) and the phase angle (of the vector) (Fig. 2.14B). The impedance of a capacitor and an inductance has pure imaginary values of opposite signs; impedance of a capacitor decreases as a function of the frequency and that of an inductor increases as a function of the frequency. The impedance of a circuit that contains a capacitor and an inductor will therefore be zero at a certain frequency (Fig. 2.14B). That frequency is known as the resonance frequency. If the circuit in question also contains a resistor, the impedance will not be zero at the resonance frequency but it will have the value of the resistance at that frequency.

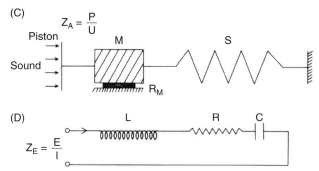

FIGURE 2.14   (A) A simple mechanical system consisting of a mass (M), elasticity (S) and friction (R). (B) Relationship between the different elements of the impedance ($Z = R + jX$) and the frequency, f, of the mechanical system in (A). (C) The mechanical system in (A) equipped with a rigid piston to form an acoustic system. (D) Electrical analogue of the mechanical system in (A) (reprinted from Møller, 1964, with permission from Taylor & Francis).

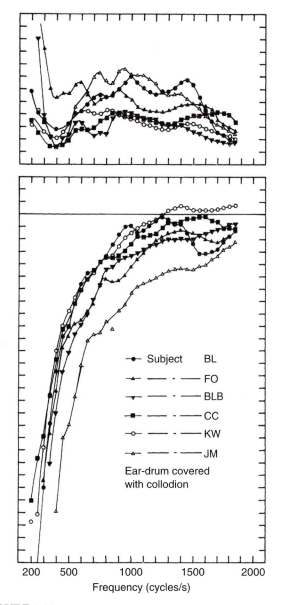

the irregularities in the impedance function results from the properties of the tympanic membrane. The properties of a triangular shaped portion of the tympanic membrane known as the *pars flaccida membrana tympani* are assumed to contribute to the irregular pattern of the acoustic impedance of the human ear (Figs 2.15 and 2.16). This part of the tympanic membrane is relatively loose and its vibrations are not transferred to the manubrium of the malleus as effectively as vibrations of other parts of the membrane. Similar irregularities are not present in the acoustic impedance of animals, such as the cat, probably because the cat's tympanic membrane does not have a pars flaccida.

4.6. Contributions of Individual Parts of the Middle Ear to its Impedance

Studies of the contribution of the different parts of the middle ear to its overall impedance have been done in animal experiments where the middle ear can be altered experimentally [89]. The possibilities of manipulating the human middle ear are naturally much more limited than what is the case in animals but the use of pathologies for such studies can provide useful information about the function of the middle ear. The immobilization of the ossicular chain as it occurs in patients with otosclerosis has been used in development of electrical and mathematical models of the human middle ear [167].

The properties of the tympanic membrane have been studied by measuring the ear's impedance when the manubrium is prevented from vibrating. When the malleus is immobilized the vibrations of the tympanic membrane are not transferred to a motion of the malleus and the measured acoustic impedance is that of the tympanic membrane itself. In the cat the acoustic impedance of the tympanic membrane with the malleus immobilized is very high for frequencies below 3 kHz (Fig. 2.17) [89] indicating that it functions in a similar way as a rigid piston for those frequencies. These results do not provide information regarding whether or not the equivalent area of this "piston" is different for different frequencies.

Comparing the ear's acoustic impedance with the vibration velocity of the malleus for constant sound pressure at the tympanic membrane provides information about the ability of the tympanic membrane to convert sound into vibration of the manubrium of malleus (Fig. 2.18).

The two curves in Fig. 2.18, showing the acoustic impedance and the inverse velocity of the malleus in the cat, are parallel for low frequencies (up to approximately 2 kHz) but deviate above 2 kHz, indicating that the tympanic membrane functions in a similar

FIGURE 2.15 The acoustic impedance measured in the ear canal and transformed to the estimated plane of the tympanic membrane, in six individuals with no known ear disorders (reprinted from Møller, 1961, with permission from the American Institute of Physics).

middle-ear diseases. Measurements of the acoustic impedance in the same individual show a high degree of reproducibility (Fig. 2.16) [95]. The variations in the impedance obtained in different individuals are therefore a result of permanent individual differences.

This individual variation has several causes. When the tympanic membrane in humans was covered with a thin layer of collodion, the individual variations in the acoustic impedance became smaller and the small irregularities in the curves of the acoustic impedance decreased indicating that the individual variation and

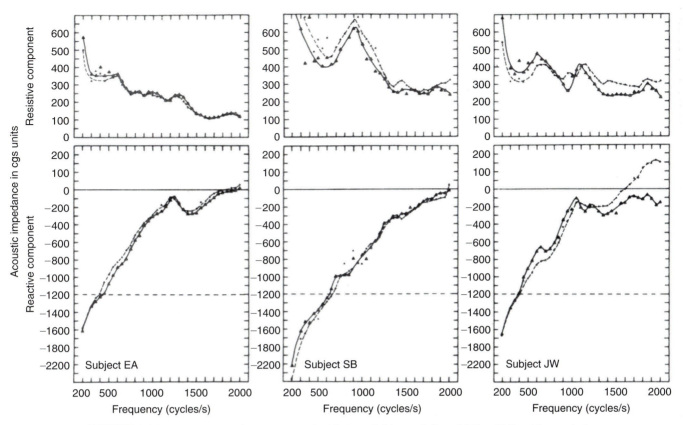

FIGURE 2.16   Acoustic impedance measured with 2 weeks' interval (from Møller, 1960, with permission from the American Institute of Physics).

way as a rigid piston for frequencies only up to approximately 2 kHz. (The inverse vibration velocity is expressed in arbitrary units and the two curves were made to superimpose at low frequencies.) This means that the effective area of the tympanic membrane changes with the frequency above 2 kHz.

The results of experiments obtained in the cat may not be directly applicable to the human ear because the tympanic membrane in humans has a more complex pattern of vibration and it may be less stiff than that of the cat. Studies of the human tympanic membrane done in cadaver ears [64] showed that the tympanic membrane has a smaller effective area at high frequencies than it has at lower frequencies.

Experiments in cats and rabbits show that severing the connection between the incus and the stapes (the incudo-stapedial joint) reduces the resistive component of the ear's acoustic impedance below 4 kHz to very small values (Fig. 2.19) [89], suggesting that the real component (friction) of the ear's acoustic impedance is mainly contributed by the cochlea. Elimination of the friction component of the middle ear makes the resonance of the middle ear more pronounced.

Below 4 kHz the reactive (imaginary) component of the ear's acoustic impedance was only little altered by disconnecting the cochlea, indicating that the cochlea contributes little elasticity and mass to the middle ear. The effect on the ear's acoustic impedance from interrupting the incudo–stapedial join is more complex for frequencies above 4 kHz than below (Fig. 2.19) [89] as has been observed by other investigators [144].

Animal experiments have shown that the reactive component of the ear's acoustic impedance for frequencies below 3 kHz decreases after opening of the middle-ear cavity [89]. This is because the middle-ear cavities add stiffness to the middle ear.

The air pressure in the middle-ear cavity is normally kept close to the ambient pressure by the occasional opening of the Eustachian tube that connects the middle-ear cavity with the pharynx. When the air pressure is not the same on both sides of the tympanic membrane, the function of the middle ear changes causing a decrease in sound conduction to the cochlea and the ear's acoustic impedance changes [89. 153].

The effect is more pronounced at low frequencies than at high frequencies and it is largest for a negative

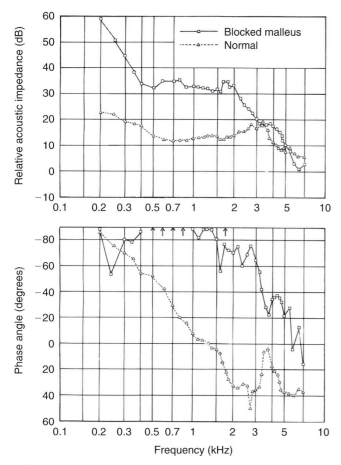

FIGURE 2.17 The acoustic impedance at the tympanic membrane measured in a cat, before (dashed lines and triangles) and after that the ossicular chain was immobilized (solid lines and squares) (reprinted from Møller, 1965, with permission from Taylor & Francis).

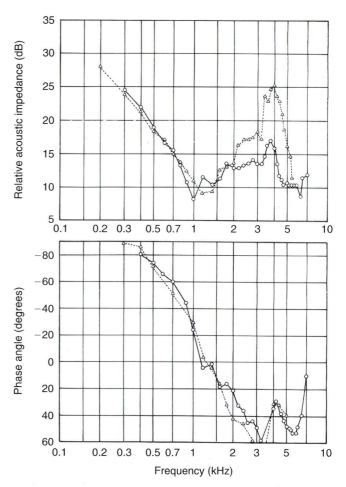

FIGURE 2.18 Comparison of the acoustic impedance at the tympanic membrane with the inverse velocity of the malleus for constant sound pressure at the tympanic membrane in a cat. The impedance is given in decibels relative to 100 cgs units and the inverse vibration velocity is given in arbitrary decibel values. Circles = accoustic impedance at the tympanic membrane; triangles = sound pressure at the tympanic membrane divided by the veloicty of the malleus (reprinted from Møller, 1963, with permission from the American Institute of Physics).

BOX 2.8

ACOUSTIC PROPERTIES OF THE TYMPANIC MEMBRANE

If the tympanic membrane functions in the same way as a (ideal) piston, the mechanical force that acts on the manubrium of malleus is proportional to the sound pressure at the tympanic membrane. The ratio between the vibration velocity of the malleus and the sound pressure will then be equivalent to the velocity of the manubrium divided by the force that acts on the membrane, thus the inverse impedance (namely, admittance). This means that measurement of the vibration velocity of the malleus (for constant sound pressure) is a measure of the ability of the tympanic membrane to convert sound into vibration of the malleus, thus a measure of the function of the tympanic membrane. (The velocity of the vibration is the first derivative of the amplitude and the velocity for sinusoidal vibrations at constant sound pressure level can be computed from the vibration amplitude by multiplying it with the frequency, which is the same as adding 6 dB/octave to the amplitude when the amplitude is expressed in dB.)

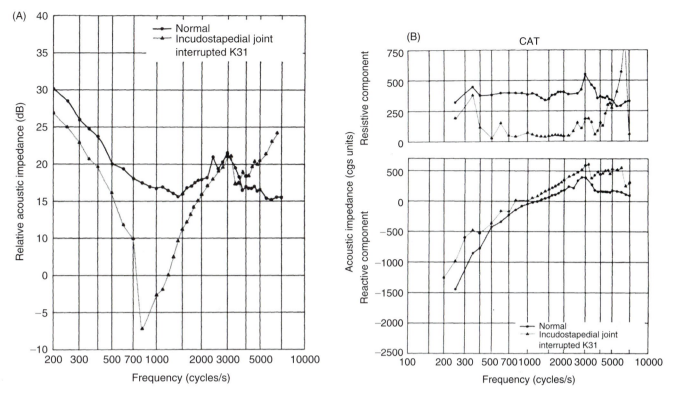

FIGURE 2.19   (A) Effect of interrupting the incudo-stapedial joint on the acoustic impedance of the ear of a cat. Absolute value of the impedance (given in decibels relative to 100 cgs units). (B) Effect of interrupting the incudo-stapedial joint on the acoustic impedance of the ear of a cat. The same data as in (A) with the real and the imaginary parts of the impedance shown separately (reprinted from Møller, 1965, with permission from Taylor & Francis).

pressure in the middle-ear cavity (corresponding to a positive pressure in the ear canal). Animal experiments have shown that the ear's acoustic impedance is lowest when the air pressure in the middle-ear cavity is the same as that in the ear canal [89]. The ear's acoustic impedance increases both when the pressure is increased and when it is decreased (Fig. 2.21) but not exactly in the same way. While both positive pressure and negative pressure in the middle-ear cavity cause the stiffness of the middle ear to increase, negative pressure in the middle-ear cavity reduces the resistive component of the ear's acoustic impedance more than

---

BOX 2.9

EFFECT OF THE BONY SEPTUM IN THE CAT'S MIDDLE-EAR CAVITY

The cat has a bony septum separating the middle-ear cavity in two compartments that communicate by a small hole in the septum. The reactive component of the acoustic impedance of the cat's ear changes rapidly as the frequency is changed around 4 kHz because of the resonator. Comparison of the acoustic impedance of the cat's ear before and after removal of that septum confirms that this hole together with the cavities act as a Helmholz resonator, which makes the effect of the middle-ear cavities in the cat different from that in other animals such as the rabbit, which does not have a similar septum in the middle ear. Removing the bony septum of the middle ear makes the middle-ear cavity act as a simple stiffness component similar to that in the rabbit, which has a middle ear that has a single middle-ear cavity adding stiffness [89]. The middle-ear cavity in humans is different from that of these animals in that it is much larger and it contains many air cells.

EARLY STUDIES OF THE EFFECT OF THE AIR PRESSURE IN THE MIDDLE-EAR CAVITY

Some of the earliest published studies of the effect of a difference in the static pressure in the ear canal were published by von Békésy [6: 95–126] who used psychoacoustic methods (loudness balance) (Fig. 2.20) and showed that the effect from a pressure difference between the two sides of the tympanic membrane on the sound transmission through the middle ear is largest at low frequencies.

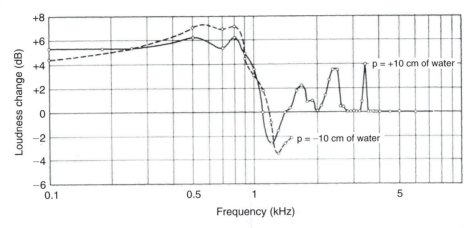

FIGURE 2.20 The effect on sound transmission through the middle ear from static air pressure of 10 cm H$_2$O measured by loudness matching. The attenuation is given in positive dB values (reprinted from Békésy, 1933, with permission from McGraw Hill).

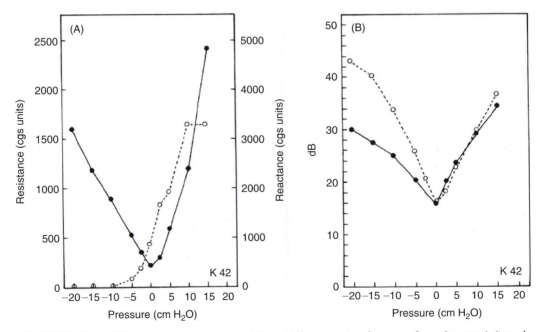

FIGURE 2.21 (A) Effect of static air pressure in the middle ear cavity of a cat on the ear's acoustic impedance (resistive [open circles] and reactive components [filled circles] shown separately) (reprinted from Møller, 1965, with permission from Taylor & Francis). (B) Effect on static air pressure in the middle ear cavity of a cat. Comparison between the change in the ear's acoustic admittance and change in its transmission. The admittance is given in dB relative to 100 cgs units and the transmission is given in arbitrary dB values (reprinted from Møller, 1965, with permission from Taylor & Francis).

positive pressure does (Fig. 2.21A). Positive pressure in the middle-ear cavity causes the acoustic admittance to decrease by approximately the same amount as the decrease in the transmission of sound through the middle ear (Fig. 2.21B). Negative pressure in the middle ear cavity causes a larger decrease in transmission than the same amount of positive pressure. That may be explained by the fact that negative pressure in the middle-ear cavity reduces the resistive component of the ear's acoustic impedance (Fig. 2.21A). Since the resistive component of the ear's acoustic impedance mainly originates in the cochlea (Fig. 2.19) a reduction of the resistive component of the ear's acoustic impedance indicates that negative pressure causes the cochlea to become decoupled from the middle ear explaining why a negative pressure causes a larger decrease in the transmission of sound to the cochlea than a positive pressure.

Measurement of changes in the ear's acoustic impedance when the air pressure in the sealed ear canal is varied is known as tympanometry. Tympanometry has found widespread clinical usage as a diagnostic tool because it provides a non-invasive way to determine the pressure in the middle-ear cavity. The use of tympanometry for that purpose is based on the finding that the ear's impedance changes as a function of the difference between the air pressure in the ear canal and the tympanic cavity and that the impedance has its lowest value when the pressure is the same in the ear canal as it is in the middle-ear cavity (see Fig. 2.21B) [89].

When tympanometry is used clinically, changes as a function of air pressure in the ear canal are usually expressed in acoustic admittance (also known as immittance).

Tympanometry also provides information about the function of the middle ear in general. Usually the acoustic impedance (or admittance) is measured at a single frequency but the variation in the ear's impedance as a result of air pressure in the ear canal is different for different frequencies (Fig. 2.22). Some investigators have made use of that fact to gain more diagnostic information from tympanometry [20].

The middle-ear muscles normally contract as an acoustic reflex (see Chapter 8). Contraction of the tensor tympani muscle pulls the manubrium of malleus inward, increasing the stiffness of the middle ear and displacing the tympanic membrane inward. The stapedius muscle pulls the stapes in a direction that is perpendicular to the piston-like motion of the stapes in response to sound causing a sliding movement in the incudo-stapedial joint.

Animal studies have shown that contraction of the tensor tympani muscle causes the tympanic membrane to move inward, the sound transmission through the middle ear to decrease and the ear's acoustic impedance to increase (Fig. 2.23) [102]. Contraction of the stapedius muscle also changes the sound transmission through the middle ear and it changes the ear's acoustic impedance but it causes little or no movement of the tympanic membrane. When both muscles were brought

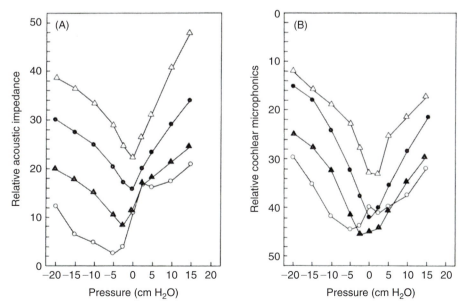

FIGURE 2.22 Acoustic impedance (A) and cochlear microphonics at constant sound pressure at the tympanic membrane (B) as a function of air pressure in the middle-ear cavity of a cat for different frequencies: Open triangles = 0.5 kHz; filled circles = 1 kHz; filled triangles = 2 kHz; open circles = 3 kHz (reprinted from Møller, 1965, with permission from Taylor & Francis).

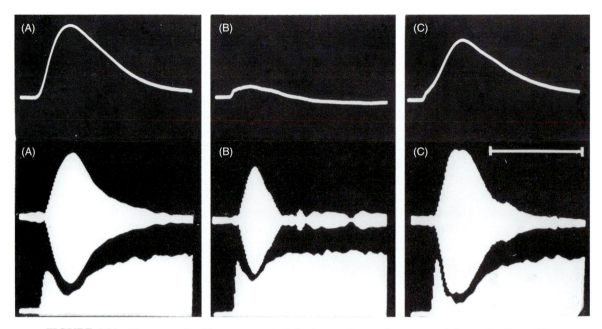

FIGURE 2.23 Upper graphs: The movement of the tympanic membrane caused by contraction of the tensor tympani muscle (A), the stapedius muscle (B), and both muscles together (C) recorded by measuring the change in the air pressure in the sealed ear canal. The tensor tympani muscle and the stapedius muscle were brought to contractions independently by electrical stimulation of these muscles independently (or rather the nerve that innervates the muscle). Middle graphs: Change in the acoustic impedance or the ear measured at 0.8 kHz. Lower graphs: Change in the CM recorded from the round window, for 800 Hz stimulation (reprinted from Møller, 1965, with permission from Taylor & Francis).

to contraction simultaneously, the movement of the tympanic membrane was smaller than it is when the tensor tympani was brought to contract alone (Fig. 2.23) but the change in transmission and the ear's acoustic impedance was larger than when these muscles were brought to contract one at a time. Thus, contraction of

the stapedius muscle impedes the motion of the tympanic membrane induced by contraction of the tensor tympani muscle.

The tensor tympani muscle contracts during swallowing when the Eustachian tube is opening and it has been suggested that contractions of the tensor tympani

---

BOX 2.11

EARLY STUDIES OF THE EFFECT OF CONTRACTION OF THE MIDDLE-EAR MUSCLES

It was probably Hallpike (1935) that first showed experimental evidence that contraction of the middle-ear muscles caused a change in the sound transmission through the middle ear. Several investigators [42, 152, 154] have used recordings of the cochlear microphonic potential from the round window of the cat and observed the change in this potential when the middle-ear muscles were brought to contract in response to a loud sound presented to the opposite ear to elicit the acoustic middle-ear reflex (see Chapter 8).

The displacement of the tympanic membrane by contraction of the tensor tympani muscle can be recorded by

measuring the change in the air pressure in the sealed ear canal. Kato (1913) was probably the first to report on studies of contractions of the middle ear muscles by recording the displacement of the tympanic membrane by measuring changes in the air pressure in the sealed external ear canal in animal experiments. At about the same time Mangold (1913) used a similar method in humans and elicited a contraction of the middle-ear muscles by presenting a loud sound to the opposite ear. Similar methods were later used by other investigators [84, 142].

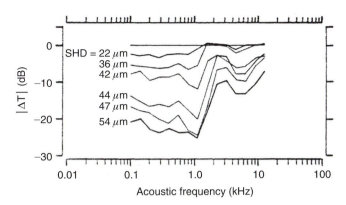

FIGURE 2.24 Change in transmission in one middle ear as a function of frequency for six different sound intensities (expressed in stapes displacement in millimeters) (reprinted from Pang and Peake, 1986, with permission from Springer).

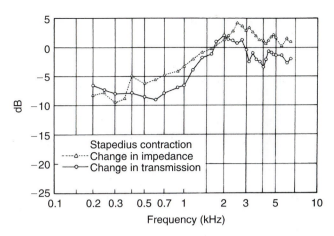

FIGURE 2.25 Change in sound transmission through the middle ear in a cat as a result of contraction of the stapedius muscle (solid lines and circles), together with the concomitant change in the ears acoustic admittance (dashed lines and triangles) (reprinted from Møller, 1965, with permission from Taylor & Francis).

improve air exchange in the tympanic cavity by displacing a small quantity of air in the middle-ear cavity whenever it contracts. If the air is not replaced, the content of oxygen in the air will decrease because oxygen is absorbed at the mucosal surface in the middle-ear cavity.

Contraction of the stapedius muscle decreases the sound conduction through the middle ear. The contraction causes a gradual decrease of transmission as a function of the stapes displacement (Fig. 2.24) [107]. The attenuation is largest in the low frequency range but during strong contractions sound transmission is

also reduced in the high frequency range. The attenuation caused by contraction of the stapedius muscle is approximately 8 dB in the cat for frequencies below 1 kHz (Fig. 2.25) [89]. Comparisons of the change in the acoustic impedance and the change in the transmission properties of the middle ear (Fig. 2.25) support the hypothesis that contraction of the stapedius muscle causes some kind of "decoupling" between the middle ear and the cochlear fluid.

3

Physiology of the Cochlea

1. ABSTRACT

1. The cochlea separates sounds according to their frequency (spectrum) so that different spectral components of sounds activate different populations of auditory nerve fibers.
2. Sensory transduction occurs in inner hair cells.
3. Outer hair cells are active elements that act as "motors" that reduce the influence of friction on the motion of the basilar membrane. This action of the outer hair cells increases the vibration amplitude of the basilar membrane for low sound intensities (by approximately 50 dB) and increases its frequency selectivity.
4. The location of the maximal response shifts toward the base of the cochlea.
5. The role of outer hair cells in increasing the frequency selectivity of the basilar membrane is greatest at low sound intensities.
6. The non-linear action of the cochlea provides amplitude compression of sounds before initiation of nerve impulses in the auditory nerve. Without that, it was not possible to code sounds in the auditory nerve in the large range of intensities that are covered by hearing.
7. The cochlea can generate different kinds of sounds. These sounds are conducted "backwards" by the middle ear, setting the tympanic membrane in motion and thereby generating sounds that can be recorded by a sensitive microphone placed in the ear canal. This is known as otoacoustic emission (OAE).

8. There are several kinds of OAE:
   i. Transient evoked otoacoustic emission (TEOAE) is elicited by a transient sound and generated by reflection of the traveling wave on the basilar membrane.
   ii. Spontaneous otoacoustic emission (SOAE) is a sustained sound that is generated without any sound being applied to the ear.
   iii. Distortion product otoacoustic emission (DPOAE) is a measure of non-linear distortion in the cochlea. DPOAE is elicited by applying two tones to the ear, and measuring the amplitude of a difference tone (usually the $2f_2$-f_1 tone).
9. The olivocochlear efferents influence the function of outer hair cells and by that the OAE is affected.

2. INTRODUCTION

Sensory cells in the cochlea transform sound into a code of nerve impulses in the auditory nerve and that conveys the information to the brain about sounds that reach the ear within the audible range. In addition the cochlea separates sounds according to their spectrum (frequency) so that different populations of hair cells become activated by sounds of different frequency (spectrum). Besides that, the cochlea compresses the amplitude of sounds and thereby makes it possible to accommodate the large dynamic range of natural sounds.

Interplay between theoretical and experimental work has been extremely successful in unraveling the intricate functions of the cochlea both with regard to the frequency analysis in the cochlea and with regard to sensory transduction. The more knowledge that is accumulated about the function of the cochlea the more it becomes evident that the cochlea is a far more complex organ than envisioned by early investigators. Many features not included in the earlier hypotheses have been added as a result of the extensive experimental work.

An example of how new information has totally revised the conception of the function of the cochlea was the discovery that the two groups of sensory cells, inner and outer hair cells, have fundamentally different functions. While the inner hair cells convert the vibration of the basilar membrane into a neural code in the individual fibers of the auditory nerve, the outer hair cells act as "motors" that compensate for the loss of energy in the cochlea and thereby improve the ear's sensitivity and sharpens its frequency selectivity for weak sounds.

It has been questioned whether the frequency selectivity of the cochlea is indeed the basis for our ability to detect changes in the frequency of a pure tone, as small as only a few hertz. The results of recent studies have also cast doubt about the role of spectral analysis in the ear as the basis for discrimination of complex sounds such as speech sounds and instead emphasizing the role of the temporal coding of sounds such as vowels and it is now believed that the main role of frequency selectivity of the basilar membrane is to divide sounds into different spectral bands before the information is processed by the auditory nervous system. The mammalian ear can process sounds the spectrum of which covers 10 octaves and that would not be possible without separation of the spectrum into suitable sized pieces so that the temporal information in different frequency bands can be coded independently in the discharge pattern of auditory nerve fibers (discussed in Chapters 5 and 6).

3. FREQUENCY SELECTIVITY OF THE BASILAR MEMBRANE

Sound analysis in the cochlea is normally equated with spectral analysis that is ascribed to the interplay between the dynamic properties of the basilar membrane and that of the surrounding fluid. Helmholtz (1863) was the first to formulate and prove that the ear performs spectral analysis of sounds. Before that, Ohm (1843) suggested that the ear could separate a sound into its frequency components. These earlier hypotheses were inspired by the finding that any complex waveform (such as natural sounds) can be

divided into a sum of a series of sinusoidal waveforms. Fourier analysis is the mathematical technique of separating a complex waveform such as natural sounds into a series of sine waves. Helmholtz suggested that the basilar membrane performed such spectral analysis and he believed that it was accomplished because the basilar membrane functioned as a series of resonators that were tuned to different frequencies covering the audible range, a function similar to that of the strings of a piano.

Although it was already hypothesized 150 years ago that the cochlea is involved in frequency analysis of sounds it was the fundamental research by von Békésy[1] that brought experimental proof that the cochlea actually does perform spectral analysis of sounds. He presented experimental evidence that a tone of a certain frequency caused the highest vibration amplitude at a certain point along the basilar membrane. This means that each point along the basilar membrane is tuned to a certain frequency and a frequency scale can be laid out along the cochlea with high frequencies located at the base and low frequencies at the apex of the cochlea (Fig 3.1).

Von Békésy [6] convincingly demonstrated that sounds set up a traveling wave motion along the basilar membrane and this traveling wave motion is the basis for the frequency selectivity and not resonance of the basilar membrane as proposed by Helmholz (1883). He concluded that the motion of the basilar membrane becomes a traveling wave motion because the stiffness of the basilar membrane decreases from the base of the cochlea to its apex. Other investigators had earlier suggested other kinds of wave motion along the basilar membrane. Ewald's hypothesis that sounds give rise to standing waves on the basilar membrane is dated back to 1898.

During the time when our understanding of the function of the cochlea steadily increased, theoretical work by investigators such as Ranke (1950) and Zwislocki (1948) were important in guiding work of experimentalists by asking relevant questions. Experimental studies of the vibration of the basilar membrane that could confirm the various hypotheses about the function of the basilar membrane as a spectrum analyzer have been hampered by the extremely small amplitude of the vibration of the basilar membrane. Until the early 1970s the only data about the vibration of the basilar membrane and its frequency selectivity that were available were obtained in studies

[1]Georg von Békésy did his fundamental work on the function of the ear between 1928 and 1956. His early work was published in the German language and all his work is has been translated into English and published in journal articles. His work is also collected in a book [6].

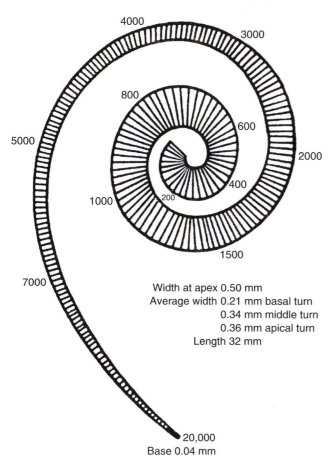

FIGURE 3.1 Schematic drawing of the basilar membrane of the human cochlea showing that the width of the basilar membrane increases from the base of the cochlea to its apex. High frequencies are represented in the basal end of the cochlea and lower frequencies toward the apex (from Stuhlman, 1943, with permission from John Wiley & Son).

more frequency selective for low intensity sounds than for high intensity sounds [120]. The cause for the non-linearity was not discovered until 1983 [12].

3.1. Traveling Wave Motion

Sounds set the cochlear fluid into motion and the motion of the cochlear fluid in turn sets the basilar membrane into motion. The mechanical properties of the basilar membrane and how they vary along the membrane determine which kind of wave motion a sound gives rise to. The traveling wave motion on the basilar membrane is a result of the gradual decrease in the stiffness of the basilar membrane from the basal portion of the basilar membrane toward the cochlear apex. The energy that is transferred to the basal portion of the basilar membrane propagates as a traveling wave motion toward the cochlear apex. As the wave travels along the basilar membrane toward less stiff parts of the basilar membrane, the propagation velocity of the wave decreases (Fig. 3.2) and consequently the wavelength of the motion decreases. (The wavelength is the distance between two identical points of the wave that travels along the basilar membrane.) When the wave motion slows, energy piles up, first causing the vibration amplitude to increase [77]. The increase in amplitude is counteracted by frictional losses of energy and when the wavelength of the traveling wave reaches small values, these losses increase rapidly and the wave propagation comes to a halt and the traveling wave

done in human cadaver ears by a single investigator [6]. The results obtained showed that the basilar membrane was broadly tuned. This work was mostly done in the 1930s when limitations in technology made it necessary to use extremely high sound levels to observe the motion of the basilar membrane.

Other investigators took these results, obtained at these extreme high sound intensities, to represent auditory frequency selectivity in the entire intensity range of hearing because it was assumed that the basilar membrane functioned as a linear system that allowed such extrapolations of these experimental findings. (Readers who are interested in details about the development of hypotheses and experimental studies of the cochlea as a spectrum analyzer are referred to extensive literature on the matter [22, 23, 54, 55, 151, 152].) It was not until the beginning of the 1970s that it became evident that the motion of the basilar membrane is non-linear and that it was

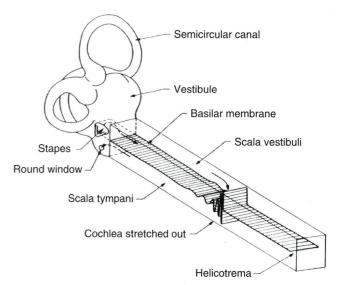

FIGURE 3.2 Schematic illustration of the traveling wave motion along the basilar membrane. The cochlea is shown schematically as a straight tube (reprinted from Zweig et al., 1976, with permission from the American Institute of Physics).

becomes extinguished. The location on the basilar membrane where that occurs is a function of the frequency of the sound and that is the basis for the basilar membrane's ability to separate sounds according to their frequency (spectrum). A high frequency tone travels a short distance and a low frequency tone travels a longer distance. The distance that the wave motion travels on the basilar membrane before being extinct is a direct function of the frequency of the sound that sets the basilar membrane into motion.

Zwislocki [166] suggested that the tectorial membrane together with the hairs (stereocilia) of the hair cells form a mass-stiffness resonator that is coupled to the basilar membrane and thereby contributes to its frequency selectivity properties. Using a mechanical model composed of a steel reed with a mass on top Zwislocki and Kletsky [170] showed that this hypothesis was plausible. When the reed was set into up and down vibrations, the mass would exhibit lateral movements when the frequency of the vibrations was equal to the resonance frequency of the reed-mass combination. Later, the results of these theoretical and model studies were confirmed in animal experiments [169] showing that the mechanical properties of the tectorial membrane and the stereocilia of the outer hair cells contribute to the frequency selectivity of the basilar membrane by forming resonators along the basilar membrane. These resonators are tuned to different frequencies at different locations along the basilar membrane because the mass of the tectorial membrane varies along the membrane and the length (and thereby the stiffness) of the hairs also varies along the basilar membrane. These resonators together with the traveling wave motion are the bases for the frequency selectivity of the cochlea.

It is interesting that resonators were again introduced as a contributor to the frequency selectivity but in a different way than Helmholtz's resonance theory stated; the tectorial membrane and hairs of hair cells and not the fibers of the basilar membrane forms the resonators. Also, the tectorial membrane is assumed to act in conjunction with the traveling wave motion in creating the frequency selectivity of the cochlea and therefore the traveling wave motion is not the only mechanism of frequency selectivity.

3.2. Basilar Membrane Frequency Tuning Is Non-linear

The frequency tuning of the basilar membrane, as it was known from von Békésy's experiments (Fig. 3.3) was much too broad to explain psychoacoustic data on frequency discrimination in the auditory system.

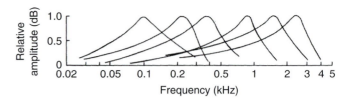

FIGURE 3.3 The vibration amplitude of the basilar membrane at different locations along the membrane of the cochlea in a human cadaver, shown as a function of frequency (reprinted from von Békésy, 1942, and article no. 42 in von Békésy, 1960, with permission from McGraw Hill).

That resulted in many attempts to explain how the broad frequency tuning of the basilar membrane could be sharpened. When technology in the beginning of the 1970s had advanced to a level that made it possible to measure the vibration of the basilar membrane

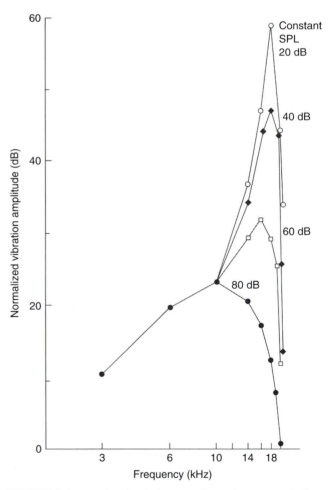

FIGURE 3.4 Amplitude of vibration of a single point on the basilar membrane in an anesthetized guinea pig in response to pure tones of 4 different intensities, at 20 dB intervals, as a function of the frequency. The curves were shifted so that they would have coincided if the cochlea had been a lines system (adapted from Johnstone et al., 1986).

for sounds in the upper physiologic intensity range (90–70 dB SPL) [120] it became evident that the frequency selectivity of the basilar membrane was greater at low sound levels than at high levels. The vibration of the basilar membrane is thus non-linear. Subsequent studies in the guinea pig using a larger range of sound intensities further elucidated the non-linearity of the basilar membrane vibration [61] (Fig. 3.4).

3.3. Frequency Tuning of the Basilar Membrane

Rhode [119] showed that the frequency selectivity of the basilar membrane deteriorates after death (Fig. 3.5) and that is an indication that metabolic energy is necessary to maintain the high degree of frequency selectivity of the basilar membrane. This explains why the tuning of the basilar membrane von Békésy obtained was so broad. (He studied cadaver ears and the sound levels used in those studies were probably above 145 dB SPL.) It is also seen from Fig. 3.5 that the frequency to which a point of the basilar membrane is tuned to shifts towards lower frequencies after death.

That the sharp tuning of auditory nerve fibers deteriorates after oxygen deprivation was shown by Evans as early as 1975 [33] (see Fig. 6.7 and Chapter 6), thus an indication that the sharp tuning of auditory nerve fibers depends on metabolic energy. At the time these results were published it was believed that the high degree of frequency selectivity of single auditory nerve fibers was a result of sharpening (by a "second filter") of the tuning of the basilar membrane, believed to be neural in nature. (At that time it was believed that the basilar membrane was broadly tuned because the only data that were available were those obtained in cadavers by von Békésy.) However, this "second filter" was never found. The question about the sharp tuning was resolved by the results of Rhode's study showing that the frequency selectivity of the basilar membrane deteriorates after death (Fig. 3.5) [119], thus indicating that metabolic energy was necessary to maintain sharp frequency tuning not only of the responses from auditory nerve fibers but also of the basilar membrane vibration. How this was accomplished was explained much later when it was discovered that the outer hair cells are active elements that make the tuning of the basilar membrane non-linear and sharpens the tuning of the basilar

---

BOX 3.1

MEASUREMENTS THAT SHOW THE BASILAR MEMBRANE TO BE NON-LINEAR

Rhode's [120] measurements of the vibration of the basilar membrane in an anesthetized animal at different sound intensities showed that the sharpness of the tuning decreased with increasing sound intensity and the frequency to which a certain point of the basilar membrane was tuned shifted when the sound intensity was changed. These studies were the first to show that the way the basilar membrane vibrates depends on the intensity that is used to set it into vibration. When technology advanced it became possible to measure the vibration of the basilar membrane at even lower sound intensities, down to near threshold of hearing (Fig. 3.4) [61].

Before Rhode's [120] study was published, electrophysiological studies of cochlear tuning by Honrubia and Ward [51] had demonstrated that the location of the maximal vibration amplitude shifts towards the base of the cochlea when the stimulus intensity is increased from low to high. These investigators showed that maximal

cochlear microphonic (CM) response recorded from the scala media shifted more than 4 mm towards the base of the cochlea when the sound intensity was increased from 60 to 100 dB thus demonstrating that the location of the maximal vibration amplitude shifts towards the base of the cochlea when the sound intensity is increased. When the sound level is varied the frequency to which a certain point is tuned may shift by several octaves. These investigators were probably the first to show results that indicated that the tuning of the basilar membrane is affected by sound intensity. Later studies showed that the shift in frequency tuning of the basilar membrane as a function of sound intensity was somewhat less and the shift was different in the low frequency range compared with the high frequency range [93, 94]. Harrison and Evans [48] arrived at essentially similar results.

This non-linearity of the basilar membrane vibration also provides compression of the amplitude of the basilar membrane vibration relative to the sound stimulus.

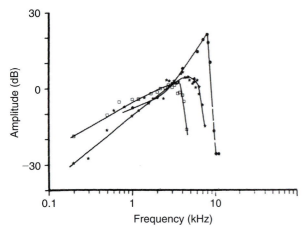

FIGURE 3.5 The amplitude of the displacement of the basilar membrane in a monkey obtained in a similar way as the results shown in Fig. 3.4. The top curve shows the results when the monkey was alive (anesthetized), and the two other curves show results obtained 1 h after the death of the monkey and 7 h after the death (reprinted from Rhode, 1973, with permission from Elsevier).

membrane [12]. That the frequency selectivity of the basilar membrane is dependent on metabolic energy was confirmed in studies where intracellular recordings from inner hair cells were made in the guinea pig [13].

3.4. Role of the Outer Hair Cells in Basilar Membrane Motion

The discovery in 1983 by Brownell [12] that outer hair cells act as "motors" that compensate for frictional losses of energy in the cochlea at low sound intensities brought our understanding of the function of the cochlea forward. It changed the concept of how the cochlea functions in a fundamental way and it explained the results of several different kinds of experimental results. The active role of the outer hair cells as "motors" that compensate for the energy losses in the propagation of the traveling wave on the basilar membrane increases the sensitivity and the frequency selectivity of the ear and it explains why metabolic energy is necessary to maintain the normal sensitivity and frequency selectivity of the ear. It also explains why the frequency tuning of the basilar membrane in cadavers and in living animals are different (Fig. 3.5) [119] and it explains why metabolic insult, induced by oxygen deprivation, causes the threshold of auditory nerve fibers to increase and the tuning to become wider (Fig. 6.7, Chapter 6) [33].

The role of outer hair cells in the function of the cochlea is related to their motility [2, 13], meaning that outer hair cells can contract and expand. Two kinds of motility of outer hair cells have been

demonstrated: One is a fast change in length of a maximum of approximately 5%, and the other is a slow change that can be much larger. The fast change can follow sound frequencies and is believed to sharpen the cochlear frequency selectivity and it probably also causes amplitude compression. The slow change occurs over seconds and it may (slowly) change the sensitivity of the ear. These changes in length, fast and slow, can be elicited in different ways, one being by sound that reaches the ear. The fast change, first observed in outer hair cells that were isolated from the cochlea (of guinea pigs), can be elicited by passing electrical current through the hair cell [2, 13]. Slow changes in length of hair cells occur in response to changes in the concentration of potassium ions in the surrounding fluid.

It is believed that the fast motility of outer hair cells can be elicited by receptor currents evoked by sound stimulation. Thus, sound stimulation that causes motion of the outer hair cells generates receptor currents that can cause (further) motion of the outer hair cells. This is assumed to be the basis for the positive feed back that explains how the outer hair cells become "motors" that amplify the vibration of the basilar membrane. If too strong, this positive feedback may cause self-sustained vibrations that result in generation of sound that can be measured in the ear canal (as otoacoustic emission [OAE]).

The most obvious implication of the active process mediated by the outer hair cells is increased sensitivity and greater acuity of frequency analysis for weak sounds. The active process of the cochlea is assumed to account for approximately 50dB of the ear's sensitivity, and total loss of the function of outer hair cells causes hearing loss of approximately that amount. This is why injuries that affect mainly outer hair cells, such as from exposure to loud noise, do not cause elevation of the hearing threshold that exceeds 50 dB (see Chapter 9).

The widening of the cochlear tuning with increasing sound intensity may serve to adapt the auditory frequency analyzer to perform optimally over a large range of sound intensities. Sharp spectral filters such as the cochlear filters at low sound intensities improve the signal-to-noise ratio and may thus benefit the detection of weak sounds and sounds in noise. Broader filters such as cochlear filters at higher sound intensities are better suited for processing of fast changes in the amplitude of a sound, which may be important for the discrimination of complex sounds. Changes in the width of the cochlear filters with changes in sound intensity may thus make the cochlea function in an optimal way both for weak sounds where detection is important, as well as for louder sounds where discrimination between different sounds is important requiring good temporal resolution.

The non-linearity of the cochlea also causes a shift in the location of the maximal amplitude of the deflection of the basilar membrane in response to stimulation with pure tones. This shift had already been demonstrated in 1968 in studies of the cochlea by Honrubia and Ward [51] and later in studies of tuning of single auditory nerve fibers [93, 94] and in recordings from cochlear hair cells [168]. That the location of maximal deflection of the basilar membrane is not only a function of the frequency (spectrum) of sounds but also depends on their intensity has implications for theories about the physiologic basis for discrimination of frequency and it explains the one-half octave shift in noise induced hearing loss (see Chapter 6).

It has been known for a long time that neural activity in the olivocochlear efferent fibers affects the response of auditory nerve fibers [45, 148, 155]. This can partly be explained by the efferent influence on the output of inner hair cells. That efferent neural activity controls the function of outer hair cells implies that it can change the mechanical properties of the basilar membrane. That function is mediated by the medial olivocochlear fibers (see Fig. 5.15B, Chapter 5),

which terminate directly on the outer hair cells (see Fig. 1.12, Chapter 1). Activity in these fibers releases transmitter substances that can affect the mechanical properties of the outer hair cells and by that the mechanical properties of the basilar membrane.

3.5. Epochs of Research in Cochlear Mechanics

Dallos [22] offered a perspective on the development of theories on cochlear function and experimental work by dividing the era of research on the function of the cochlea as a spectrum analyzer into three epochs. The first period was characterized by Helmholtz's theories (1863), that postulated that a series of lightly damped mechanically tuned resonant elements were located along the basilar membrane (reviewed by Wever in 1949 [151]). The second epoch from late 1940 to early 1970 was dominated by von Békésy's [6] experimental demonstration of the spectral analysis in the cochlea as being the result of a traveling wave motion along the basilar membrane (Fig. 3.2). The third epoch starting in the 1970s runs until the present. That period is dominated by the finding that the traveling wave motion is boosted

---

BOX 3.2
_____

OLIVOCOCHLEAR BUNDLE AND COCHLEAR NON-LINEARITY

Studies of the function of the olivocochlear bundle have been done in animals by electrical stimulation of the olivocochlear bundle where it comes close to the surface of the floor of the fourth ventricle. The fact that the medial olivocochlear efferent fibers that terminate on outer hair cells can be activated by sound stimulation of the opposite ear makes it possible to study the function of outer hair cells on the sensitivity of the ear and on basilar membrane tuning in humans (and in animals) [149, 150]. Studies have demonstrated that such activation of the efferent system by contralateral sound stimulation can change basilar membrane tuning and otoacoustic emission [19, 88].

Another important feature of the active cochlea is the non-linear conversion of sound into vibration of the basilar membrane. The action of the outer hair cells to amplify the motion of the basilar membrane provides compression of the range of sound intensities (automatic gain control). The range of sound intensities that the ear can

handle is enormous (approximately one to one million) and it is possible because of the compression of the amplitude of the vibration of the basilar membrane that occurs before the sound is coded in the discharge pattern of cochlear nerve fibers. This compression (automatic gain control) makes it possible to code the large range of sound intensities in the discharge pattern of auditory nerve fibers. The neural transduction in the cochlea together with the non-linearity of the motion of the basilar membrane are the physiological basis for this amplitude compression that makes it possible for the auditory system to process sound over a range of approximately 100 dB.

It has been estimated that a 10 dB increase of the sound at the tympanic membrane only results in an increase of 2.5 dB of the vibration of the basilar membrane [17]. This compression of the amplitude scale takes place before transduction into a neural code and functions in a similar way as the automatic gain control that is often incorporated in artificial communication systems.

by active processes of the outer hair cells that inject energy into the system [12], compensating for frictional losses. The result is greater sensitivity of the ear and sharper tuning of the basilar membrane for sounds of low intensities. The finding that the frequency to which a point along the basilar membrane is tuned depends on the sound intensity has consequences regarding the role of the place hypothesis for frequency discrimination (see p. 100).

Recognition of the role of a resonant system consisting of the tectorial membrane and the hairs of hair cells by Zwislocki and co-workers [169] may be regarded as a fourth epoch in our understanding of cochlear micromechanics [22]. This resonator, together with the traveling wave properties of the basilar membrane are the bases for the frequency selectivity of each small segment of the basilar membrane.

In summary, as we understand it now, cochlear frequency selectivity is a result of a combination of at least three different mechanisms: (1) The traveling wave motion on the basilar membrane; (2) the active function of the outer hair cells that inject energy into the motion of the basilar membrane and causes the basilar membrane motion to become non-linear; and (3) the resonance of the tectorial membrane and its attachment (mainly the stereocilia of the outer hair cells).

4. SENSORY TRANSDUCTION IN THE COCHLEA

Sensory transduction in the cochlea has been the subject of many studies and probably more is known about the sensory transduction in the ear than that in any other sensory system. Early research efforts were directed to understand how the two groups of hair cells, the inner and the outer hair cells, converted the vibrations of the basilar membrane into a neural code of single auditory nerve fibers. That so few nerve fibers terminated on the outer hair cells was puzzling before it was understood that the outer and inner hair cells have fundamentally different functions. It is now assumed that only inner hair cells participate in the transduction of motion of the basilar membrane into a neural code in the auditory nerve. The connections between inner hair cells and auditory nerve fibers are in many ways similar to other synapses but its ability to transmit timing information may be better than what it is in other synapses.

4.1. Excitation of Hair Cells

Hair cells are type II receptors (see [101]) that consist of stereocilia that are sensitive to displacement (bending).

Bending of the stereocilia causes change in the intracellular potential of the hair cells which becomes less negative when the stereocilia are bent in one direction and more negative when bent in the opposite direction. The molecular basis for sensory transduction has been studied intensively during the past decade, and it will be discussed below (p. 50). It was shown many years ago that bending of the stereocilia of hair cells in the lateral line organ of fish in the direction toward the kinocilium depolarize the hair cells [36] and is thus excitatory. Deflections in the opposite direction hyperpolarizes the cells (Fig. 3.6A) [35, 36]. Cochlear hair cells are assumed to function in a similar way but they do not have kinocilia. A basal body is located at the place of the kinocilium. Studies of mammalian hair cells have later confirmed the results of the studies of hair cells of the lateral line organ (Fig. 3.6A, B & C) [124]. Deflection of the hairs toward the location of the basal body (toward the tallest row of stereocilia) has been regarded to be excitatory.

When low frequency tones (for instance, 0.3 kHz) were used as stimuli in early studies of the function of hair cells as mechano-transducers, it appeared that hair cells responded to the amplitude of the displacement of the basilar membrane. More recently when tones of higher frequencies have been used it has become evident that inner hair cells may be excited by either the amplitude of the basilar membrane displacement or by the velocity of the basilar membrane motion [133, 171]. The visco-elastic coupling between the tectorial membrane and the stereocilia makes hair cells sensitive to the velocity of the motion of the basilar membrane because a visco-elastic coupling transmits the change in displacement, thus the velocity. Since the velocity of the motion increases proportionally with the frequency, low frequency motion of the basilar membrane is less effective than high frequency motion in displacing the stereocilia.

4.2. Which Phase of a Sound Excites Hair Cells (Rarefaction or Condensation)?

If deflection of the stereocilia of inner hair cells toward the basal body is excitatory, motion of the basilar membrane toward the scala vestibuli would be excitatory. An outward movement of the stapes (caused by the rarefaction phase of sounds) would cause the basilar membrane to be deflected towards the scala vestibuli. This concept was supported by early recordings of the compound action potential from the round window of the cochlea in response to click stimulation [109]. However, these results were challenged by later studies [70, 133, 171], which

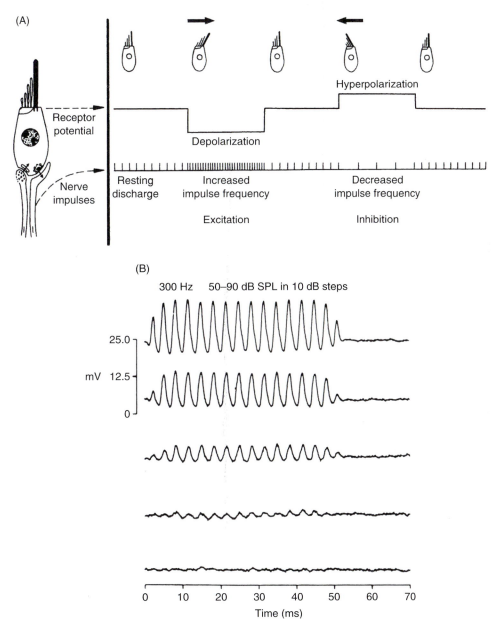

FIGURE 3.6   (A) Schematic illustration of excitation of hair cells from the lateral line organ of a fish. The picture shows how intracellular potentials are affected by bending of the stereocilia of hair cells in the lateral line organ of fish (reprinted from Flock, 1965, with permission from Cold Spring Harbor Laboratory Press). (B) Waveform of the voltage recorded intracellularly (receptor potentials) from inner hair cells in a guinea pig at different sound intensities (reprinted from Russell and Sellick, 1983, with permission from Cold Spring Harbor Laboratory Press).

(Continued)

revealed a more complex relationship between the acoustical waveform of a sound and the excitation of cochlear hair cells and inner hair cells were shown to respond to the motion of the basilar membrane in both directions (up and down) or only in one direction. Studies of the response of single auditory nerve fibers have shown that nerve fibers that are tuned to high frequencies respond when the basilar membrane was deflected toward the scala tympani while nerve fibers that are tuned to low frequency sounds respond to motion of the basilar membrane either toward scala vestibuli or towards scala tympani [123].

These results can be understood by considering that the force that acts on the stereocilia can be generated either by the shearing motion between the reticular lamina and the tectorial membrane, or by the flow of

(C)

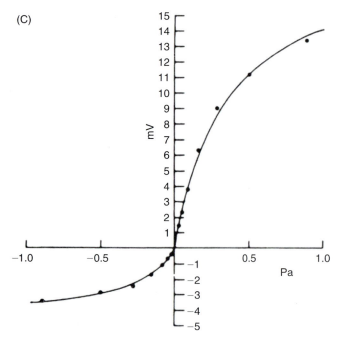

FIGURE 3.6 *(Continued)* (C) Voltage of the response seen in Fig. 3.9B. (reprinted from Russell and Sellick, 1983, with permission from The Physiological Society).

endolymph that is caused by displacement of the tectorial membrane.

Studies by Ruggero [123] confirmed that excitation in the cochlea is complex and that it can best be described in the following way: Auditory nerve fibers tuned to low frequencies respond to low intensity sounds when the basilar membrane moves at its highest speed in the direction toward the scala vestibuli, whereas nerve fibers that are tuned to high frequencies respond to low frequency sounds of moderate intensity (60–80 dB SPL) when the basilar membrane moves at its highest velocity towards. Other studies of the stimulation of the inner hair cells have shown further evidence that the motion of the stereocilia is more complex than that of the basilar membrane and that the deflection of the stereocilia of inner hair cells, that is the basis for excitation of auditory nerve fibers, is not a direct function of the motion of the basilar membrane [87, 165].

4.3. Molecular Basis for Sensory Transduction

The discovery that stereocilia contains actin, a protein found in muscles, was the beginning of a series of developments of understanding of the extremely complex machinery that is responsible for the function of hair cells as receptors and as active elements ("motors") that increases the sensitivity of the ear [37]. The subsequent discovery of the active role of outer

hair cells [12, 13] explained many questions regarding the function of the ear. Prestin is another protein that plays an important role in the motility of outer hair cells [25]. Prestin is a membrane protein that performs a direct voltage-to-force conversion. It is found in outer hair cells and its action is different from the motors that are based on enzymatic activity.

The basic function of mechanoreceptors assume a specialized membrane where mechanical deformation cause ionic conduction to change and that subsequently changes the membrane potential of the hair cells. This specialized membrane is located on the stereocilia and displacement of the cilia opens specific ionic channels that are located at or near the tips of the stereocilia [54]. The inflow of ions results in the release of one or more neurotransmitters and that is the way sensory transduction in hair cells occurs. Both potassium and calcium ions (K^+ and Ca^{2+}) are involved in this process. The first step in the cascade that eventually results in a process controlling the discharge of auditory nerve fibers is a change in the conductivity of the mechanically gated ion channel. Opening of this channel causes the membrane potentials to become less negative (depolarization), which in turn leads to opening of voltage-gated Ca^{2+} ion channels. The influx of Ca^{2+} leads to release of a neurotransmitter at the base of the hair cells. Activation of hair cells causes an increased flux of K^+ from endolymph through the hair cells and then into the endolymph, from where the potassium ions are taken up and secreted back into perilymph. This means that sensory activation causes potassium ions to be cycled within the cochlea (Fig. 3.7) [147].

Many studies have shown that the membranes of cochlear hair cells have voltage gated calcium channels that control release of transmitter substances [40]. The role of Ca^{2+} ion channels has been studied in animals using the calcium channel blocker nimodipine and it has been found that Ca^{2+} ion channels are

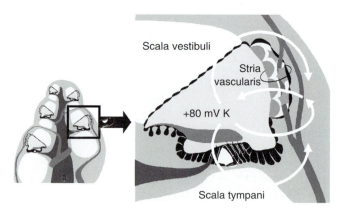

FIGURE 3.7 Schematic drawing of ionic flow in the cochlea (reprinted from: Wangemann, 2002, with permission from Elsevier).

important for the final stage of cochlear transduction, either for the release of transmitters from inner hair cells or for the postsynaptic spike generation [162]. The summating potential (SP) has a neural component and when that has been removed, the SP is not to be affected by administration of nimodipine, an L-type Ca^{2+} ion channel blocker involved in neural transduction in the cochlea [162]. These studies also suggested that there are nimodipine sensitive channels in outer hair cells thus indicating a role of L-type Ca^{2+} ion channels in the function of outer hair cells. Evidence has also been presented that indicates that T-type Ca^{2+} channels are not involved in any stage of transduction in hair cells.

Other studies have found glutamate to be a neural transmitter involved in afferent neural transmission [59] but also aspartate may have a role in afferent transmission in the cochlea. Studies in rats have shown that a nicotinic acetylcholine receptor is expressed in hair cells and it has been hypothesized that it mediates efferent inhibition via calcium activated potassium channels [40], especially in outer hair cells [52].

The role of potassium ion (K^+) channels in transduction in hair cells depends on the frequency of stimulation [32]. The K^+ channels respond cycle-by-cycle to the waveform of low frequency stimulation. This variation in membrane conductivity cause distortion that makes the K^+ channels in IHC to be involved in the change in the membrane potential. This change in the membrane potential is known as the receptor potential that is associated with low frequencies. For high frequency stimulation the K^+ channels affect the size of the DC response.

Recent advances in the understanding of the transduction processes that occurs in hair cells pave the way for understanding some forms for congenital hearing loss [76]. Congenital hearing loss occurs in approximately 1 in 1000 live births and the main part of the impairment of hearing is assumed to be caused by defects in the function of hair cells. Studies have shown evidence that encoding of the gap junction protein connexin-26 is responsible for many occurrences of congenital hearing loss [68].

Connexins are a group of proteins that are important in communication between the components inside the cells using gap junctions and it is important for the exchange of electrolytes, second messengers and metabolites [68, 132, 147]. There are expressions of this protein in the stria vascularis, the basement membrane, the limbus and the spiral prominence of the human cochlea. Loss of connexin-26 disrupts many fundamental processes in hair cells. The gene coding for connexin-26 has been identified which is important for genetic counseling. Mutation of the connexin gene may be involved in the pathophysiology of some forms of progressive adult deafness and understanding of the processes that underlies this transformation opens new prospects for understanding of some forms of genetically related hearing impairment and for providing genetic counseling. The understanding of the molecular mechanisms of sensory transduction in the cochlea is also important for diagnostics and for development of effective therapeutic means for genetically related hearing disorders.

Cochlear hair cells are susceptible to insults from overstimulation, chemicals, aging and unknown factors. In most situations it is outer hair cells that degenerate or become injured. These changes are irreversible but measures have been identified that can prevent or reduce the destruction of hair cells. During the past two decades great advantages have been made in understanding the mechanisms involved in destruction of hair cells and it has become evident that these processes are complex and subsequent changes in the function of the auditory nervous system may contribute to the observed hearing impairment (see Chapters 9 and 10). The fact that hearing loss from noise exposure (noise induced hearing loss [NIHL]) depends not only on the intensity and duration of the noise exposure but also on previous exposures [15, 86] are signs that the central nervous system is involved in the processes that injure hair cells from noise exposure. There are signs that the pathology is more complex than just mechanical stress on hair cells. Age related hearing impairment has earlier been assumed to be caused by injuries to hair cells but studies show indications that the degree of hearing impairment depends on exposure to sounds [157] indicating that the processes that injure hair cells are complex. Also genetic causes of hearing impairment and deafness have complex pathologies (see Chapter 9).

The mechanisms of hair cell damage from administration of aminoglycoside antibiotics has been studied extensively [39, 127, 160] and it is now believed that such injury to hair cells can be viewed as an abnormal initiation of apoptosis (programmed cell death).

The apoptosis of hair cells from aminoglycoside antibiotics starts when the antibiotic enters the hair cells and accumulates in the cells. This triggers a cascade of events in the cell that eventually lead to the death of the cell. The damage from aminoglycoside antibiotics is initiated by reactive oxygen species [39. 127] and the orderly series of events that follows activate caspases, which trigger a series of protelytic events that degrade the various proteins in the cell's nuclei and cytoplasm, leading to "cell suicide" [38, 117]. In vitro studies have shown that caspase inhibitors can promote survival of hair cells from aminoglycoside induced apoptosis [81].

The effect of the oxidative stress can be reduced by increasing the level of agents that neutralize oxygen free radicals. Such therapy using adenoviral vectors seems to reduce the effect of administration of ototoxic antibiotics [62]. Animal experiments have shown that administration of agents that neutralize free oxygen radicals can decrease the hearing loss after administration of ototoxic antibiotics (Kanamycin) [114]. Ionized calcium (Ca^{2++}), which is a second messenger, has also been implicated in apoptosis from ototoxic antibiotics [82]. The molecular biology of this process begins to become understood [82] leading to development of ways to inhibit this process [132] and thereby rescue hair cells from destruction. Such advances in understanding of the processes that control hair cell death may also lead to ways to induce regeneration of hair cells, thus curing hearing loss and deafness. Hair cells in the bird and reptiles can regenerate spontaneously after injuries to hair cells [21, 125] but mammalian hair cells do not have that ability.

A better understanding of the series of events that occur when hair cells dies also can help develop ways to regenerate hair cells and several approaches in that direction has been tried, including the use of stem cells. Some success has been achieved regarding regenerating hair cells in mammals [58, 82], but it is not clear whether such regenerated cells will establish the contacts with auditory nerve fibers that are necessary for establishing a functioning cochlea. Stem cells can differentiate into a variety of cell types and there are indications that supporting cells may have the properties of stem cells, thus providing the basis for regeneration of cochlear and vestibular hair cells [82]. However, despite the enormous progress in our

understanding of these cellular processes regarding the life and death of hair cells we are still far away from any practical clinical therapy for lost hair cells but many of the factors that now are known to cause destruction of cochlear hair cells are linked to activities that can be modified or reduced.

4.4. Endocochlear Potential

The endocochlear potential (EP) is present as a potential difference between the perilymphatic and endolymphatic fluid spaces. The EP was studied by von Békésy [6] who found that the endolymphatic space had a potential of approximately +80 mV relative to the tissue surrounding the cochlea, and that the scala vestibuli had a potential of approximately 5 mV (Fig. 3.8). That means that the difference between the electrical potential in the perilymphatic and the endolymphatic space is approximately 75 mV. The EP is higher (80–120 mV) near the base of the cochlea than what it is in higher turns where it is 50–80 mV. These studies were done in the guinea pig but similar values were found in the cat.

The stria vascularis (see Fig. 3.9) plays an important role in the function of the cochlea. It is a multi-layered epithelium that is a part of a barrier around the endolymphatic space. On the apical side it faces the endolymph and on the basal side it faces the spiral ligament. Strial marginal cells and vestibular dark cells secrete potassium ions into the endolymph (Fig. 3.9) [147]. The secretion of potassium ions by the strial marginal cells is necessary for maintaining the endolymphatic potential [146].

The proper rate of secretion of potassium ions is required not only for maintaining the endolymphatic

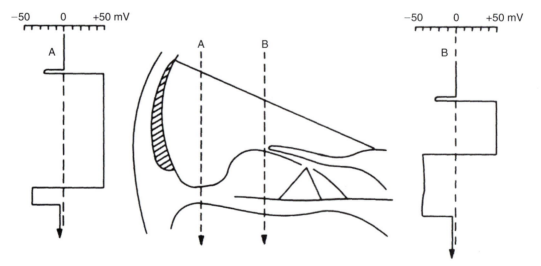

FIGURE 3.8 Electrical potentials recorded in different locations of the cochlea (reprinted from von Békésy, 1952, with permission from Nature Publishing Group).

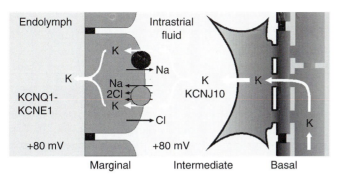

FIGURE 3.9 Schematic drawing of ionic flow in the stria vascularis (reprinted from Wangemann, 2002, with permission from Elsevier).

potentials but also for maintaining the normal ionic composition of the endolymph, and its volume. Multiple control mechanisms regulate the rate of K$^+$ secretion. Since it may be suspected that abnormal volume of endolymphatic fluid is involved in Ménière's disease and possibly in sudden deafness, studies of the control mechanisms for the endolymphatic fluid may have a direct clinical implication. Evidence has been presented that stress can trigger attacks of Ménière's disease [131] and it is interesting to note that the secretion of K$^+$ ions in the strial marginal cells of the stria vascularis seems to be stimulated by beta1 adrenergic receptors and inhibited by M3 and M4 muscarinic receptors [146]. This means that stress elicited sympathetic activity that increase the level of adrenergic substances may be involved in upsetting the ionic balance in the stria vascularis and thereby perhaps explain how stress can be involved in triggering attacks of the Ménière's disease.

4.5. Cochlea as a Generator of Sound

It was reported anecdotally many years ago that the ears of some animals (dogs) sometimes emitted sounds that could be heard by an observer but it was not until Kemp [63] published his study on cochlear echoes that sound generation by the cochlea was described scientifically. Since then, many papers have been published on the subject and it has become evident that the cochlea can generate several different kinds of sounds, now commonly known as otoacoustic emission (OAE). The types of otoacoustic emission include linear reflection at a point along the basilar membrane, and non-linear distortion in the cochlea [130]. There are two kinds of linear reflection: Spontaneous emission and reflection emission. The spontaneous emission is generated at a certain location along the basilar membrane and travels in both directions from there. The wave that travels towards the apex is extinguished but the wave that travels in

the opposite direction reaches the oval window and sets the middle ear bones and the tympanic membrane in motion creating a sound in the ear canal. Non-linear distortion is the source of distortion emission. Reflection emission and distortion emission are evoked emissions. Both stimulus-frequency otoacoustic emissions (SFOAEs) and DPOAEs are assumed to be caused by nonlinearities in the cochlea that act as "sources" of backward-traveling waves.

All these kinds of otoacoustic emissions are normally very weak sounds and sensitive microphones and recording equipment are required to study such sounds. In rare cases, however, SOAE can be heard by an observer.

When a transient sound is presented to the ear, reflected sound, first known as the cochlear echo (or Kemp [63] echo, after the person who described it), occurs with a latency of 5–15 ms. The cochlear echo, or TEOAE (Fig. 3.10) is assumed to be caused by reflection of the traveling wave at some point along the basilar membrane [130]. It is not entirely clear why that happens but any inhomogeneity in the basilar membrane may cause reflection of the traveling wave. Normally the traveling wave propagates smoothly without any reflection along the basilar membrane and all energy is dissipated before the wave reaches the cochlear apex. A slight inhomogeneity at a

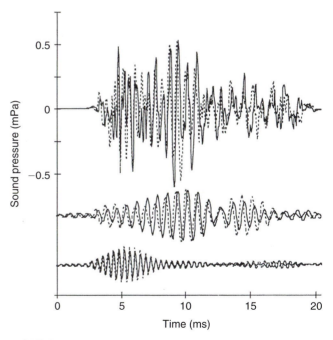

FIGURE 3.10 Click evoked cochlear echo (transient evoked otoacoustic emission [TEOAE]) recorded in standing position (solid lines) and in Trendelenburg position (supine with head lowered, dotted lines). Top traces are unfiltered, bottom two traces are filtered with different settings (reprinted from Büki et al., 1996, with permission from Springer Verlag).

certain location along the basilar membrane, however, will cause some of the energy to be reflected and that energy will travel in the opposite direction, namely, toward the base of the cochlea [130]. Since the outer hair cells contribute to the mechanical properties of the basilar membrane, loss of outer hair cells or injury to these hair cells at a certain location along the basilar membrane could cause such discontinuities. Two mechanisms, linear reflection versus non-linear distortion, can explain the mechanisms of generations of two broad classes of emissions – reflection-source and distortion-source emissions. The reflected wave is assumed to be amplified by the cochlear amplifier and that is the way the active process in the cochlea become involved in generation of TEOAE [130]. When a reflected wave reaches the basal region of the cochlea it will set the cochlear fluid into motion and that will cause the stapes to vibrate. This vibration is conducted (backwards) by the ossicular chain so that the tympanic membrane generates sound in the ear canal.

The reflected sound from a transient sound such as a click sound appears in the ear canal with a certain delay of 5–10 ms after the presentation of the click sound. The delay is the travel time on the basilar membrane to the location of the inhomogeneity and back again to the base of the cochlea. The main components of the reflected sound in response to broad band clicks is usually an oscillation with a narrow spectrum (Fig. 3.10) indicating that it originates from a narrow segment of the basilar membrane. The frequency of the oscillation is different for different individuals but stable over many years in a certain individual.

BOX 3.3

OTOACOUSTIC EMISSION

Detailed analysis of the TEOAE reveals that the frequency contents of the TEOAE is related to the stimulus that elicited the TEOAE in a complex way and the TEOAE contains energy at frequencies not represented in the stimulus. Normally, broad band clicks are used to elicit TEOAE but when the spectrum of the stimulus clicks are limited, for instance, by high pass filtering it emerges that the TEOAE contains frequency components outside (below) the range of the stimulus sounds [161]. There is reason to believe that the amplitude of the different frequency components of the TEOAE reflects the physiologic condition of the areas of the cochlea that are tuned to these frequencies. Intermodulation distortion between the spectral components of the stimulus may generate some of the spectral components of the TEOAE.

The TEOAE could be explained without involvement of the active properties of outer hair cells by reflection of the wave motion on the basilar membrane as was done by Kemp [63]. However, the TEOAE is largest for low stimulus intensities and its amplitude grows in a non-linear fashion when the stimulus intensity is increased indicating that the cochlear echo is caused by active processes in the cochlea. That the TEOAE is generated by active processes in the cochlea was further supported by Wilson [158] who showed that in some individuals, the TEOAE did not die away with time but persisted for long periods.

Self-oscillation of the outer hair cells generates continuos sounds by the cochlea without any external sound eliciting it, and that is known as spontaneous otoacoustic emission (SOAE). SOAE has the character of a pure tone and that indicates that SOAE may be produced by a narrow segment of the basilar membrane where the outer hair cells oscillate.

Distortion product otoacoustic emission (DPOE) is used extensively for diagnostic purposes. These emissions are generated by presenting two tones simultaneously to the ear. Because of the non-linearities of the cochlea, several distortion products are generated and the emissions therefore contain components of other frequencies than the stimulus tones. The component with the frequency of $2f_1-f_2$ has the largest amplitude. (The f_1 and f_2 are the two tones that are presented.) Since these combination tones can be affected by stimulation of the olivocochlear (efferent) bundle [116] it confirms that hair cells are actively involved in generation of distortion products. The amplitude of this cubic distortion product (Fig. 3.11) is a measure of the non-linearity. Since the non-linearity is caused by the (normal) function of the outer hair cells, the amplitude of the cubic distortion product becomes a measure of the function of the outer hair cells. When outer hair cells are injured or destroyed, the cubic distortion product has a lower amplitude than normal or it may be absent.

Aspirin affects the mechanical properties (tuning) of the basilar membrane as evidenced from measuring cubic distortion product of the OAE [11]. Aspirin has an acute effect and a long-term effect depending on the dosage. Acetylsalicylate (aspirin) reduces the outer hair cell motility immediately after administration of large amounts of the drug. Other studies in animals seem to indicate that long term administration of acetylsalicylate may enhance outer hair cell motility [53].

BOX 3.4

ASPIRIN AFFECTS OTOACOUSTIC EMISSION

Using recording of the cubic distortion product measured in the ear canal of humans it was shown that administration of 3.84 g aspirin for two days resulted in a downshift of the center frequency of the produced distortion product measured in the ear canal in 3 of 8 participants [11]. There was no change in the bandwidth of the cochlear filter.

Acetylsalicylate has many effects, the best known being inhibition of synthesis of prostaglandins which has been related to the drug's anti-inflammatory effect. Acetylsalicylate also affect blood circulation and that may contribute to the effect on the function of the cochlea.

4.6. Efferent Control of Hair Cells

Efferent control of hair cells is mediated through the olivocochlear bundle and acts in fundamentally different ways on inner and outer hair cells. The efferent connections to outer hair cells (see p. 91) can control the function of the hair cells. Efferent activity in olivocochlear fibers also controls the activity in the dendrites that leave the inner hair cells before it reaches the spike generator at the first node of Ranvier.

Lateral efferent fibers terminate on auditory (afferent) nerve fibers where these leave the inner hair cells (see Chapter 1) and that explains how stimulation of these efferent fibers can modulate (decrease) the excitability of the afferent fibers. Efferent activity can thus control the output of inner hair cells before activating the spike generator at the first node of Ranvier. Activity in the efferent nerve fibers releases a transmitter substance that alters the excitability of afferent nerve fibers where they leave the inner hair cells. It is not known in detail what the transmitter substances are. Evidence has been presented that acetylcholine plays an important role but also other known neurotransmitters may be involved [115].

The physiological effect of the efferent innervation has been confirmed in experiments in animals and it has been shown that electrical stimulation of the olivocochlear efferent bundle decreases sound evoked activity in single auditory nerve fibers [35, 115].

Since the efferent nerve fibers terminate on the cell bodies of the outer hair cells, activity in these fibers can affect the function of the outer hair cells and thus controls the motility of these cells [19, 88]. Studies have shown that stimulation of the olivocochlear bundle can change the way the basilar membrane vibrates.

4.7. Autonomic Control of the Cochlea

There is anatomical evidence of a considerable adrenergic innervation of the sensory cells of the cochlea (Chapter 1) [27, 136]. Densert [27] described the anatomical basis for autonomic influence on the function of the cochlea and showed that there are adrenergic nerve endings located close to the hair cells. Secretion of norepinephrine from these nerve endings may thereby affect the sensitivity of the hair cells. Autonomic influence on the cochlea includes the adrenergic control of blood flow in the cochlea [74, 105]. Other adrenergic influence on the function of the cochlea involves the finding that the rate of K^+ secretion in strial marginal cells is stimulated by beta1-adrenergic receptors and inhibited by M3 and/or M4 muscarinic adrenergic receptors [146]. However, little experimental evidence of the effect on the function of the cochlea from adrenergic innervation has been presented. It has been shown that norepinephrine is present in the cochlea but it does not seem to be liberated from sympathetic terminals [118]. Electrical stimulation of the stellate ganglion that gives rise to the adrenergic innervation of the cochlea has only little effect on the click evoked AP recorded

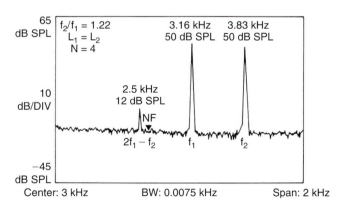

FIGURE 3.11 Illustration of DPOAE from a normal human ear, elicited by two tones of the same intensity (50 dB SPL), with frequencies of 3.16 and 3.83 kHz. The $2f_1$-f_2 component (2.5 kHz) has an intensity of 12 dB SPL (reprinted from Lonsbury-Martin and Martin, 1990, with permission from Lippincott).

from the round window of the cochlea [75, 111]. Electrical stimulation of the stellate ganglion, however, also affects the blood flow in the cochlea [74] and that may be at least partly responsible for the (small) changes in the cochlear potentials. Sympathectomy of the ear reduces temporary threshold shift caused by exposure to loud sounds [9] but only bilateral sympathectomy had this protective effect [50]. This means that sympathetic innervation of the cochlea may mediate protection against noise induced (permanent) hearing loss (see Chapter 9). Sympathectomy is an effective treatment of certain forms of tinnitus, such as that which occurs in Ménière's disease [108] and it has therefore been suggested that the sympathetic nervous system may modulate (increase) the sensitivity of cochlear hair cells (see Chapter 10).

5. AUTOREGULATION OF BLOOD FLOW TO THE COCHLEA

The vascular supply to the cochlea is complex. The blood supply to the cochlea comes through the labyrinthine artery, which is an end-artery and the inner ear is without collateral blood supply [3]. In order for the cochlea to function properly, the blood supply to the cochlea must be kept within relatively narrow limits and arterial pulsation must be kept low [105]. Pulsation of the blood flow in the cochlea could excite hair cells, which would result in constantly hearing ones own pulses. That the labyrinthine artery consists of many parallel arteries reduces blood pulsation because narrow vessels act as low pass filters and thus help to supply the cochlea with a smooth flow of blood.

Autoregulation of cerebral blood flow is an important mechanism for maintaining constant perfusion of the brain, independent of fluctuations in systemic blood pressure. In the brain, autoregulation is maintained by controlling the width (lumen) of arterioles. It is not known if there is a similar regulation of cochlear blood flow. The labyrinthine artery consists of many small arterioles, which could be the anatomical basis for such an autoregulation. However, cochlear flood flow is affected by catecholamines (such as epinephrine) and that speaks against autoregulation.

4

Sound Evoked Electrical Potentials in the Cochlea

1. ABSTRACT

1. Three different sound evoked potentials can be recorded from the cochlea: the cochlear microphonics (CM); the summating potential (SP); and the action potential (AP).
2. The CM follows the waveform of a sound and its amplitude increases with increasing stimulus intensity in a linear fashion up to a certain intensity above which it reaches a plateau. CM recorded from the round window is mainly generated by outer hair cells in the basal portion of the cochlea.
3. The SP is generated by cochlear hair cells. It is the most variable of the cochlear potentials and it may depend on the pressure in the cochlea in a systematic way.
4. The AP is the compound action potentials of the auditory nerve. Recorded from the round window of the cochlea in small animals it has two negative peaks (N_1 and N_2). The N_1 is generated in the most peripheral portion of the auditory nerve and the N_2 is generated by the cochlear nucleus. The latency of the AP decreases with increasing stimulus intensity while at the same time its amplitude increases. The latency of the AP is shorter in response to high frequency sounds than to low frequency sounds of the same intensity.
5. Cochlear potentials recorded from the human ear are known as the electrocochleogram (ECoG) that comprises all the sound evoked cochlear potentials (SP, CM, and AP).
6. The endolymphatic potential (EP) is a steady potential that is not evoked by sound.

2. INTRODUCTION

Recordings of sound evoked potentials from the cochlea in animal experiments have played an important role for understanding the function of the cochlea. In animals, sound evoked cochlear potentials are commonly recorded from an electrode placed on or near the round window of the cochlea or by electrodes placed inside the cochlea. In humans evoked potentials from the cochlea, known as electrocochleographic (ECoG) potentials, are recorded from electrodes placed on the cochlear capsule or in the ear canal near the tympanic membrane. ECoG potentials have found some use as a diagnostic tool to assess pathologies of the ear (see p. 229).

3. ELECTRICAL POTENTIALS IN THE COCHLEA

Three distinctly different kinds of sound evoked potentials can be identified in recordings from the cochlea namely the cochlear microphonics (CM), the summating potential (SP), and the action potential (AP).[1] The CM and the AP were discovered first and followed by the SP. The CM and the SP are generated

---

[1] The notion "AP" stands for action potentials and it was first used by the researchers of the cochlear electrophysiology many years ago and it has been used ever since. Although the initial negative peak N_1 is the same as that of the compound action potentials (CAP) of a nerve, we will use "AP" for the compound action potentials recorded from the cochlea.

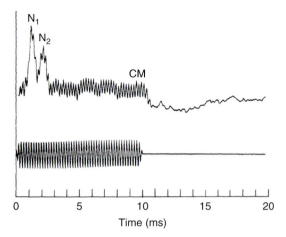

FIGURE 4.1   Response recorded from the round window of the cochlea of a rat to 5 ms long bursts of a 5-kHz tone with a rapid onset showing CM and AP (N_1, N_2). The SP is represented by the baseline shift during the tone burst. The sound is shown below. A negative potential is shown as an upward deflection (reprinted from Møller, 1983, with permission from Elsevier).

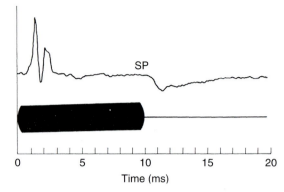

FIGURE 4.2   Response recorded from the round window of the cochlea of a rat in response to a 20-kHz tone burst to show AP and SP (reprinted from Møller, 1983, with permission from Elsevier).

by cochlear hair cells, while the AP is generated by the auditory nerve. When recorded from the round window of the cochlea in small animals, potentials generated in the cochlear nucleus contribute to the AP. The endolymphatic potential (EP) is a steady potential that is not evoked by sound. The EP is generated by the ionic differences between the different compartments of the cochlea.

All three sound evoked potentials (CM, AP, and SP) can be recorded simultaneously from an electrode placed at the round window when an appropriate sound stimulus is used (Fig. 4.1). The typical AP response that can be recorded from the cochlea in small animals has two negative peaks, N_1, N_2. The N_1 appears approximately 1.5 ms after the onset of a high frequency tone burst stimulus. The CM in response to pure tones appears as a sinusoidal oscillation that is present throughout the duration of the stimulus. The SP occurs as a deflection of the baseline during a tone burst. The CM and SP occur without any noticeable latency.

Depending on the stimulus, each one of these potentials may dominate the recording. The response to a transient sound such as a click sound will be dominated by the AP response. When a high frequency tone is used as stimulus, the low-pass filter that is a part of commonly used physiologic amplifiers may attenuate the recorded CM so that it is not visible in the recording and only the SP and the AP will appear (Fig. 4.2). Since the AP only appears at the onset of a tone burst (and to some extend at the offset of a tone), the potentials that occur after a few milliseconds will almost entirely be the CM and SP.

When click sounds are used as stimuli the polarity of the CM reverses when the polarity of the sound is reversed (from rarefaction to condensation clicks), while the polarities of the SP and the AP are nearly independent of the click polarity. The latency of the AP may be slightly different when elicited by rarefaction clicks compared with condensation clicks.

3.1. Cochlear Microphonics

The cochlear microphonics (CM) were first recorded in the 1930s in animal experiments where an electrode was placed in contact with the exposed round window. When a person spoke into the animal's ear and the amplified CM was passed on to a loudspeaker, an observer could hear the speech sounds as if a microphone had been connected to the amplifier. That drew attention in the beginning of the era of auditory physiology because it gave the impression that the ear functioned in a way similar to a microphone. It also gave the potential its name cochlear microphonics. Between 1930–1970 much research effort was devoted to studies of the CM recorded from the round window of an experimental animal [23, 152]. The relationship between the amplitude of the CM and the sound intensity was studied in detail and so was the harmonic distortion of the CM elicited by pure tones. It was found that the amplitude of the cochlear microphonic potentials recorded from the round window of a cat was a linear function of the intensity of the stimulus sound up to a certain intensity where the amplitude no longer increased when the sound level is increased [152].

The dramatic demonstration of the ear as a microphone gave a much too simplified impression of the function of the ear. Nevertheless, recordings of the CM became important in studies of the function of the cochlea and studies of CM have probably produced more journal articles than studies of any other single phenomenon of hearing.

The CM recorded from an electrode at the round window is the sum of potentials generated by a large population of hair cells. The CM recorded from an electrode placed at the round window of the cochlea is generated mostly by the outer hair cells that are located in the basal portion of the cochlea [23]. The CM recorded in that way has an initial positive deflection in response to a rarefaction click [66, 67]. A pair of electrodes placed inside the cochlea will record from a small portion of the sensory epithelium and such recordings are more useful in studies of the function of the cochlea. The technique of recording from pairs of fine wires placed inside the cochlear capsule near a specific location of the basilar membrane were introduced early in the history of cochlear electrophysiology [140]. These investigators connected the two electrodes to the two inputs of a differential amplifier. Such recordings show frequency selectivity in accordance with the tuning of the basilar membrane. The guinea pig has been used frequently for such studies because its cochlea protrudes into the middle ear space, making it possible to gain access to all the turns of the cochlea. Later the technique developed by Tasaki has been used in many investigations to study the difference in the electrical potentials in the scala media, scala tympani and scala vestibuli (for a review, see Dallos [23]).

3.2. Summating Potential

The summating potential (SP) is, as the name indicates, a summation of sound evoked potentials. The SP appears as a potential that follows the envelope of a sound, and it can therefore be readily demonstrated when tone bursts are used as stimuli. The SP may be regarded as the extracellular potential of hair cells. It is mainly a distortion product of sound stimulation caused by the basilar membrane not being deflected the same amount in both directions but it also has a neural component. The amplitude of the SP is to some extent a measure of the asymmetry of the motion of the basilar membrane.

While the SP can be recorded by placing an electrode on the round window the best way to record the SP is by placing recording electrodes inside the cochlear capsule. Recorded from the round window, the SP is dependent on many factors and it is highly sensitive to impairment of the function of the cochlea. The amplitude and even the polarity of the SP is affected by the pressure in the scala media or rather by the distension of the Reissner's membrane and presumed deflection of the basilar membrane.

The SP can be identified in recordings from humans (it is a part of the ECoG) and it has found some use in diagnosis of disorders associated with inner ear hydrops and possibly distension of the Reissner's membrane. However, the large individual variation in the SP is an obstacle in its use for diagnosis of disorders of the ear (Ménière's disease, see Chapter 10).

Recent development of methods for selective destruction of inner and outer hair cells in animals has made it possible to study the contribution from these two groups of hair cells to the SP independently. Carboplatin is used to selectively destroy inner hair cells and ototoxic antibiotics such as Kanamycin is used to destroy (mostly) outer hair cells. Recordings from the round window in such studies showed that both inner and outer hair cells contributed to the SP but inner hair cells to a greater extent than outer hair cells [28]. It was concluded that the amplitude of the SP recorded from the round window is roughly proportional to the ratio between inner and outer hair cells. The relative contributions from the two groups of hair calls depend on the sound level used to elicit the SP and how it is recorded. A noticeable variability in the SP was observed.

3.3. Action Potential

The action potential (AP) has been recorded in small animals from electrodes placed at the round window and inside the cochlear capsule as well as from electrodes placed on the surface of the cochlear capsule. When recorded in small animals the AP has two negative peaks, N_1 and N_2. (The basic properties of the AP were studied in the 1960s and 1970s [109, 141].) The waveform is similar when recorded from these different locations but the amplitude is largest when recorded from inside the cochlea.

The N_1 of the AP is the electrical activity generated in the auditory nerve, probably at the location where impulse activity in the fibers of the auditory nerve is initiated (the first node of Ranvier). It therefore reflects the neural transduction in the cochlea. The AP is the most important of the cochlear potentials for studies of the normal function of the cochlea and for studies of pathologies of the cochlea.

The N_1 peak of the AP has a slightly shorter latency than the responses that can be recorded from single auditory nerve fibers [99] (Fig. 4.3) and that supports the hypothesis that the N_1 peak is generated by the distal part of the auditory nerve.

The latency of the N_2 peak is similar to that of the peak in the histogram of latencies of responses from single cells in the cochlear nucleus (Fig. 4.4) [99]. This supports the hypothesis that the N_2 peak is generated

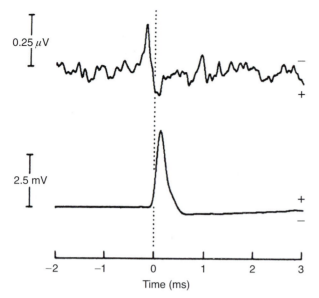

0.25 μV

2.5 mV

-2 -1 0 1 2 3
Time (ms)

FIGURE 4.3 Illustration of how single discharges of the auditory nerve contribute to the AP recorded from the round window in a cat. The record was obtained by averaging the response from the round window with the averager triggered by single discharges of a fiber of the auditory nerve. No stimuli were applied to the ear. The record represents 12,600 discharges. The characteristic frequency of the fiber was 21 kHz and its spontaneous rate was 30 discharges per second (reprinted from Kiang et al. 1976, with permission from University Park Press).

by the cochlear nucleus and conducted passively to the recording site at the round window. Removal of the cochlear nucleus in the rat (by suction) eliminated the second peak of the AP recorded from the round window (Fig. 4.5) [99] further supporting the hypothesis that the N_2 peak of the AP recorded from the round window is generated by the cochlear nucleus.

The latency of the cochlear AP depends on the intensity and the spectrum of the stimulus sound [90] as illustrated by the responses to band-pass filtered clicks (Fig. 4.6). At a given stimulus intensity the latency of the N_1 (and the N_2) is shortest when the energy of the stimulus clicks is concentrated to high frequencies. This is because high frequency sounds produce their maximal deflection of the basilar membrane near the base of the cochlea and thus a shorter travel time than what is the case for low frequency sounds that travels further towards the apex of the cochlea. The latency of the AP is more dependent on the stimulus intensity when the stimuli have their energy in the low frequency range than when the energy of the click stimuli is located in the high frequency range.

The decrease in latency with increasing stimulus intensity has several causes. The N_1 component of the AP depends on the function of the hair cells as well as the synaptic transmission between the hair cells and the auditory nerve fibers. A nerve fiber discharges when the excitatory postsynaptic potential (EPSP) has exceeded a certain threshold value. The higher the stimulus intensity the steeper is the rise of the

BOX 4.1

RELATIONSHIP BETWEEN THE ACTION POTENTIAL AND DISCHARGES OF SINGLE AUDITORY NERVE FIBERS

Kiang and co-workers [65] recorded responses from the round window of the cochlea in small animals and discharges from a single auditory nerve fiber at the same time. These investigators averaged the response from the round window with the signal averager triggered by the recorded discharges from an auditory nerve fiber. After a sufficient number of responses were averaged, the recording from the round window of the cochlea revealed an AP response that resulted from the discharges in a single auditory nerve fiber. Kiang and co-workers [65] identified a component of the averaged round window response that occurred 0.25–0.5 ms before the discharge

they recorded from the single auditory nerve fiber (Fig. 4.3), and which triggered the signal averager. They named that component the N_0 and they interpreted their results to support the assumption that the AP recorded from the round window is a summation of discharges of the peripheral portion of the auditory nerve. The recorded potentials had very small amplitudes because they were generated by only one single nerve fiber. They used either spontaneous activity of the auditory nerve fiber or activity elicited by stimulation with pure tones with a frequency equal to the fiber's characteristic frequency for these studies.

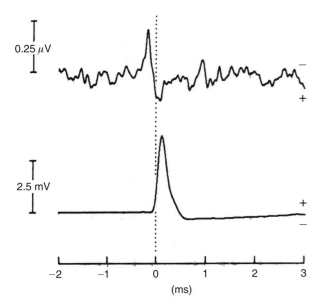

FIGURE 4.4 The AP recorded from the round window of a rat compared with the distribution of latencies of single nerve fibers of the auditory nerve (A) and the distribution of latencies of single nerve cells of the cochlear nucleus (B) (reprinted from Møller, 1983, with permission from Elsevier).

EPSP and it thus takes a shorter time to reach the threshold of firing when the stimulus intensity is high [96]. An additional source of the decrease in latency with increasing sound intensity is related to the non-linear function of the cochlea [100] (see Chapter 3), which causes the maximal deflection of the basilar membrane to shift toward the base of the cochlea when the sound intensity is increased [94, 99]. A shift of the maximal displacement of the basilar membrane toward the base of the cochlea decreases the time it takes for a displacement of the basilar membrane to the location where it has its maximal deflection. This shortening of the distance that the wave travels on the basilar membrane contributes to the decrease in the latency with increasing stimulus intensity of the recorded AP.

Since the N_1 of the AP is the summation of neural discharges of many nerve fibers, its amplitude is largest when many nerve fibers discharge simultaneously. Transient sounds such as click sounds cause many nerve fibers to discharge within a narrow time interval and the AP is therefore best recorded in response to transient sounds. The activation of hair cells in the (high frequency) basal portion of the basilar membrane has the highest degree of synchronization and the AP therefore has its largest amplitude when elicited by clicks that contain much energy at high frequencies, or by high frequency tone bursts with a fast onset.

The amplitude of the N_1 peak of the AP increases with increasing stimulus intensity. The commonly used stimuli for studies of the AP are click sounds that are generated by applying 100 μS rectangular waves to an earphone. The amplitude of the N_1 of the AP in response to such stimuli rises more slowly from threshold to approximately 50 dB above the threshold than it does above 50 dB above the threshold. A curve that shows the amplitude of the N_1 peak as function of the stimulus intensity thus has two segments with different slopes (Fig. 4.7). While the amplitude of the AP in response to condensation clicks increases monotonically as a function of the stimulus intensity, the response to rarefaction clicks has a clear two-segment relationship between amplitude and stimulus intensity. This two-segment increase in amplitude was earlier regarded as a sign of two different excitatory

---

BOX 4.2

GENERATION OF COCHLEAR AP

The peak in a histogram of the distribution of latencies of single auditory nerve fibers coincides with the latency of the AP obtained in the same animal species (rat) [99] supporting the hypothesis that the AP is generated by the auditory nerve.

Two different hypotheses about the generation of the N_2 peak have been presented. One hypothesis postulates that the N_2 component of the AP is generated by a second firing of auditory nerve fibers while the first firing contributes to the N_1 peak. The other hypotheses state that the N_2 peak is generated in the cochlear nucleus and conducted to the recording site by passive conduction. This hypothesis seems plausible in view of the small distance between the cochlea and the cochlear nucleus in (small) animals and it is further supported by comparing the AP response in the rat and cells in the cochlear nucleus in the same animal species (Fig 4.4).

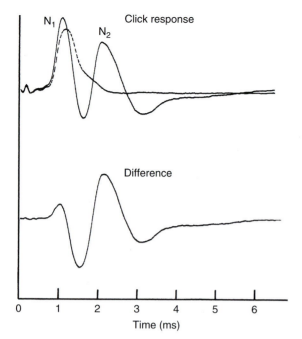

FIGURE 4.5 Top curves: responses recorded from the round window of the cochlea in response to clicks at approximately 40 dB above the threshold before (solid line) and after (dashed line) removal of the cochlear nucleus. Lower curve: the difference between the two top curves (reprinted from Møller, 1983, with permission from Elsevier).

mechanisms in the cochlea, each operating in a different intensity range [110]. However, it was shown later that the shape of the stimulus response curves of amplitude of the N_1 peak is related to the spectrum of the stimulus sounds and the peculiarities in the response to clicks with a duration of 100 µS are absent when clicks with a wider spectrum were used as stimuli (Fig. 4.8) [91]. Thus, the two-segment amplitude function is only present in response to low-pass filtered clicks. When the spectrum of the stimulus clicks was varied by varying the duration of the rectangular wave[2] it became apparent that only stimuli that were generated by 100 µS rectangular waves caused the two-segment stimulus-response curves (Fig. 4.8). The choice of stimuli based on 100 µS rectangular waves was an unfortunate one but it is still the most

[2]When expressed in dB the spectrum of a sound that is generated by an earphone is the sum of the spectrum of the electrical signal that is used to drive the earphone and the frequency transfer function of the earphone. The spectrum of a rectangular wave is flat up to a certain frequency above which it falls off towards high frequencies (Fig. 4.9). It has a dip at a frequency that is 1/T, where T is the duration of the rectangular wave. Thus, the spectrum of a rectangular wave the duration of which is 100 µS has a dip (null) at 10 kHz. This means that independent of the characteristics of the earphone, the spectrum of a click sound generated by applying a 100 µS rectangular wave to an earphone will have a dip at 10 kHz.

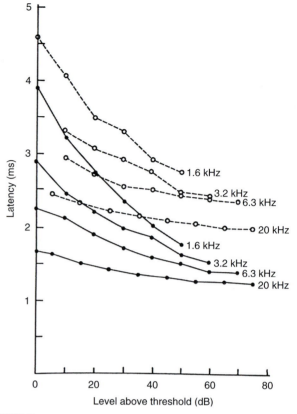

FIGURE 4.6 Latency of the N1 (solid lines) and N2 (dashed lines) peaks of the response recorded from the round window of a rat to one-third octave band-pass filtered clicks. The center frequencies of the band-pass filter setting used are given by legend numbers (reprinted from Møller, 1983, with permission from Elsevier).

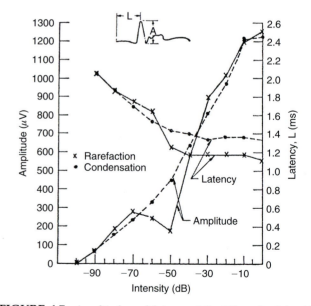

FIGURE 4.7 Amplitude and latency of the N1 peak of the AP recorded from the round window of the cochlea of a cat to condensation and rarefaction clicks. The intensity is given in decibels with an arbitrary reference (reprinted from Peake et al., 1962, with permission from the American Institute of Physics).

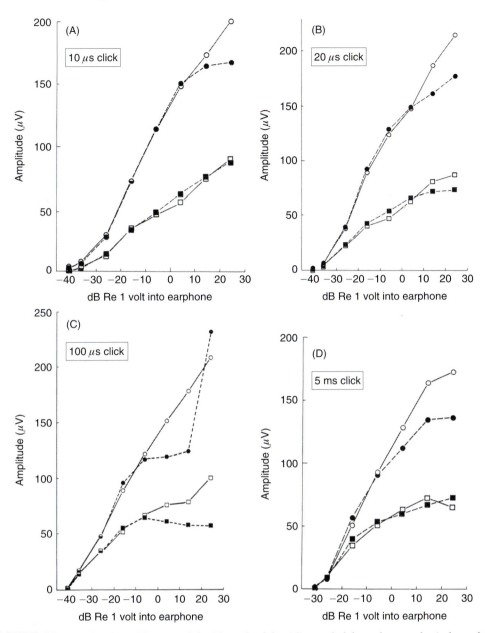

FIGURE 4.8   Amplitude and latency of the N_1 peak of the AP recorded from the round window of the cochlea of a rat to click sound generated by applying rectangular waves of different duration to the sound transducer. Circles = N_1; squares = N_2; open symbols and solid lines = rarefaction clicks; closed symbols and dashed lines = condensation clicks. The duration of the clicks is given by legends (reprinted from Møller, 1986, with permission from Taylor & Francis).

used stimuli for research and for clinical studies of auditory evoked potentials (see Chapters 7).

Cochlear frequency selectivity (frequency tuning curves) can be studied from recordings of the AP in animals [24] by using (weak) tone bursts to elicit the AP (test tone) and another tone to mask the response. The two tone bursts may either overlap each other (simultaneous masking) or the masker may be presented a brief time before the test sound that elicits the AP

response (forward masking). The test sound is set to the frequency at which tuning is to be determined and the frequency of the masker is varied in a frequency range above and below the test frequency while its amplitude is adjusted so that it reduces the amplitude of the evoked AP by a certain amount (20% or 30%). AP tuning curves are similar to frequency tuning curves obtained using recordings from single auditory nerve fibers (Fig. 4.12 [47]).

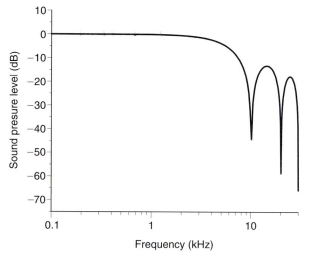

FIGURE 4.9 Spectrum of a rectangular wave with duration of 100 μS.

3.4. Electrocochleographic Potentials

Auditory evoked potentials recorded from the cochlear capsule or the round window in humans are used as a clinical test and became known as electrocochleographic (ECoG) potentials. The ECoG (Fig 4.13) consists of the CM, SP, and AP [30]. When the ECoG was first introduced, it was common to use tone bursts or band-pass filtered clicks to elicit the response, but more recently, broad band click sounds have been the most common stimuli. Band-pass filtered clicks have well-defined spectra and such sounds are often generated by applying short impulses to 1/3 octave band-pass filters [31]. The clinical value of the ECoG using tones of different frequencies or band-pass filtered clicks over the use of broad band clicks was never convincingly proven and tones and filtered clicks did not gain widespread

---

BOX 4.3

RESPONSES TO CONDENSATION AND RAREFACTION CLICKS

Studies of the response to low-pass filtered clicks confirm that the difference between the response to condensation and rarefaction clicks is more pronounced when the spectrum of the stimulus sound is limited to the low frequency range (Fig. 4.10). When low-pass filtered clicks are added to broadband clicks, the response to condensation and rarefaction clicks are different only for high intensity stimulus sounds. Thus adding low frequency

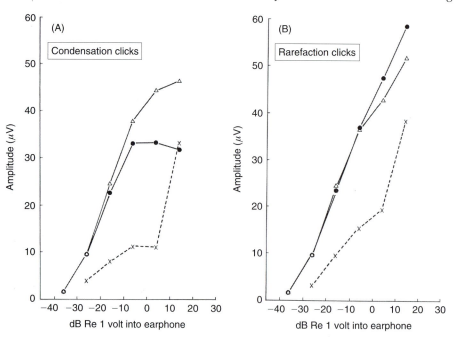

FIGURE 4.10 Amplitude and latency of the N_1 peak of the AP recorded from the round window of the cochlea of a rat to click sounds generated by applying rectangular waves of 20 μS duration to the ear (triangles and solid lines), and by low-pass filtered 10 μS rectangular waves (2.3 kHz cut off) (dashed lines and crosses). The solid lines and filled circles show the response obtained when these two sounds were presented simultaneously (reprinted from Møller, 1986, with permission from Taylor & Francis).

BOX 4.3   *(cont'd)*

components to broad band clicks reduces the amplitude of the N_1 of the AP response to clicks of high intensity (Fig. 4.10) [91] for condensation clicks, but it increases the amplitude of the N_1 peak of the AP response to rarefaction clicks. It is thus the low frequency component of a transient sound that causes the irregular (two-segment) increase in the amplitude of the N_1 peak of the AP when recorded from the round window in response to transient sounds.

When low frequency sounds (for instance, low-pass filtered clicks) are used as stimuli and recordings are made from the round window, the CM may interfere with the AP because the low frequency sounds have a longer duration than broad band click sounds. Broad band noise mask the AP leaving the CM unchanged. The AP and the CM can therefore be studied in separation by using masking with broad band noise. Appropriate masking can make it possible to study the CM in isolation and when such "clean" CM response is subtracted from the unmasked response (containing both CM and AP), a "clean" AP response without contamination with the CM is obtained (Fig. 4.11) [91]. This is a better method to eliminate the CM response than the commonly used method of reversing the polarity of every other stimulus (alternating condensation and rarefaction clicks) because the AP may be different in response to condensation and rarefaction clicks.

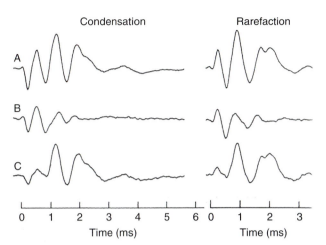

FIGURE 4.11   Row A: response from the round window of the cochlea of a rat to 2.3 kHz low-pass filtered clicks. Row B: response to the same stimulus to which broad band noise is added to mask the AP response showing the "clean" CM. Row C is the difference between the masked (B) and the unmasked responses (A) to show the "clean" AP (reprinted from Møller, 1986, with permission from Taylor & Francis).

---

clinical use, probably because of the greater complexity in generating such stimuli compared to wide band clicks.

The neural component of the ECoG recorded from the human ear is mainly generated by the distal end of the auditory nerve in the cochlea with less contributions from the cochlear nucleus than the AP in animals. This is because the longer distance between the cochlea and the cochlear nucleus in humans compared with that in the small animals used in studies of the responses from the cochlea (see Chapter 1).

Recording of the ECoG from the surface of the cochlear capsule (otic capsule) involves piercing the tympanic membrane with the recording electrode, thus an invasive approach (transtympanic ECoG). This procedure requires certain skills and also involves risks of infections, etc. To avoid these disadvantages, investigators [18] developed extratympanic methods for recording ECoG using an electrode placed deep in

the ear canal or at the tympanic membrane (Fig. 4.14) [159]. The amplitude of such recordings is lower than those recorded from the surface of the cochlear capsule and their waveform is different.

Recordings of ECoG are used clinically. In one application, the SP component is used in diagnosis of cochlear hydrops (see Chapter 9).

Frequency tuning curves can be obtained in humans using recordings of ECoG potentials and the same masking technique as described above (p. 63) [29]. This is useful because it can be obtained in individuals with normal hearing as well as in individuals with hearing loss and it can thus be used to study the effect of cochlear injury on its frequency selectivity. As in animal experiments, studies in humans have used either forward masking or simultaneous masking. In humans, forward masking yielded slightly narrower tuning curves than simultaneous masking (Fig. 4.15).

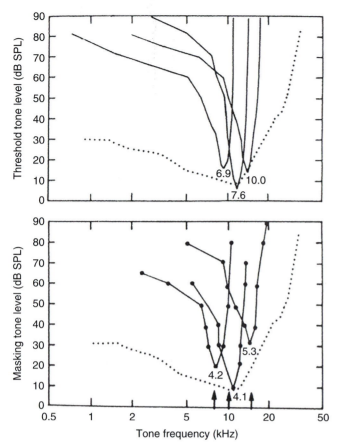

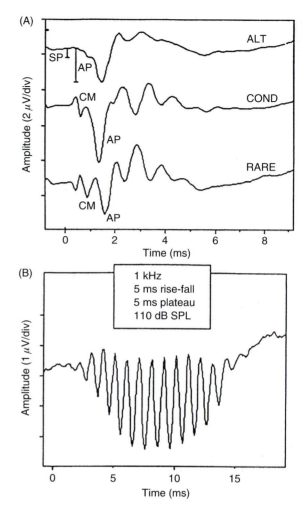

FIGURE 4.12 Frequency tuning curves. Upper graph: Frequency threshold curves for three normal guinea pig cochlear nerve fibers with characteristic frequency 9.4, 11.6, and 14 kHz. Lower graph: AP tuning curves obtained from the same animal as the curves in the upper graph. The test frequencies were 8, 10, and 15 kHz and using simultaneous masking (reprinted from Harrison et al., 1981, with permission from the American Institute of Physics).

FIGURE 4.13 (A) Typical ECoG recording from the promontorium of an individual with normal hearing in response to click stimuli (indicated by an arrow). Note that negativity is shown as a downward deflection. (B) ECoG response to tone bursts (1-kHz) (reprinted from Winzenburg et al., 1993, with permission from Elsevier).

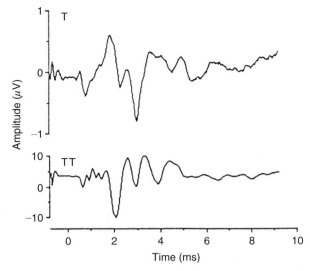

FIGURE 4.14   Extratympanic ECoG recorded from the surface of the ear canal (T) compared with transtympanic ECoG (TT) (reprinted from Winzenburg et al., 1993, with permission from Elsevier).

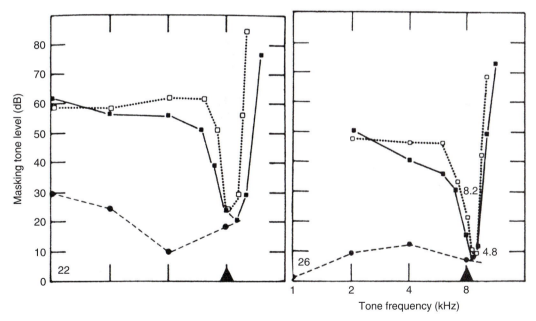

FIGURE 4.15   AP tuning curves obtained in an individual guinea pig with nearly normal hearing using simultaneous masking (solid lines) and using forward masking (dotted lines) (reprinted from Harrison et al., 1981, with permission from the America Institute of Physics).

SECTION I REFERENCES

1. Anson BJ and Donaldson JA. *Surgical anatomy of the temporal bone and ear*. Philadelphia: W.B. Saunders Co., 1973.

2. Ashmore JF. A fast motile response in guinea pig outer hair cells: the cellular basis of the cochlear amplifier. *J Physiol* 388: 323–347, 1987.

3. Axelsson A. Comparative anatomy of cochlear blood vessels. *Am J Otolaryngol* 9: 278–290, 1988.

4. Axelsson A and Ryan AF. Circulation of the inner ear: I. Comparative study of vascular anatomy in the mammalian cochlea. In: *Physiology of the ear*, edited by Jahn AF and Santos–Sacchi J. New York: Raven Press, 1988, p. 295–315.

5. Békésy von G. D–C resting potentials inside the cochlear partition. *J Acoust Soc Am* 24: 72–76, 1952.

6. Békésy von G. *Experiments in hearing*. New York: McGraw Hill, 1960.

7. Berger EH and Killion MC. Comparison of the noise attenuation of three audiometric earphones, with additional data on masking near threshold. *J Acoust Soc Am* 86: 1392–1403, 1989.

8. Blauert J. Binaural localization. *Scand Audiol (Stockholm) Suppl* 57: 7–26, 1982.

9. Borg E. Protective value of sympathectomy of the ear in noise. *Acta Physiol Scand* 115: 281–282, 1982.

10. Brodel M. *Three unpublished drawings of the anatomy of the human ear*. Philadelphia: W.B. Saunders, 1946.

11. Brown AM, Williams DM, and Gaskill SA. The effect of aspirin on cochlear mechanical tuning. *J Acoust Soc Am* 93: 3298–3307, 1993.

12. Brownell WE. Observation on the motile response in isolated hair cells. In: *Mechanisms of hearing*, edited by Webster WR and Aitken LM. Melbourne: Monash University Press, 1983, p. 5–10.

13. Brownell WE, Bader CR, Bertrand D, and Ribaupierre de Y. Evoked mechanical responses of isolated cochlear hair cells. *Science* 227: 194–196, 1985.

14. Büki B, Avan P, and Ribari O. The effect of body position on transient otoacoustic emission. In: *Intracranial and Intralabyrinthine fluids*, edited by Ernst A, Marchbanks R and Samii M. Berlin: Springer Verlag, 1996, p. 175–181.

15. Canlon B, Borg E, and Flock A. Protection against noise trauma by pre-exposure to a low level acoustic stimulus. *Hear Res* 34: 197–200, 1988.

16. Carlborg B, Konradsson KS, and Farmer JC. Pressure relation between labyrinthine and intracranial fluids: experimental study in cats. In: *Intracranial and intralabyrinthine fluids*, edited by Ernst A, Marchbanks R and Samii M. Berlin: Springer–Verlag, 1996, p. 63–72.

17. Chatterjee M and Zwislocki JJ. Cochlear mechanisms of frequency and intensity coding. II. Dynamic range and the code for loudness. *Hear Res* 124: 170–181, 1998.

18. Coats AC. On electrocochleographic electrode design. *J Acoust Soc Am* 56: 708–711, 1974.

19. Collet L, Kemp DT, Veuillet E, Duclaux R, Moulin A, and Morgon A. Effect of contralateral auditory stimuli on active cochlear micro–mechanical properties in human subjects. *Hear Res* 43: 251–262, 1990.

20. Colletti V. Multifrequency tympanometry. *Audiology* 16: 278–287, 1977.

21. Corwin JT and Cotanche DA. Regeneration of sensory hair cells after acoustic trauma. *Science* 240: 1772–1774, 1988.

22. Dallos P. The active cochlea. *J Neuroscience* 12: 4575–4585, 1992.

23. Dallos P. *The auditory periphery: biophysics and physiology*. New York: Academic Press, 1973.

24. Dallos P and Cheatham MA. Compound action potential tuning curves. *J Acoust Soc Am* 59: 591–597, 1976.

25. Dallos P and Fakler B. Prestin, a new type of motor protein. *Nat Rev Mol Cell Biol* 3: 104–111, 2002.

26. Davis H, Benson RW, Covel WP, Fernandez C, Goldstein R, Y. K, Legouix JP, McAuliffe DR, and Tasaki I. Acoustic trauma in guinea pig. *J Acoust Soc Am* 25: 1180–1189, 1953.

27. Densert O. Adrenergic innervation in the rabbit cochlea. *Acta Otolaryngol (Stockh)* 78: 345–356, 1974.

28. Durrant JD, Wang J, Ding DL, and Salvi RJ. Are inner or outer hair cells the source of summating potentials recorded from the round window? *J Acoust Soc Am* 104: 370–377, 1998.

29. Eggermont JJ. Compound action potential tuning curves in normal and pathological human ears. *J Acoust Soc Am* 62: 1247–1251, 1977.

30. Eggermont JJ. Electrocochleography. In: *Handbook of sensory physiology*, edited by Keidel W and Neff W. New York: Springer–Verlag, 1976, p. 625–705.

31. Eggermont JJ, Spoor A, and Odenthal DW. Frequency specificity of toneburst electrocochleography. In: *Electrocochleography*, edited by Ruben RJ, Elberling C and Salomon G. Baltimore: University Park Press, 1976, p. 215–246.

32. Emst van MG, Giguere C, and Smoorenburg GF. The generation of DC potentials in a computational model of the organ of Corti: effects of voltage–dependent K+ channels in the basolateral membrane of the inner hair cell. *Hear Res* 115: 184–196, 1998.

33. Evans EF. Normal and abnormal functioning of the cochlear nerve. *Symp Zool Soc Lond* 37: 133–165, 1975.

34. Feddersen WE, Sandel TT, Teas DC, and Jeffress LA. Localization of high frequency tones. *J Acoust Soc Am* 29: 988–991, 1957.

35. Fex J. Auditory activity in centrifugal and centripetal cochlear fibers in cat. *Acta Physiol Scand* 55: 5–68, 1962.

36. Flock A. Transducing mechanisms in lateral line canal organ. *Cold Spring Harbor Symp Quant Biol* 30: 133–146, 1965.

37. Flock A and Cheung HC. Actin filaments in sensory hairs of inner ear receptor cells. *J Cell Biol* 75: 339–343, 1977.

38. Forge A and Li L. Apoptotic death of hair cells in mammalian vestibular sensory epithelia. *Hear Res* 139: 97–115, 2000.

39. Forge A and Schacht J. Aminoglycoside antibiotics. *Audiol Neurotol* 5: 3–22, 2000.

40. Fuchs PA. Synaptic transmission at vertebrate hair cells. *Curr Opin Neurobiol* 6: 514–519, 1996

41. Fullerton BC, Levine RA, Hosford Dunn HL, and Kiang NYS. Comparison of cat and human brain stem auditory evoked potentials. *Hear Res* 66: 547 570., 1987.

42. Galambos R and Rupert AL. Action of the middle–ear muscles in normal cats. *J Acoust Soc Amer* 31: 349–355, 1959.

43. Goode RL, Ball G, and Nishihara S. Measurement of umbo vibration in human subjects – Method and possible clinical applications. *Am J Otol* 3, 1993.

44. Guinan JJ and Peake WT. Middle-ear characteristics of anesthetized cats. *J Acoust Soc Am* 41: 1237–1261, 1967.

45. Guinan Jr JJ and Gifford ML. Effects of electrical stimulation of efferent olivocochlear neurons on cat auditory-nerve fibers: II. Spontaneous rate. *Hear Res* 33: 115–128, 1988.

46. Gyo K, Aritomo H, and Goode RL. Measurement of the ossicular vibration ratio in human temporal bones by use of a video measuring system. *Acta Otolaryng* 103: 87–95, 1987.

47. Harrison RV, Aran J-M, and Erre JP. AP tuning curves in normal and pathological human and guinea pig cochlea. *J Acoust Soc Am* 69: 1374–1385, 1981.

48. Harrison RV and Evans EF. Reverse correlation study of cochlear filtering in normal and pathological guinea pig ears. *Hear Res* 6: 303–314, 1982.

49. Hawkins JE. Auditory physiologic history: A surface view. In: *Physiology of the ear*, edited by Jahn AF and Santos–Sacchi J. New York: Raven Press, 1988.

50. Hildesheimer M, Sharon R, Muchnik C, Sahartov E, and Rubinstein M. The effect of bilateral sympathectomy on noise induced temporary threshold shift. *Hear Res* 51: 49–53, 1991.

51. Honrubia V and Ward PH. Longitudinal distribution of the cochlear microphonics inside the cochlear duct (guinea pig). *J Acoust Soc Am* 44: 951 958, 1968.

52. Housley GD and Ryan AF. Cholinergic and purinergic neurohumoral signalling in the inner ear: a molecular physiological analysis. *Audiol Neurootol* 2: 92–110, 1997.

53. Huang ZW, Luo Y, Wu Z, Tao Z, Jones S, and Zhao HB. Paradoxical enhancement of active cochlear mechanics in long–term administration of salicylate. *J Neurophysiol* 93: 2053–2061, 2005.

54. Hudspeth A. How the ear's works work. *Nature* 341: 97–404, 1989.

55. Hudspeth AJ. Mechanical amplification of stimuli by hair cells. *Curr Opin Neurobiol* 7: 480–486, 1997.

56. Hudspeth AJ and Corey DP. Sensitivity, polarity, and conductance change in the response of vertebrate hair cells to controlled mechanical stimuli. *Proc Natl Acad Sci* 74: 2407–2411, 1977.

57. Hughes GB. *Textbook of otology*. New York: Thieme–Straton, 1985.

58. Izumikawa M, Minoda R, Kawamoto K, Abrashkin KA, Swiderski DL, Dolan DF, Brough DE, and Raphael Y. Auditory hair cell replacement and hearing improvement by Atoh1 gene therapy in deaf mammals. *Nat Med* 11: 71–76, 2005.

59. Jager W, Goiny M, Herrera–Marschitz M, Flock A, Hokfelt T, and Brundin L. Sound-evoked efflux of excitatory amino acids in the guinea-pig cochlea in vitro. *Exp Brain Res* 121: 425–432, 1998.

60. Johnson A and Hawke M. The nonauditory physiology of the external ear canal. In: *Physiology of the ear*, edited by Jahn AF and Santos-Sacchi J. New York: Raven Press, 1988, p. 41–58.

61. Johnstone BM, Patuzzi R, and Yates GK. Basilar membrane measurements and the traveling wave. *Hear Res* 22: 147–153, 1986.

62. Kawamoto K, Sha SH, Minoda R et al. Antioxidant gene therapy can protect hearing and hair cells from ototoxicity *Mol Ther* 9: 173–181, 2004.

63. Kemp DT. Stimulated acoustic emissions from within the human auditory system. *J Acoust Soc Am* 64: 1386–1391, 1978.

64. Khanna SM and Tonndorf J. Tympanic membrane vibrations in cats studied by time-averaged holography. *J Acoust Soc Am* 51: 1904–1920, 1972.

65. Kiang NY–S, Moxon EC, and Kahn AR. The relationship of gross potentials recorded from the cochlea to single unit activity in the auditory nerve. In: *Electrocochleography*, edited by Ruben RJ, Elberling C, and Salomon G. Baltimore: University Park Press, 1976, p. 95–115.

66. Kiang NY–S, Moxon EC, and Levine PA. Auditory–nerve activity in cats with normal and abnormal cochleas. In: *Sensorineural hearing loss*, edited by Wolstenholme GEW and Knight J. London: CIBA Foundation, 1970.

67. Kiang NY-S and Peake WT. Components of electrical responses recorded from the cochlea. *Ann Otol Rhinol and Laryng* 69: 448–458, 1960.

68. Kikuchi T, Kimura RS, Paul DL, Takasaka T, and Adams JC. Gap junction systems in the mammalian cochlea. *Brain Res Brain Res Rev* 32: 163–166, 2000.

69. Killion MC, Wilber LA, and Gudmundsen GI. Insert earphones for more interaural attenuation. *Hearing Instruments* 36: 34–36, 1985.

70. Konishi M and Nielsen DW. The temporal relationship between motion of the basilar membrane and initiation of nerve impulses in auditory nerve fibers. *J Acoust Soc Am* 53: 325, 1973.

71. Kurokawa H and Goode RL. Sound pressure gain produced by the human middle ear. *Otolaryngol Head Neck Surg* 113: 349–355, 1995.

72. Lang J. *Clinical anatomy of the head*. New York: Springer–Verlag, 1983.

73. Lang J. Facial and vestibulocochlear nerve, topographic anatomy and variations. In: *The cranial nerves*, edited by Samii M and Jannetta P. New York: Springer–Verlag, 1981, p. 363–377.

74. Laurikainen EA, Kim D, Didier A, Ren T, Miller JM, Quirk WS, and Nuttall AL. Stellate ganglion drives sympathetic regulation of cochlear blood flow. *Hear Res* 64: 199–204, 1993.

75. Lee AH and Møller AR. Effects of sympathetic stimulation on the round window compound action potential in the rat. *Hear Res* 19: 127–134, 1985.

76. Lefebvre PP and Van De Water TR. Connexins, hearing and deafness: clinical aspects of mutations in the connexin 26 gene. *Brain Res Brain Res Rev* 32: 159–162, 2000.

77. Lighthill J. Biomechanics of hearing sensitivity. *J Sound Vibration* 113: 1–13, 1991.

78. Lim DJ. Effects of noise and ototoxic drugs at the cellular level in the cochlea. *Am J Otolaryngol* 7: 73–99, 1986.

79. Lonsbury-Martin BL and Martin GK. The clinical utility of distortion-product otoacoustic emissions. *Ear and Hear* 11: 144–154, 1990.

80. Marchbanks RJ. Hydromechanical interactions of the intracranial and intralabyrinthine fluids. In: *Intracranial and intralabyrinthine fluids*, edited by Ernst A, Marchbanks R, and Samii M. Berlin: Springer–Verlag, 1996.

81. Matsui J, Haque A, Huss D et al. Caspase inhibitors promote vestibular hair cell survival and function after aminoglycoside treatment in vivo. *J Neurosci* 23: 6111–6122, 2003.

82. Matsui JI and Cotanche DA. Sensory hair cell death and regeneration: two halves of the same equation. *Curr Opin Otolaryngol Head Neck Surg* 21: 418–425, 2004.

83. Melloni BJ. The internal ear. *What's new?* 57: 14–19, 1957.

84. Mendelson E. A sensitive method for registration of human intratympanic muscle reflexes. *J Appl Physiol* 11: 499–502, 1957.

85. Merchant SN, Ravicz ME, Puria S, Voss SE, Wittemore KR, Peake WT, and Rosowski JJ. Analysis of middle ear mechanics and application to diseased and reconstructed ears. *Amer J Otol* 18: 139–154, 1997.

86. Miller JM, Watson CS, and Covell WP. Deafening effects of noise on the cat. *Acta Oto Laryng Suppl* 176: 1–91, 1963.

87. Mountain DC and Cody AR. Multiple modes of inner hair cell stimulation. *Hear Res* 132: 1–14, 1999.

88. Mountain DC, Geisler CD, and Hubbard AE. Stimulation of efferents alters the cochlear microphonic and sound-induced resistance changes measured in the scala media of the guinea pig. *Hear Res* 3: 231–240, 1980.

89. Møller AR. An experimental study of the acoustic impedance of the middle ear and its transmission properties. *Acta Otolaryngol (Stockh)* 60: 129–149, 1965.

90. Møller AR. *Auditory physiology*. New York: Academic Press, 1983.

91. Møller AR. Effect of click spectrum and polarity on round window N_1N_2 response in the rat. *Audiology* 25: 29–43, 1986.

92. Møller AR. Effect of tympanic muscle activity on movement of the eardrum, acoustic impedance, and cochlear microphonics. *Acta Otolaryngol (Stockh)* 58: 525–534, 1965.

93. Møller AR. Frequency selectivity of phase–locking of complex sounds in the auditory nerve of the rat. *Hear Res* 11: 267–284, 1983.

94. Møller AR. Frequency selectivity of single auditory nerve fibers in response to broadband noise stimuli. *J Acoust Soc Am* 62: 135–142, 1977.

95. Møller AR. Improved technique for detailed measurements of the middle ear impedance. *J Acoust Soc Am* 32: 250–257, 1960.

96. Møller AR. Latency of unit responses in the cochlear nucleus determined in two different ways. *J Neurophysiol* 38: 812–821, 1975.

97. Møller AR. Network model of the middle ear. *J Acoust Soc Am* 33: 168–176, 1961.

98. Møller AR. Noise as a health hazard. *Ambio* 4: 6–13, 1975.

99. Møller AR. On the origin of the compound action potentials (N_1N_2) of the cochlea of the rat. *Exp Neurol* 80: 633–644, 1983.

100. Møller AR. Origin of latency shift of cochlear nerve potentials with sound intensity. *Hear Res* 17: 177–189, 1985.

101. Møller AR. *Sensory systems: anatomy and physiology.* Amsterdam: Academic Press, 2003.

102. Møller AR. The acoustic impedance in experimental studies on the middle ear. *Int Audiology* 3: 123–135, 1964.

103. Møller AR. The middle ear. In: *Foundation of modern auditory theory*, edited by Tobias JV. New York: Academic Press, 1972, p. 133–194.

104. Møller AR. Transfer function of the middle ear. *J Acoust Soc Am* 35: 1526–1534, 1963.

105. Nakashima T, Naganawa S, Sone M, Tominaga M, Hayashi H, Yamamoto H, Liu X, and Nuttall AL. Disorders of cochlear blood flow. *Brain Res Brain Res Rev* 43: 17–28, 2003.

106. Nordlund B. Physical factors in angular localization. *Acta Otolaryng (Stockh)* 54: 76–93, 1962.

107. Pang XD and Peake WT. How do contractions of the stapedius muscle alter the acoustic properties of the ear? In: *Lecture notes in biomathematics.* Berlin: Springer Verlag, 1986, p. 360.

108. Passe EG. Sympathectomy in relation to Ménière's disease, nerve deafness and tinnitus. A report of 110 cases. *Proc Roy Soc Med* 44: 760–772, 1951.

109. Peake WT, Goldstein MH, and Kiang NYS. Responses of the auditory nerve to repetitive stimuli. *J Acoust Soc Am* 34: 562–570, 1962.

110. Peake WT and Kiang NYS. Cochlear responses to condensation and rarefaction clicks. *Biophysical J* 2: 23–34, 1962.

111. Pickles JO. An investigation of sympathetic effects on hearing. *Acta Oto Laryngol* 87: 67–71, 1979.

112. Pickles JO, Comis SD, and Osborne MP. Cross-links between stereocilia in the guinea pig organ of Corti, and their possible relation to sensory transduction. *Hear Res* 15: 103–112, 1984.

113. Pierson LL, Gerhardt KJ, Rodriguez GP, and Yanke RB. Relationship between outer ear resonance and permanent noise-induced hearing loss. *Am J Otolaryngol* 15: 37–40, 1994.

114. Pierson MG and Møller AR. Prophylaxis of kanamycin-induced ototoxity by a radioprotectant. *Hear Res* 4: 79–87, 1981.

115. Puel JL. Chemical synaptic transmission in the cochlea. *Prog Neurobiol* 47: 449–476, 1995.

116. Puel JL and Rebillard G. Effect of contralateral sound stimulation on distortion product 2F1–F2: Evidence that the medial efferent system is involved. *J Acoust Soc Am* 87: 1630–1635, 1990.

117. Raff M. Cell suicide for beginners. *Nature* 396: 119–122, 1998.

118. Rarey KE, Ross MD, and Smith CB. Quantitative evidence for cochlear, non-neuronal norepinephrine. *Hear Res* 5: 101–108, 1981.

119. Rhode WS. An investigation of post-mortem cochlear mechanics using the Mossbauer effect. In: *Basic mechanisms in hearing*, edited by Møller AR. New York: Academic Press, 1973.

120. Rhode WS. Observations of the vibration of the basilar membrane in squirrel monkeys using the Mossbauer technique. *J Acoust Soc Am* 49: 1218–1231, 1971.

121. Rosowski JJ. Models of external- and middle-ear function. In: *Auditory computation*, edited by Hawkins HL, McMullen TA, Popper AN, and Fay RR. New York: Springer Verlag, 1996.

122. Rosowski JJ. The effects of external- and middle-ear filtering on auditory threshold and noise-induced hearing loss. *J Acoust Soc Am* 90: 124–135, 1991.

123. Ruggero MA. Responses to sound of the basilar membrane of the mammalian cochlea. *Current Opinion in Neurobiology* 2: 449–456, 1992.

124. Russell IJ and Sellick PM. Low frequency characteristic of intracellularly recorded receptor potentials in guinea-pig cochlear hair cells. *J Physiol* 338: 179–206, 1983.

125. Ryals BM and Rubel EW. Hair cell regeneration after acoustic trauma in adult Coturnix quail. *Science* 240: 1774–1776, 1988.

126. Santi P. Cochlear microanatomy and ultrastructure. In: *Physiology of the ear*, edited by Jahn AF and Santos-Sacchi J. New York: Raven Press, 1988, p. 173–199.

127. Sha SH and Schacht J. Stimulation of free radical formation by aminoglycoside antibiotics. *Hear Res* 128: 112–118, 1999.

128. Shaw EAC. The external ear. In: *Handbook of sensory physiology*, edited by Keidel WD and Neff WD. New York: Springer-Verlag, 1974, p. 455–490.

129. Shaw EAC. Transformation of sound pressure level from the free field to the eardrum in the horizontal plane. *J Acoust Soc Am* 56: 1848–1861, 1974.

130. Shera CA and Guinan Jr. JJ. Evoked otoacoustic emissions arise by two fundamentally different mechanisms: a taxonomy for mammalian OAEs. *J Acoust Am*, 1999.

131. Soderman AC, Moller J, Bagger-Sjoback D, Bergenius J, and Hallqvist J. Stress as a trigger of attacks in Meniere's disease. A case-crossover study. *Laryngoscope* 114: 1843–1848, 2004.

132. Sohl G, Odermatt B, Maxeiner S, Degen J, and Willecke K. New insights into the expression and function of neural connexins with transgenic mouse mutants. *Brain Res Brain Res Rev* 47: 245–259, 2004.

133. Sokolich WG, Hamernick RP, Zwislocki JJ, and Schmiedt RA. Inferred response polarities of cochlear hair cells. *J Acoust Soc Am* 59: 963–974, 1976.

134. Song BB, Sha SH, and Schacht J. Iron chelators protect from aminoglycoside-induced cochleo- and vestibulotoxity in guinea pig. *Free Rad Biol Med* 25: 189–195, 1998.

135. Spoendlin H. Structural basis of peripheral frequency analysis. In: *Frequency analysis and periodicity detection in hearing*, edited by Plomp R and Smoorenburg GF. Leiden: A.W. Sijthoff, 1970, p. 2–36.

136. Spoendlin H and Lichtensteiger W. The adrenergic innervation of the labyrinth. *Acta Otolaryngol (Stockh)* 61: 423–434, 1966.

137. Spoendlin H and Schrott A. Analysis of the human auditory nerve. *Hear Res* 43: 25–38, 1989.

138. Stuhlman O. *An introduction to biophysics.* New York: Wiley, 1943.

139. Svane-Knudsen V and Michelsen AC. The impulse response vibration of the human ear drum. In: *Lecture notes in biomathematics.* Berlin, 1985, p. 21–27.

140. Tasaki I, Davis H, and Legouix JP. The space-time pattern of the cochlear microphonics (guinea pig), recorded by differential electrodes. *J Acoust Soc Am* 24: 502–518, 1952.

141. Teas DC, Eldredge DH, and Davis H. Cochlear responses to acoustic transients: an interpretation of whole-nerve action potentials. *J Acoust Soc Am* 32: 1438–1459, 1962.

142. Terkildsen K. Movements of the eardrum following intraaural muscle reflexes. *Arch Otolaryngol* 66: 484–488, 1957.

143. Tomita H, Yamada K, Ghadami M, Ogura T, Yanai Y, Nakatomi K, Sadamatsu M, Masui A, Kato N, and Niikawa N. Mapping of

the wet/dry earwax locus to the pericentromeric region of chromosome16. *Lancet* 359: 2000–2002, 2002.

144. Tonndorf J, Khanna SM, and Fingerhood B. The input impedance of the inner ear in cats. *Ann Otol Rhin Laryng* 75: 752–763, 1966.

145. Tos M. *Manual of middle ear surgery: Vol 2 Mastoid surgey and reconstructive procedures.* Stuttgart: Thieme Medical Publishers, 1995.

146. Wangemann P. Adrenergic and muscarinic control of cochlear endolymph production. *Adv Otorhinolaryngol* 59: 42–50, 2002.

147. Wangemann P. K+ cycling and the endocochlear potential. *Hear Res* 165: 1–9, 2002.

148. Warr WB and Guinan JJ. Efferent innervation of the organ of Corti: two separate systems. *Brain Res Brain Res Rev* 173: 152–155, 1979.

149. Warren EH and Liberman MC. Effects of contralateral sound on auditory–nerve responses. I. Contributions of cochlear efferents. *Hear Res* 37: 89–104, 1989.

150. Warren EH and Liberman MC. Effects of contralateral sound on auditory–nerve responses. II. Dependence on stimulus variables. *Hear Res* 37: 105–121, 1989.

151. Wever EG. *Theory of hearing.* New York: John Wiley & Sons, 1949.

152. Wever EG and Lawrence M. *Physiological acoustics.* Princeton, NJ: Princeton University Press, 1954.

153. Wever EG, Lawrence M, and Smith KR. The middle ear in sound conduction. *Arch of Otolaryng* 48: 12–35, 1948.

154. Wever EG and Vernon JA. The effect of the tympanic muscle reflexes upon sound transmission. *Acta Otolaryngol (Stockh)* 45: 433–439, 1955.

155. Wiederhold ML and Kiang NYS. Effects of electrical stimulation of the crossed olivocochlear bundle on single auditory-nerve fibers in the cat. *J Acoust Soc Am* 48: 950–965, 1970.

156. Wiggers HC. The function of the intraaural muscles. *Amer J Physiol* 120: 771–780, 1937.

157. Willott JF, Turner JG, and Sundin VS. Effects of exposure to an augmented acoustic environment on auditory function in mice: roles of hearing loss and age during treatment. *Hear Res* 142: 79–88, 2000.

158. Wilson JP. Evidence for cochlear origin for acoustic re-emissions, threshold fine structure and tonal tinnitus. *Hear Res* 2: 233–252, 1980.

159. Winzenburg SM, Margolis RH, Levine SC, Haines SJ, and Fournier EM. Tympanic and transtympanic electrocochleography in acoustic neuroma and vestibular nerve section surgery. *Am J Otol* 14: 63–69, 1993.

160. Wu WJ, Sha SH, and Schacht J. Recent advances in understanding aminoglycoside ototoxicity and its prevention. *Audiol Neurootol* 7, 2002.

161. Yates GK and Withnell RH. The role of intermodulation distortion in transient-evoked otoacoustic emissions. *Hear Res* 136: 49–64, 1999.

162. Zhang SY, Robertson D, Yates G, and Everett A. Role of L-type Ca(2+) channels in transmitter release from mammalian inner hair cells I. Gross sound-evoked potentials. *J Neurophysiol* 82: 3307–3315, 1999.

163. Zweig G, Lipes R, and Pierce JR. The cochlear compromise. *J Acoust Soc Am* 59: 975–982, 1976.

164. Zwislocki JJ. An analysis of the middle ear function. Part II: Guinea pig ear. *J Acoust Soc Am* 35: 1034–1040, 1963.

165. Zwislocki JJ. Are nonlinearities observed in firing rates of auditory-nerve afferents reflections of a nonlinear coupling between the tectorial membrane and the organ of Corti? *Hear Res* 22: 217–222, 1986.

166. Zwislocki JJ. Five decades of research on cochlear mechanics. *J Acoust Soc Am* 67: 1679–1685, 1980.

167. Zwislocki JJ. Some impedance measurements on normal and pathological ears. *J Acoust Soc Am* 29: 1312–1317, 1957.

168. Zwislocki JJ. What is the cochlear place code for pitch? *Acta Otolaryngol (Stockh)* 111: 256–262, 1991.

169. Zwislocki JJ and Cefaratti LK. Tectorial membrane: II. Stiffness measurements in vivo. *Hear Res* 42: 211–228, 1989.

170. Zwislocki JJ and Kletsky EJ. Micromechanics in the theory of cochlear mechanics. *Hear Res* 2: 505–512, 1980.

171. Zwislocki JJ and Sokolich WG. Velocity and displacement responses in auditory nerve fibers. *Science* 182: 646, 1973.

II

THE AUDITORY NERVOUS SYSTEM

The auditory nervous system is the most complex of all sensory pathways. Studies of the function of the auditory nervous system have earlier been regarded to be mainly of academic interest but recent studies have revealed that many of the pathologies that earlier were regarded to be located to the ear are in fact caused by changes in the function of the auditory nervous system. The complexity of the anatomy of the auditory nervous system is considerable and the discussions in Chapter 5 present the main aspects of the anatomy of the auditory system.

Most of our knowledge about the function of the auditory nervous system is based on studies of single nerve cells and the impulse traffic in single nerve fibers. However, single elements of the nervous system do not play a similar role for the function of the nervous system as single elements do for the function of manmade systems such as computers and telephone systems. Comparing the nervous system to a telephone system, as was common in the beginning of the last century, provides not only an oversimplified picture of the nervous system but also an incorrect one because the function of the nervous system is based on the function of large groups of elements with complex and numerous interconnections. Studies of the response pattern of single nerve cells therefore provide a view of the working of the nervous system through a narrow and distorted window.

Now, the nervous system is often compared with computers. While computers are more complex systems than old-fashioned telephone systems and thus offer a closer similarity with the nervous system, the analogy remains a great oversimplification. Just comparing the number of elements illustrates how incompatible such a comparison is. It has been estimated that the human central nervous system has at least 100 billion nerve cells. Compare that to the "brain" of modern personal computers, which have approximately 50 million active elements (transistors), thus there are more than 1,000 times more elements in the brain than in a microprocessor.

Many nerve cells have hundreds and even thousands of inputs while the transistors in a computer have only a few inputs. The nervous system has a vast number of interconnections between nerve cells and that has no analogy in modern computers, even in the most advanced computer systems. The number of connections in the nervous system is unknown but estimated to be several trillions. In the nervous system, the existence of a specific element and its connections are not known and only the probabilities of their existence may be known. Similar probabilistic systems are unknown in the world of manmade systems.

The working of the central nervous system is extremely difficult to fathom. We can essentially only study the working of one or perhaps a few neural elements at a time and we know little about the connections of the elements we study. It is unknown which features of the function of a single element in the nervous system are important and that adds uncertainty to the interpretation of the results of studies of the nervous system.

Plasticity is prevalent in the nervous system and its ability to self organize is unique. Self-organizing computer programs are emerging but appear fundamentally simple compared with the plasticity of the central nervous system. Neural plasticity may be involved in disorders of the nervous system and that will be discussed in Section III of this book.

Understanding the basic anatomy and the normal function of the auditory nervous system is a prerequisite to understand and diagnose the pathologies of the auditory nervous system. In this section we will discuss the basic anatomy (Chapter 5) and the normal function of the auditory nervous system (Chapter 6). This section also contains a chapter on the evoked potentials that can be recorded from the nervous system (Chapter 7). Both near-field and far-field potentials will be described. This section also includes a chapter on the acoustic middle-ear reflex (Chapter 8).

5

Anatomy of the Auditory Nervous System

1. ABSTRACT

1. The ascending auditory pathways are more complex than the ascending pathways of other sensory systems such as the somatosensory system or the visual system. Two separate pathways, the classical and the non-classical, have been identified.

2. The classical ascending pathways are also known as the tonotopic system, and the non-classical or adjunct systems are known as the extralemniscal, diffuse or polysensory system.

3. In the classical ascending auditory system, three main relay nuclei are located between the auditory nerve and the primary auditory cerebral cortex: (1) the cochlear nucleus (CN); (2) the central nucleus of the contralateral inferior colliculus (ICC); and (3) the lateral portion of the ventral division of the contralateral medial geniculate body (MGB). All information is interrupted in these nuclei by synaptic transmission. The neurons of the vertical MGB project to the primary auditory cerebral cortex (AI) and some other auditory cortical areas.

4. Bifurcations of auditory nerve fibers and fibers of the auditory pathways are abundant and this is the basis for parallel processing in the auditory system. Consequently the number of nerve fibers that connect the different nuclei increases from the periphery to the cerebral cortex.

5. Some fibers of the ascending pathways send collaterals to neurons in nuclei such as those of the superior olivary complex and the nucleus of the lateral lemniscus. Some fibers are also interrupted in these nuclei.

6. Several fiber tracts connect neurons in nuclei of the two sides of the ascending auditory pathways. The lowest level of such crossover occurs at the cochlear nuclei but the most prominent connections occur between the nuclei of the superior olivary complex, the inferior colliculus and the auditory cerebral cortices (but not the MGB).

7. The non-classical ascending auditory pathways branch off the classical pathways at several levels, the most prominent being the central nucleus of the inferior colliculus.

8. The non-classical pathways project, via the dorsal portion of the MGB, to cortical areas other than the AI such as secondary auditory cortices (AII) and the anterior auditory field (AAF). That means that these pathways bypass the processing that occurs in the AI cortex.

9. Cells in the dorsal MGB also provide subcortical connections to structures such as the amygdala ("low route"), which can be reached from the classical pathways only through a long chain of neurons in the cerebral cortex including neurons in the association cortices.

10. Some neurons in the non-classical pathways receive input from other sensory systems (somatosensory and visual) in addition to auditory input.

11. Separation of information of different kinds occurs in the association cortices ("stream segregation") where different kinds of information are processed in anatomically different populations of neurons.

12. Descending pathways travel in parallel with the ascending pathways. Although two or three separate systems can be identified it may be more appropriate to regard the descending pathways as being reciprocal to the ascending pathways, extending from the auditory cortex to cochlear hair cells.

13. The most peripheral part of the descending pathways projects from nuclei in the superior olivary complex (SOC) to the hair cells of the cochlea. This is the best known part of the descending pathways.

2. INTRODUCTION

The auditory nervous system consists of ascending and descending systems. Two ascending systems, known as the classical and non-classical auditory systems, have been identified. The term "non-classical pathways" is used in this book to describe pathways that use the dorsal nuclei of the thalamus and that project to secondary cortices rather than primary cortices. Other investigators have used terms such as the extralemniscal system, the adjunct, diffuse or non-specic system or the polysensory system. These terms are generally synonymous. Some investigators, however, have divided the non-classical pathways into two separate systems, the diffuse system and the polysensory system [252].

The classical auditory pathways are known as the tonotopic system because they have distinct frequency tuning and the neurons are organized anatomically according to the frequency to which they are tuned. The non-classical auditory system may be analogous to the pain pathways of the somatosensory system. The non-classical pathways consist of complex structures that connect to many other brain areas but the anatomy of these pathways is not completely known. An abnormal activation of the non-classical auditory pathways may explain the similarities between hyperactive hearing disorders (tinnitus) and central neuropathic pain [187].

Two descending systems, the corticofugal system and the olivocochlear system, have been identified but it may be more appropriate to regard the descending pathways as being reciprocal pathways to the ascending pathways. Descending auditory pathways are organized mostly parallel to the ascending pathways extending from the cerebral cortex to the cochlear hair cells. One part of the descending pathways, the olivo-cochlear system, connects nuclei of the superior olivary complex (SOC) with hair cells of the cochlea and it is the best known of the descending pathways.

3. CLASSICAL ASCENDING AUDITORY PATHWAYS

The classical ascending auditory system is more complex than the ascending pathways of other sensory systems. It has three main nuclei with several additional nuclei where some of the ascending fibers are interrupted by synaptic connections (Fig. 5.1A & B). The auditory nerve extends from the organ of Corti to the cochlear nucleus (CN) where each nerve fiber makes contact with neurons in all three main divisions of the CN. From the cochlear nucleus, fibers cross over to the opposite side in three fiber tracts that connect to the central nucleus of the contralateral inferior colliculus (ICC) (Fig. 5.1C). Fibers from the ICC project to the medial geniculate body (MGB). The fibers from the MGB project to the primary auditory cortex (AI), and the posterior auditory field (PAF) (Fig. 5.2).

There are connections between the two sides of the brain at several levels of the classical ascending auditory pathways (Fig. 5.1C). These connections are important for directional hearing. This anatomical organization is the basis for the parallel and hierarchical neural processing that occurs in the classical ascending auditory pathways.

3.1. Auditory Nerve

The auditory nerve (AN) in man has approximately 30,000 fibers. The auditory nerve is part of the eighth cranial nerve (CN VIII), which also includes the (superior and inferior) vestibular nerve. The AN has two types of fibers known as type I and type II. The nerve fibers of the AN are bipolar cells that have their cell bodies in the spiral ganglion located in the modiolar region of the cochlea (for details see Ryugo [257]). The peripheral portions of the type I fibers of the AN terminate on the inner hair cells (see Chapter 1) and the central portions terminate on the cells of the cochlear nucleus (CN). The type I nerve fibers are known as the radial fibers and they are thought to carry all the auditory information from the organ of Corti to higher centers of the central nervous system. The average diameter of myelinated (type I) cochlear nerve fibers in the internal auditory meatus in children is

(A)

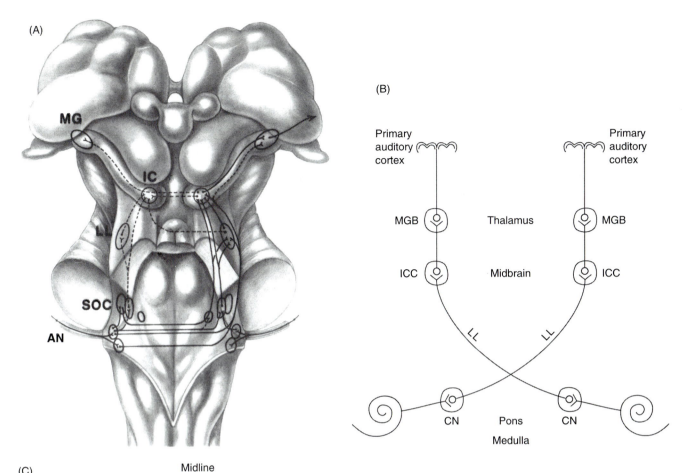

(B)

(C)

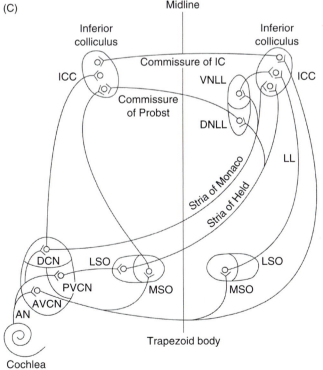

FIGURE 5.1   (A) Schematic drawing of the anatomical locations of the ascending auditory pathways. AN = auditory nerve; SOC = superior olivary complex; LL = lateral lemniscus; IC = inferior colliculus; MG = medial geniculate body (from Møller, 1988). (B) Schematic diagram showing the main nuclei and fiber tracts of the classical ascending auditory system pathways. CN = cochlear nucleus; LL = lateral lemniscus; ICC = inferior colliculus; MGB = medial geniculate body (reprinted from Møller, 2003, with permission from Elsevier). (C) More detailed drawing of the ascending auditory pathways from the ear to the IC. AVCN = anterior ventral cochlear nucleus; PVCN = posterior ventral cochlear nucleus; DCN = dorsal cochlear nucleus; LSO = lateral superior olive; MSO = medial superior olive; SH = stria of Held (intermediate stria); LL = nucleus of the lateral lemniscus; DNLL = dorsal nucleus of the lateral lemniscus; VNLL = ventral nucleus of the lateral lemniscus; ICC = central nucleus of the inferior colliculus (reprinted from Møller, 2003, with permission by Elsevier).

Classical auditory
pathways

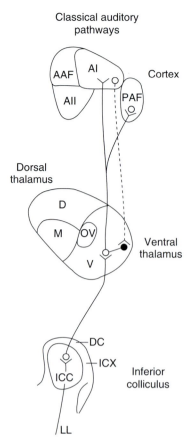

FIGURE 5.2 Schematic drawing of the classical ascending pathways from the ICC to the ventral division of the MGB, and the connections from the MGB to the auditory cerebral cortex. V = ventral division; D = dorsal division; M = medial division of the MGB; AAF = anterior auditory field; AI = primary auditory field; AII = secondary auditory field; PAF = posterior auditory field; OV = ovoid part of medial geniculate body; ICX = external inferior colliculus; DC = dorsal cortex (of inferior colliculus); and LL = lateral lemmiscus (reprinted from Møller, 2003, with permission from Elsevier).

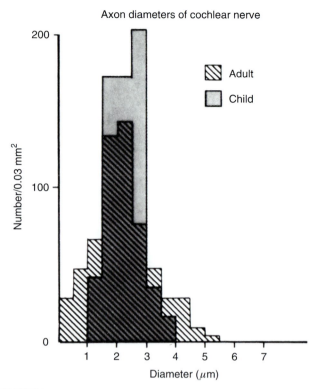

FIGURE 5.3 Distribution of diameters of myelinated auditory nerve fibers in humans. Results obtained in an adult are compared with that found in a child (reprinted from Spoendlin and Schrott, 1989, with permission from Elsevier).

2.5 μm [281] (Fig. 5.3). The diameter of the myelinated fibers in the osseous spiral lamina is approximately half that of the fibers of the AN in the internal auditory meatus. At any cross-section of the central portion of the AN, the variations in the diameters of these type I nerve fibers are small, which implies that the variations in the conduction velocity of different auditory nerve fibers are small. The information that is carried in different auditory nerve fibers will therefore arrive at the cochlear nucleus with very small time differences, ensuring a high degree of temporal coherence of the nerve impulses that arrive at the cochlear nucleus. Evidence has been presented that such coherence is important for discrimination of complex sounds such

as speech sounds (discussed in Chapter 6). The variation in fiber size and thus in conduction velocity, increases with age (Fig. 5.3) [281] and that may explain some of the hearing problems that are present at increasing age and which are not directly related to elevation in the pure tone threshold (see Chapter 9).

The type II fibers form the outer spiral fibers innervate outer hair cells. These fibers constitute only approximately 1% of the total population of nerve fibers in the auditory nerve. Type II nerve fibers project mostly to the dorsal cochlear nucleus (DCN), but their function is unknown. The cell bodies of the majority of ganglion cells are unmyelinated in humans, but the individual variation is large [226].

Most of our knowledge about the morphology of the auditory nerve is based on studies in animals (mostly cats) and it is incompletely known to what extent these results are applicable to humans. The human spiral ganglion is different from that of the animals most studied. The human eighth cranial nerve is much longer than it is in animals such as the cat (2.5 cm [123, 124] vs. 0.8 cm [59]). One of the reasons the eighth nerve is longer in humans than in animals is the larger size of the human head. The larger subarchnoidal space in humans compared with animals commonly used for auditory research, including the monkey, also contributes to the difference between animals and humans. This difference in the size of the subarchnoidal space has not been given much attention although it may have many implications regarding the development of certain auditory disorders (see Chapter 10).

Like other cranial nerves, the eighth nerve is twisted. The auditory portion is located caudally with respect to the superior vestibular nerve in the nerve's central course [123, 272] (Fig. 5.4), dorsally in the internal auditory meatus and in its most peripheral portion it is located ventrally with respect to the vestibular nerves (Fig. 5.4).

Each individual myelinated auditory nerve fiber (type I) is covered with peripheral myelin (generated by Schwann cells) in its peripheral course, up to a point just before leaving the internal auditory meatus where the covering changes to central myelin (generated by oligodendrocytes). The transition between central and peripheral myelin is known as the Obersteiner-Redlich zone. The central portion of the nerve is different from the peripheral portion in several aspects. Thus most of the supporting structures that are present

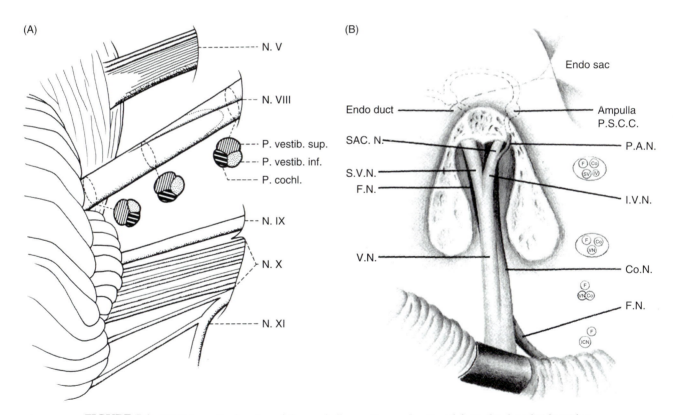

FIGURE 5.4   (A) Schematic drawing of the cerebello pontine angle viewed from the dorsal side with a cross section of the eighth cranial nerve showing the different portions of the nerve and how it rotates. N.V = fifth cranial nerve; N.VIII = eighth cranial nerve; N.IX = ninth cranial nerve; N.X = tenth cranial nerve; and N.XI = eleventh cranial nerve (reprinted from Lang, 1985, with permission from Elsevier). (B) Drawing of the anatomy of the internal auditory canal as seen from the retro mastoid approach. The posterior wall of the internal auditory meatus has been removed so that it appears as a single canal. IVN = inferior vestibular nerve; SVN = superior vestibular nerve; FN = facial nerve; VN = (entire) vestibular nerve; CoN = cochlear (auditory) nerve; SACN = saccular nerve; PAN = posterior ampulla nerve; and PSCC = posterior semicircular canal (from Silverstein et al., 1986, with permission from Otolaryngology Head and Neck Surgery).

in the peripheral portion are absent in the central portion [293]. The most central portion of the auditory nerve (approximately 1 cm) is thus similar to brain tissue. This has practical implications because it implies that the auditory nerve (and for that matter, the entire eighth cranial nerve) is at considerable risk of being injured in surgical manipulations that may occur in surgical operations in the cerebello pontine angle.

3.2. Cochlear Nucleus

The auditory nerve terminates in the cochlear nucleus (CN), which is the first relay nucleus of the ascending auditory pathways. It is located in the lower brainstem, at the junction between the medulla and the pons (the pontomedular junction) on the same side as the ear from which it receives its innervations. The CN has three main divisions, the dorsal cochlear nucleus (DCN), the posterior ventral cochlear nucleus (PVCN), and anterior ventral nucleus (AVCN) (Fig. 5.5). Before the nerve reaches the CN

each nerve fiber bifurcates and one of the two branches terminates in the AVCN and the other branch bifurcates again before terminating in the cells of the PVCN and the DCN (Fig. 5.5B). Each auditory nerve fiber thus connects to all three divisions of the CN [144]. This represents the initiation of the parallel processing that is abundant in the auditory system.

In small animals, the CN is a prominent structure of the lower brainstem, but in humans, it is comparatively small. In humans, the CN has a rostral-caudal extension of only 3 mm, but a medial-to-lateral extension of approximately 10 mm and 8 mm ventro-laterally [158]. In contrast, the cat CN is more symmetrical, with extensions of approximately 4 mm in all three planes.

The fibers from the CN mainly project to the contralateral inferior colliculus through three fiber tracts: the dorsal stria (stria of Monaco [SM]); the intermediate stria (stria of Held [SH]); and the ventral stria (trapezoid body [TB]) (Fig. 5.1C). The SM originates in the DCN. The fibers from the PVCN cross in the

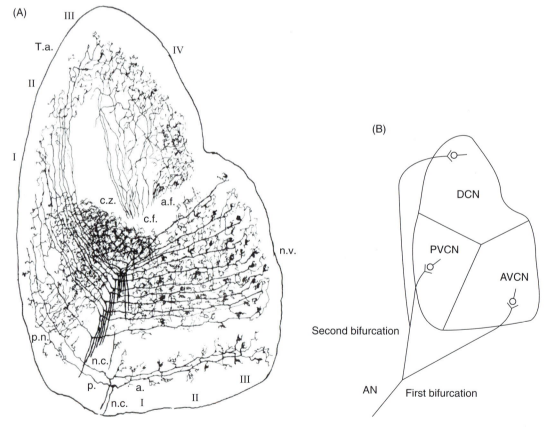

FIGURE 5.5 The cochlear nucleus. (A) Drawings of the connections of the auditory nerve with the cochlear nucleus (reprinted from Lorente de No, 1933, with permission from Lippincott Williams and Wilkins). (B) Schematic drawing of the cochlear nucleus to show the auditory nerve's (AN) connections with the three main divisions and the cochlear nucleus. DCN = dorsal cochlear nucleus; PVCN = posterior ventral cochlear nucleus; and AVCN = anterior ventral cochlear nucleus (reprinted from Møller, 2003, with permission from Elsevier).

SH and the output of the AVCN forms the TB. These three striae, after crossing to the opposite side, form the lateral lemniscus (LL), a fiber tract that projects to the central nucleus of the inferior colliculus (ICC).

Some fibers from the AVCN and PVCN do not cross the midline but ascend on the same side to reach the ipsilateral ICC (Fig. 5.1C). Fibers from the PVCN reach the dorsal nucleus of the lateral lemniscus (DNLL) and from there fibers travel to the ipsilateral ICC. The ventral cochlear nucleus also sends fibers to the facial motor nucleus and the trigeminal motor nucleus as part of the acoustic middle-ear reflex (see Chapter 8).

The two sides' cochlear nuclei are connected [148]. This is the most peripheral connection between the two side's ascending auditory pathways but its functional importance is unknown. The CN also receives input from the trigeminal somatosensory system [271, 306]. This is probably a part of the non-classical ascending auditory pathways (see p. 85).

3.3. Superior Olivary Complex

The superior olivary complex (SOC) consists of three main nuclei: the medial superior olivary nucleus (MSO); the lateral superior olivary (LSO) nucleus (Fig. 5.1C). Some of the fibers of the three striae (SM, SH, and TB) give off collaterals to nuclei of the SOC and some fibers are interrupted by synaptic contacts in one of the nuclei of the SOC before forming the LL. Nuclei of the SOC, in particular the MSO, receive input from both sides' CN.

The SOC is thus the first group of nuclei that integrate information from both ears. The nuclei of the SOC are involved in directional hearing, mainly by comparing arrival time of neural activity from the two ears (in the MSO) and intensity differences (in the LSO) (see p. 142). The nuclei of the SOC comprise some of the most complicated parts of the ascending auditory pathways and they have the largest variations between different species of mammals. The anatomical arrangements of these nuclei in humans is in many ways different from that of the commonly used experimental animals such as the cat [158].

3.4. Lateral Lemniscus and Its Nuclei

The LL is the most prominent fiber tract of the classical ascending auditory pathways (Fig. 5.1B&C). The LL is formed by the three striae that emanate from the CN. The LL is composed of fibers from all divisions of the CN. The axons of the LL cross the midline and reach the contralateral ICC. Axons from cells in the nuclei of the SOC contribute to the LL. Since fibers of the LL extend from different sources, the LL contains both second, third, and possibly, fourth order neurons although second order axons dominate.

The fibers of the LL have many collaterals some of which terminate on neurons in nuclei of the SOC, and some terminate on neurons in the dorsal and ventral nuclei of the lateral lemniscus (DNLL and VNLL). Some fibers in the LL are interrupted in the VNLL. Thus axons of the LL that originate in specific cells (octopus cells) in the contralateral PVCN do not travel directly to the ICC as other axons of the LL do, but they instead terminate in the VNLL. The DNLL receives input from both ears and is involved in binaural hearing while the VNLL mainly receives input from the contralateral ear. Some of the axons that lead away from the DNLL travel to the opposite side as the commissure of Probst and connect to neurons of the ipsilateral ICC. Neurons in the CN also connect to the ipsilateral ICC.

3.5. Inferior Colliculus

The inferior colliculus (IC) is located in the midbrain just caudal to the superior colliculus (SC). The IC is the midbrain relay nucleus where all ascending auditory information is channeled. The IC consists of the central nucleus (ICC), the external nucleus (ICX) (also known as the lateral nucleus) and the dorsal cortex of the IC (DC) (Fig. 5.6). The ICC receives its input from the LL and all the fibers of the LL are interrupted by neurons in the ICC. The ICC on one side connects to the ICC on the other side and these connections are important for directional hearing that is based on the differences in the sound intensity at the two ears.

3.6. Medial Geniculate Body

The medial geniculate body (MGB) is the thalamic auditory relay nucleus where all fibers that originate in the ICC are interrupted (see Fig. 5.2). The MGB has three distinct divisions: ventral; dorsal; and medial [160, 313]. The ventral division of the medial geniculate body includes the pars lateralis (LV) and the pars ovoidea (OV). The ventral division receives its input from the ICC.

The main output of the ICC is the brachium of the inferior colliculus (BIC), which terminate in neurons in the ventral part of the MGB but other (parallel) pathways exist. Thus Galambos [62] showed already in 1961 that the auditory cortex could be activated by sound after the brachium of the inferior colliculus (BIC) was severed. It is interesting that the number of fibers of the BIC is approximately 250,000, thus

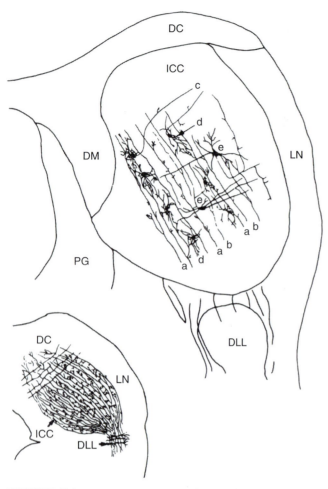

FIGURE 5.6 Schematic drawing of the inferior colliculus in frontal section. DC = dorsal cortex; DM = dorsomedial nucleus; ICC = central nucleus; LN = lateral nucleus; DLL = dorsal nucleus; PG = periaqueductal gray; a = thick lemniscal axons; b = thin lemniscal axons; c = axons of principal cells leaving the ICC; d = principal cells; and e = multipolar cells (reprinted from Ehret and Romand, 1997 *The Central Auditory Pathway*. New York: Oxford University Press; after Morest and Oliver, 1984; Oliver and Morest, 1984, with permission from Oxford Press).

approximately 10 times that of the fibers of the auditory nerve. This divergence indicates that considerable signal processing occurs in the auditory nervous system.

The ventral MGB also receives input from the thalamic reticular nucleus (RE) [310], which may exert control of the excitability of neurons in the MGB in general. The LV portion of the MGB probably also receives input from the ipsilateral ear via the ICC. There are no connections between the two sides of the MGB.

3.7. Auditory Cerebral Cortex

The auditory cortex is a complex structure where abundant connections between cells provide complex

neural processing of auditory information. Several different regions of the auditory cortex have been identified. The primary auditory cortex (AI) and the posterior auditory field (PAF) receive input from the ventral division of the MGB [35, 316] whereas the AAF receive ipsilateral input from the PO division of the MGB.

The AI is not the end station for the auditory information, and fiber tracts from the AI project to other regions of the auditory cortex and to association cortices where auditory information is integrated with other sensory information and information from other parts of the CNS. There are abundant connections between the primary auditory cortex on the two sides [28, 317].

Most descriptions of the anatomy and physiology of the auditory cortex refer to the auditory cortex in animals such as the cat, rat, guinea pig and the monkey. In the cat, rat, and guinea pig the auditory cortex is located on the surface of the cerebral cortex while in humans, the auditory cortex is located deep in the superior portion of the temporal lobe in the transverse gyrus of Hechel. The auditory cortex in humans is not visible from the surface of the brain. The anatomy of the human auditory cortex is incompletely known and different investigators place different names on the same parts of the human auditory cortices. The exact anatomical location of the different components of the human auditory cortex varies between individuals.

The axons from the MGB mainly make synaptic contacts with neurons in layer IV (Fig. 5.7B). It has been estimated that one afferent fiber from the thalamus can make contact with as many as 5,000 cortical neurons. The horizontal axons in the cortex make both inhibitory and excitatory connections with other neurons and these fibers connect functionally related neurons. The main output cells are in layers IV and V and these cells connect to cells in the MGB and the IC. These cells may integrate information from as many as 600 nearby cortical cells and may receive approximately 60,000 synapses (based on studies in monkeys [21]). There are abundant connections between primary auditory cortical areas on one side and similar areas on the other side [28, 310, 317]. It is important to keep in mind that the connections between subcortical structures and cells in the cortex as well as the connections between cells in the cortex are not "hard wired" and these connections can change as a result of expression of neural plasticity (see p. 247).

Neurons of layer III of the AI project to AII and posterior ectosylvian area (Ep) and to the contralateral AI [28, 310, 317]. The AAF receives input from the dorso-medial MGB (part of the non-classical pathways, see p. 86).

BOX 5.1

ANATOMICAL ORGANIZATION OF THE AUDITORY CEREBRAL CORTEX

The auditory cortex has six layers as do other sensory cortices (Fig. 5.7). The cells of the different layers of the cortex have specific connections (Fig. 5.7B) [155].

1. Layer I mainly contains connections between local cortical areas. This layer contains few cell bodies. Nonspecific input from the thalamus reaches neurons in layer I.

2. Neurons in layer II receive input from layer I and send connections to neurons in other layers and to other cortical areas on the same side.

3. Layer III provides the main output to other cortical areas. Neurons in layer III send connections to neurons of layer IV of cortical regions on the opposite side.

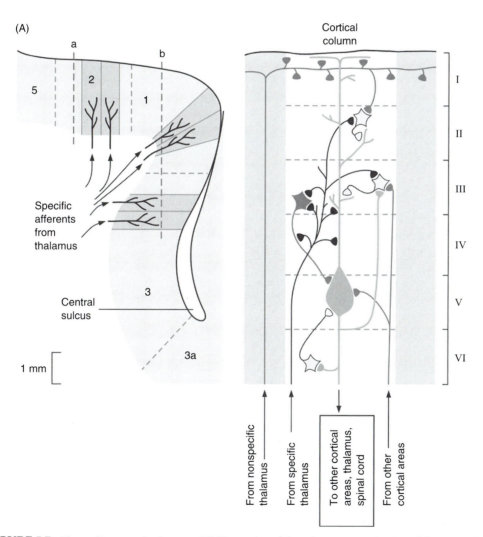

FIGURE 5.7 The auditory cerebral cortex. (A) Illustration of the columnar organization of the cortex. Left side: direction of electrode penetrations: a = parallel to a column; b = perpendicular to a column and each of the areas 1, 2, and 3 have adjacent columns, shows 3a and 5 = other areas (reprinted from Møller, 2003, with permission from Elsevier; based on Schmidt and Thews, 1983, with permission from Springer-Verlag).

(Continued)

4. Layer IV is the main receiving area where thalamic fibers from its ventral portion terminate in the internal granular layer. Subdivisions (IVa, IVb, and IVc) of layer IV have been identified in some sensory cortices.

5. The very large cells (pyramidal cells) of layer V have long axons that descend to subcortical structures.

6. Neurons of layer VI receive input from other layers and project back to the thalamus and other more peripheral nuclei of ascending sensory pathways.

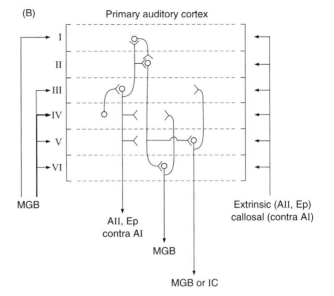

FIGURE 5.7 *(Continued)*   (B) Schematic drawing of the connections in the primary auditory cortex (reprinted from Møller, 2003, with permission from Elsevier; based on Mitani, 1985).

While neurons in the primary cortical area of hearing only respond to sound, some neurons in the other areas of the auditory cortex (AII, PAF, and AAF) also respond to other sensory modalities such as touch (somatosensory stimuli) and vision. This means that neurons in these areas receive input from other ascending sensory pathways.

The primary and secondary auditory cortices occupy only a small fraction of the neocortex and the largest portion of the neocortex is the association cortices. Association cortices receive and integrate information from different sensory systems as well as input from intrinsic sources in many parts of the CNS. This means that the association cortices perform an even higher order of processing of sensory information than the primary and secondary cortical areas. Stream segregation is a part of such complex processing (see p. 87).

The nuclei of the superior olivary complex (SOC) (Fig. 5.8) is the most peripheral level where the two sides of the ascending auditory pathways come in contact with each other. The neurons of the SOC receive input from both ears. There are also connections between the two cochlear nuclei [148], but little is known about the anatomy of these connections. Central to the SOC, several large fiber tracts connect the nuclei of the two sides. Thus the commissure of Probst connects the two DNLL with each other and it contains fibers from the DNLL on one side that

connect to the ICC on the other side. The commissure of the inferior colliculus connects the ICCs of each side.

A large fiber tract that is a part of the corpus callosum (Fig. 5.8) connects the auditory cerebral cortices on the two sides [223, 252]. Thus neurons in the auditory cortical areas are connected through diffuse fiber tracts that travel in the posterior two thirds of the corpus callosum [27]. The axons from the AAF areas cross more rostral than those from AI and axons from PAF and AII cross more caudally.

The connections between the two sides that are present at the midbrain level (ICC) and at the cerebral cortex are the main reason that sounds presented to one ear are represented in the auditory cortices on both sides. The ipsilateral connections from the CN to the IC may also play some role in providing auditory information to both sides' cortical areas. The bilateral cortical representation of sounds is the reason why it is difficult to diagnose disorders that affect the auditory cortex on one side only (see Chapter 9).

3.8. Differences between the Classical Auditory Pathways in Humans and in Animals

Most of our understanding of the anatomy and physiology of the auditory nervous system comes from studies of animals. It is therefore important to

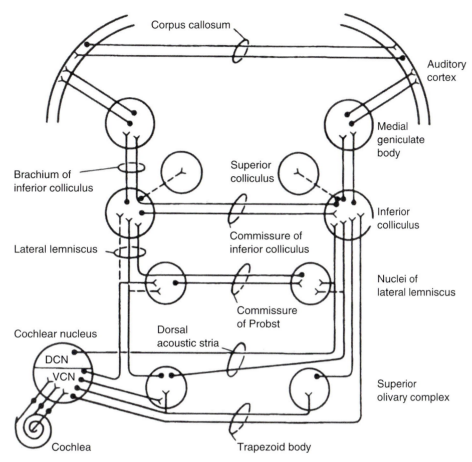

FIGURE 5.8   Schematic diagram of the ascending auditory pathways from the left cochlea showing the main nuclei and their connections including the connections between the two sides (based on Ehret and Romand, 1997, *The Central Auditory Pathway.* New York: Oxford University Press, with permission from Oxford University Press).

consider the differences between the auditory nervous system in animals and humans. The most obvious difference between the classical auditory nervous system in humans and that of commonly used experimental animals is that the auditory nerve is much longer in humans than in animals (2.5 cm [126] vs 0.8 cm in the cat [59]). The fiber tracts of the ascending auditory pathways are also in general longer in man than in small animals such as the cat [59, 126, 158] (Fig. 5.9) which has the implication that the neural travel time becomes longer in humans than in the small animals commonly used in auditory research [158, 205].

We can only speculate about the functional importance of these differences between the ascending auditory pathways in humans and that of animals that are commonly used in studies of the auditory system. The differences in the length of the auditory nerve and the length of the fiber tracts, however, are known to be important for the interpretation of auditory evoked potentials (ABR) (see Chapter 7).

4. NON-CLASSICAL ASCENDING AUDITORY PATHWAYS

The term "non-classical pathways" is used in this book for the ascending auditory pathways that are different from the classical pathways. Other investigators have used other names for these pathways such as "the diffuse system" that relates to the fact that neurons in the non-classical system are not as clearly tuned and they are not as clearly organized anatomically as those of the classical ascending pathways. The use of the term "the polysensory system" reflects the finding that the non-classical pathways receive input from other sensory systems.

Graybiel [71] described the basic anatomy of the non-classical ascending auditory system in the early 1970s. Later studies of the anatomy [3, 270, 312] have provided a general understanding of the connections in these pathways.

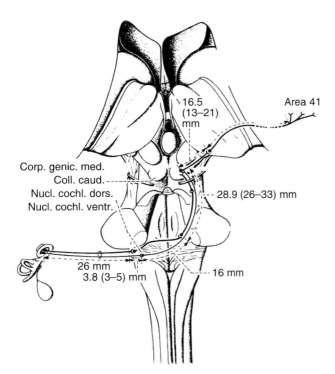

Area 41

16.5
(13–21)
mm

Corp. genic. med.
Coll. caud.
Nucl. cochl. dors.
Nucl. cochl. ventr.

28.9 (26–33) mm

26 mm
3.8 (3–5) mm

16 mm

FIGURE 5.9 Length of the main paths of the ascending auditory system in humans (modified from Lang et al., 1991, with permission from Springer-Verlag).

as the auditory reflex center as it connects to the superior colliculus (SC) to control eye movements and other motor responses to auditory stimuli that are important for directional hearing (see p. 148).

While the classical sensory pathways are interrupted by synaptic contacts with neurons in the ventral parts of the MGB of the thalamus, the non-classical sensory pathways use the dorsal and medial division of the MGB as relay (Fig. 5.10) [122]. These divisions of the MGB receive their input from the ICC and the ICX. The posterior division of the MGB (PO) receives input from the ICC and projects to the AAF cortical area. The neurons in the ventral portion of the auditory thalamus project to the primary auditory cortex but the neurons in the dorsal and medial parts of the thalamus project to secondary (AII) auditory cortex and association cortices thus bypassing the AI.

Neurons in the dorsal auditory thalamus also project to other parts of the brain such as the lateral nucleus of the amygdala thereby providing a subcortical connection to the amygdala (see p. 89). These projections have functional implications that will be discussed in Chapter 10.

Neurons in the non-classical pathways respond both to sound and to other sensory stimulations such as touch [4] and light while neurons in the classical auditory pathways up to and including the AI cortex only respond to sound stimulation. Neurons in the non-classical auditory pathways thus receive input from other sensory systems such as the somatosensory [4] and visual systems [11] (Fig. 5.10).

While early studies have shown that the non-classical pathways branch off from the classical pathways at the inferior colliculus [3] (Fig. 5.11) more recent studies indicate that the non-classical pathways branch off as early as the cochlear nucleus where neurons receive projections from the somatosensory system [270]. It is, however, the ICX of the IC and the

There are two specific differences between the classical and the non-classical auditory pathways. While the ICC is a part of the classical ascending auditory system the ICX and the DC are parts of the non-classical auditory system. The neurons of the DC deliver their output to the diffuse thalamocortical auditory system. The ICX receives input from the somatosensory system (dorsal column nuclei) and provides input to the medial portion of the MGB, and to acoustic reflex pathways (other than the acoustic middle ear reflex) (Fig. 5.10) [2]. The IC has been described and labeled

---

BOX 5.2.

ANATOMICAL DIFFERENCES BETWEEN HUMANS AND ANIMALS

The differences in the nuclei in humans and those in the animals commonly used for auditory research are greatest in the superior olivary complex [158]. There are fewer small neurons in the lateral olivary nucleus, the nucleus of the trapezoidal body in humans, compared with animals such as the cat. Small cells are also fewer in

the human cochlear nucleus than in the cat and the dorsal cochlear nucleus is much smaller and less developed in humans compared with the cat or other animals that are used in auditory research. Groups of large neurons are more developed in the human CN, the medial superior olivary (MSO) nuclei, and periolivary nuclei [158].

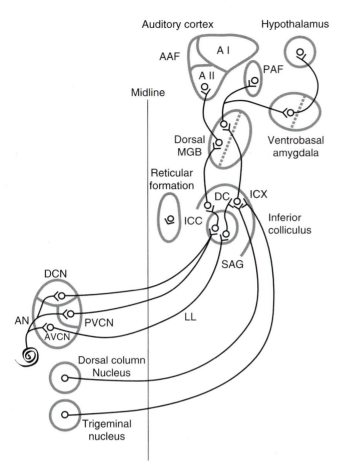

FIGURE 5.10 Simplified drawing of the non-classical ascending auditory pathways (Reprinted from Møller, 2006, with permission from Cambridge University Press).

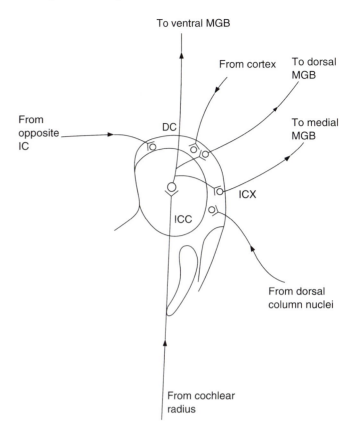

FIGURE 5.11 Schematic drawing of the connections from the ICC to the ICX and the DC, and connections from these structures to other nuclei. Also shown is the efferent input from the cerebral cortex to the ICX (From Møller, 2003, with permission from Elsevier).

DC of the IC that usually are associated with the non-classical auditory system [3, 191, 295].

In addition to receiving auditory input, neurons of the ICX also receive input from other sensory systems such as the somatosensory system (the dorsal column nuclei) [3] and from the visual and the vestibular systems [11]. The dorsal division of the MGB projects to the AII and the PAF cortical fields (Fig. 5.10) rather than the primary auditory cortex (AI) that is the target of the classical pathways. Another pathway from the IC to the primary cortex is via the posterior nucleus of the thalamus that sends axons to the AAF. The neurons of the medial division of the MGB project to the AAF, which may send collaterals to the reticular nucleus (RE) of the thalamus. The RE controls the excitability of neurons in the MGB. There neurons receive both inhibitory and excitatory input from the somatosensory system and probably also from the visual system.

There are indications that the non-classical ascending pathways are dormant in adults but active in children [220]. There are also indications that the non-classical pathways are abnormally active in connection with certain pathologies such as tinnitus [219] and hyperacusis (see p. 258) where it may cause phonophobia and perhaps depression (see Chapter 10). There are some indications that the non-classical auditory pathways may function abnormally in certain developmental disorders (autism) [218].

5. PARALLEL PROCESSING AND STREAM SEGREGATION

The information that travels in the auditory nerve is separated in different ways while being processed in the nervous system. Two fundamentally different principles of such separation have been identified. One is parallel processing, which means that the same information is processed in different populations of neurons. The other is stream segregation, which means that different kinds of information are processed in different populations of neurons. Stream segregation was first studied in the visual systems but it has later been shown to occur in other sensory systems.

5.1. Parallel Processing

Parallel processing is based on branching of the ascending auditory pathways. It begins peripherally where each auditory nerve fiber bifurcate twice to connect to neurons in each of the three main divisions of the CN (Fig. 5.5B) [159, 305] (p. 80). The fiber tract that connects the ICC with the MGB (BIC) has approximately 10 times as many nerve fibers as the auditory nerve and that is another sign of parallel processing and it means that information that is represented in the neural code in the auditory nerve is divided into many separate channels before it reaches the cerebral cortex. Another example of parallel processing is the classical and the non-classical ascending pathways.

5.2. Stream Segregation

It has been demonstrated in several sensory systems that populations of cells that process different kinds of information are anatomically segregated and that populations of cells with common properties are anatomically grouped together [70, 164, 330]. That different kinds of information are processed by different populations of cells in association cortices was first recognized in studies of the visual system where it was found that spatial and object information was processed in two anatomically separate locations (streams) in the association cortices [154, 301]. These two locations were also known as the "where" and "what" streams; "where" (spatial information) was found to be processed in a dorsal part of the cortex and a ventral stream processed the "what" (object) information (Fig. 5.12).

Recently, stream segregation was studied in the auditory system [104, 239, 248, 298] and it was shown that directional information ("where") is processed in anatomically separate locations from where object information was processed. Studies in the rhesus monkey have shown that processing of different kinds of information occurs in the lateral belt of auditory cortex where neurons in the anterior portion of this belt prefer complex sounds such as species specific communication sounds ("what") whereas neurons in the caudal portion of the belt region show the greatest spatial specificity ("where") [237, 297, 298].

Neurons in the superior temporal gyrus of the monkey (macaques) is organized in two areas with different functions. One, the most rostral stream,

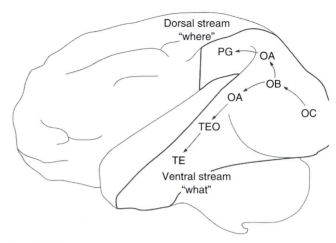

FIGURE 5.12 Illustration of the anatomical separation of information into two principal streams. Connections between the visual (striate) cortex and association cortices in the brain of the monkey (according to Mishkin et al., 1983, with permission from Elsevier).

seems to be involved in processing of object information such as that carried by complex sounds (for instance vocalization) while neurons in the other population of neurons that is located more caudally are involved in processing of spatial information.

Auditory spatial information (directional information) is not related to the location on a receptor surface as is the case for visual and somatosensory information but spatial auditory information is derived from manipulation of information from the two ears, thus computational rather than related to a receptor surface.

Speech perception is better when listening with the right ear (right ear advantage) [84] while there is no hemispheric difference with regard to identification of a speaker [118], an indication that information regarding speech perception and speech recognition is processed in different parts of the brain.

More recently, neuroimaging techniques have been used to explore the anatomical site of processing of different kinds of sounds in humans [78] and it has been shown that motion produced stronger activation in the medial part of the planum temporale, and frequency-modulation produced stronger activation in the lateral part of the planum temporale,[1] as well as an additional non-primary area lateral to Heschl's gyrus. The results of these studies were taken as indications of the existence of segregation of spatial and non-spatial auditory information. The study also

---

[1]Planum temporale: An important structure for language [78] is the posterior surface of the superior temporal gyrus of the cerebral cortex located in the temporal lobe. It is normally larger on the left side than on the right.

BOX 5.3

STREAM SEGREGATION STUDIED IN THE FLYING BAT

Other evidence of stream segregation in the auditory system comes from studies of the flying bat. Bats emit sounds and use information about the reflected sound for navigation and location of prey (echolocation). In bats, the cortical representation of distance to an object is the interval between the emission of a high frequency sound and receiving of the echo of that sound. This time difference is coded in the discharge pattern of individual neurons. Sound intervals (duration of silence) that are coded in some neurons in the auditory pathways (see p. 137) [197, 321] may therefore be regarded as spatial information because it refers to a location. Bats use low frequency sounds for communication while flying and that may be regarded as object information. Studies have shown that these two kinds of information are separated at the midbrain level (inferior colliculus [IC]) but the two streams are joined again in the cerebral cortex where the same neurons process both kinds of information [241]. Sound duration may also be coded specifically in the auditory system [24, 240]. While the coding of these kinds of sounds has been studied in animals, features like duration of sounds and duration of silent intervals are important features for discrimination of speech sounds.

---

suggested that the superior parietal cortex is involved in the spatial pathways and that it is dependent on the task of motion detection and not simply on the presence of acoustic cues for motion. These findings indicate that engagement of processing streams is dependent on the listening task.

The psychoacoustic aspects of stream segregation have been studied extensively [156, 282] and it has been related to hearing impairment (see p. 88).

5.3. Connections to Non-auditory Parts of the Brain

Auditory information can reach many parts of the brain. Naturally, auditory information can control motor systems such as extraocular muscles and neck muscles. Sound can also activate reflexes such as the acoustic middle-ear reflex and the startle reflex, and it can affect wakefulness and sounds can influence the autonomic system and the endocrine systems. The IC has often been regarded as the motor center of the auditory system although it is not involved in the acoustic middle-ear reflex (see Chapter 8) but it is involved in righting reflexes through its connection to the superior colliculus. Cells in the IC connect to many other parts of the brain with much less known function. Many of these connections are dormant in adults but the synaptic efficacy of the connections to these systems is dynamic and can be modulated by expression of neural plasticity.

Auditory information can reach the emotional brain known as the limbic system through two fundamentally different routes (Fig. 5.13) [132]. Input to the amygdala from the auditory system can evoke fear. Both the classical and the non-classical pathways provide input to the amygdala, but through very different routes. The classical pathways provide input to the amygdala through a long route involving the primary auditory cortex, secondary auditory cortex and association cortices while the non-classical pathways provide a much shorter and subcortical route to the amygdala (Fig. 5.13).

Subcortical connections from auditory pathways to limbic structures are important because the information that is mediated through such connections is probably not under conscious control. This route may be activated in certain forms of tinnitus where it can mediate fear without conscious control [219]. The non-classical pathways also have abundant projections to the reticular formation controlling wakefulness [191].

6. DESCENDING PATHWAYS

The descending pathways are at least as abundant as the ascending pathways [311, 312, 314] but much less is known about the descending pathways than the classical ascending pathways. The descending auditory pathways have often been described as two separate pathways, the corticofugal and the cortico-cochlear systems [76]. The most central part of the corticofugal system originates in the auditory cerebral cortex (Fig. 5.14A) and the cortico-cochlear system projects from the auditory cortex to the cochlear

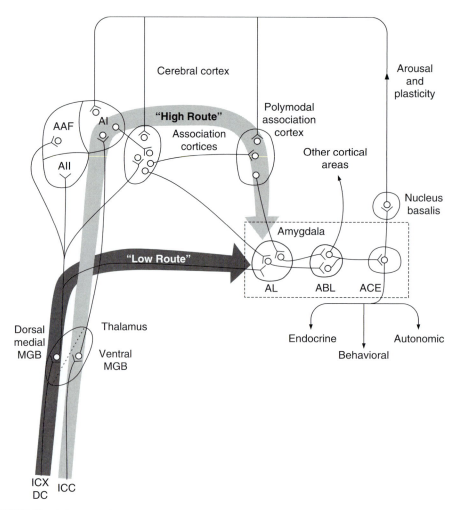

FIGURE 5.13 Schematic drawing of the connections between the classical and the non-classical routes and the lateral nucleus of the amygdala (AL), showing the "high route" and the "low route". Connections between the basolateral (ABL) and the central nuclei (ACE) of the amygdala and other CNS structures are also shown (reprinted from Møller, 2006, with permission from Cambridge University Press; based on LeDoux, 1992).

nucleus and cochlea (Fig. 5.14B). Both systems include crossed and uncrossed pathways. The descending pathways from the auditory cortices to the thalamic sensory nuclei are especially abundant [312] and extensive descending pathways reach auditory nuclei in the brainstem [314]. Instead of classifying the descending pathways separately, it seems more appropriate to regard the descending pathways as reciprocal pathways to the ascending pathways.

One large descending fiber tract originates in layers V and VI of the primary auditory cortex (Fig. 5.7B). Uninterrupted fiber tracts that originate in neurons of layer VI make synaptic connections with neurons in the MGB and neurons of layer V project to both the MGB and IC [38, 315]. The descending projections to the IC reach mainly neurons in the ICX and DC [311, 314]. The descending connections from layers V and VI may

be regarded as reciprocal innervation to the ascending connections but they are often referred to as a separate descending auditory system.

Descending pathways from the SOC reach the cochlear nucleus [76, 279], and even cochlea hair cells receive abundant efferent innervation (Fig. 5.15) [303]. The descending system that projects from SOC to the cochlea has two parts, one that projects mainly to the ipsilateral cochlea and the fibers of which travel close to the surface of the floor of the fourth ventricle (Fig. 5.15B) [72]. The other part of the olivocochlear system projects mainly to the contralateral cochlea and the fibers of that system travel deeper in the brainstem. The ipsilateral fibers originate in the lateral part (LSO) of the SOC. The system that mainly projects to the contralateral cochlea originates from medial part of the SOC (MSO). Both systems project to hair

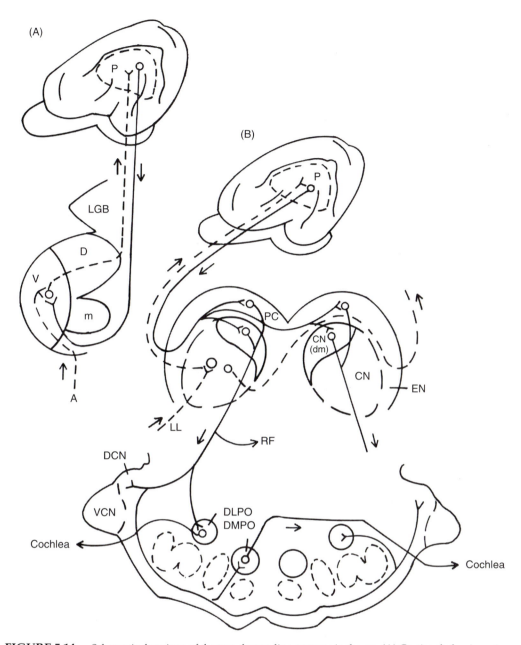

FIGURE 5.14 Schematic drawings of the two descending systems in the cat. (A) Cortico-thalamic system. (B) Cortico-cochlear and olivocochlear systems: P = principle area of the auditory cortex; LGB = lateral geniculate body; D = dorsal division of the medial geniculate body; V = ventral division of the medial geniculate body; m = medial (magnocellular) division of the medial geniculate body; PC = pericentral nucleus of the inferior colliculus; EN = external nucleus of the inferior colliculus; LL = lateral lemniscus; CN (dm) = dorsal medial part of the central nucleus of the inferior colliculus; DCN = dorsal cochlear nucleus; VCN = ventral cochlear nucleus; DLPO = dorsolateral periolivary nucleus; DMPO = dorsomedial periolivary nucleus; and RF = reticular formation (reprinted from Harrison and Howe, 1974, with permission from Springer-Verlag).

cells in the cochlea but the pathways that originate in the LSO mainly terminate on afferent fibers of inner hair cells, whereas axons of the medial system terminate mainly on the cell bodies of the outer hair cells. This description refers to the cat, and the olivocochlear system may be different in different animal species including humans.

The fact that the response of single auditory nerve fibers are affected by contralateral sound stimulation has been attributed to the efferent innervations of cochlear hair cells [304]. The finding that cochlear microphonics is affected by electrical stimulation of the efferent bundle is taken as an indication of efferent innervations of outer hair cells [163].

(A)

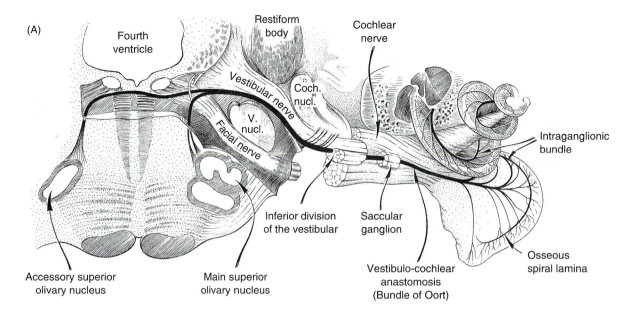

Fourth
ventricle

Restiform
body

Cochlear
nerve

Vestibular nerve

Coch.
nucl.

V.
nucl.

Facial nerve

Intraganglionic
bundle

Inferior division
of the vestibular

Saccular
ganglion

Accessory superior
olivary nucleus

Main superior
olivary nucleus

Vestibulo-cochlear
anastomosis
(Bundle of Oort)

Osseous
spiral lamina

(B)

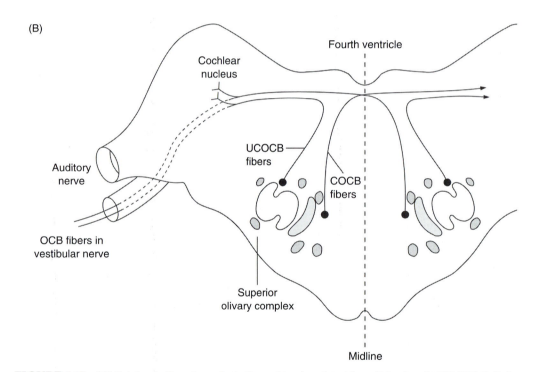

Fourth ventricle

Cochlear
nucleus

Auditory
nerve

UCOCB
fibers

COCB
fibers

OCB fibers in
vestibular nerve

Superior
olivary complex

Midline

FIGURE 5.15   (A) Origin of efferent supply to the cochlea (reprinted from Schucknecht HF, 1974 *Pathology of the Ear*. Cambridge, MA: Harvard University Press, with permission from Harvard University Press). (B) Olivocochlear system in the cat. The uncrossed olivocochlear bundle (UCOCB) and the crossed olivocochlear bundle (COCB) are shown (redrawn from Pickles, 1988, with permission from Elsevier).

6

Physiology of the Auditory Nervous System

1. ABSTRACT

1. Frequency selectivity is a prominent property of the auditory nervous system that can be demonstrated at all anatomical levels. The frequency selectivity of the basilar membrane is assumed to be the originator of the frequency tuning of auditory nerve fibers and cells in the classical ascending auditory pathways.

2. The threshold of the responses of an auditory nerve fiber is lowest at one frequency known as that fiber's characteristic frequency (CF) and a fiber is said to be tuned to that frequency. Different auditory nerve fibers are tuned to different frequencies.

3. A plot of the threshold of an auditory nerve fiber as a function of the frequency of a tone is known as a frequency threshold curve, or tuning curve.

4. Tuning curves of cells of the nuclei of the classical ascending auditory pathways have different shapes.

5. Nerve fibers of the auditory nerve, cells of auditory nuclei and those of the auditory cerebral cortex are arranged anatomically according to their characteristic frequency. This is known as tonotopical organization.

6. An auditory nerve fiber's response to one tone can be inhibited by presentation of a second tone when that tone is within a certain range of frequencies and intensities (inhibitory tuning curves).

7. Analysis of the discharge pattern of single auditory nerve fibers in response to continuous broad band noise reveals great similarity with the tuning of the basilar membrane over a large range of stimulus intensities.

8. The waveform of a tone or of complex sounds is coded in the time pattern of discharges of single auditory nerve fibers, known as "phase-locking." Phase-locking can be demonstrated experimentally in the auditory nerve for sounds with frequencies at least up to 5 kHz but may also exist at higher frequencies. The upper frequency limit for phase locking in auditory nuclei is lower than it is in the auditory nerve.

9. Convergence of input from many nerve fibers on one nerve cell improve the temporal precision of phase locking by a process similar to that of signal averaging.

10. The cochlea delivers a code to the auditory nervous system that yields information about both the (power) spectrum and the waveform (periodicity) of a sound. One of these two representations or both is the basis for discrimination of frequency.

11. The frequency selectivity of the basilar membrane is the basis for the place principle of frequency discrimination. Coding of the temporal pattern of sounds in the discharge pattern of auditory nerve fibers is the basis for the temporal principle of frequency discrimination.

12. Because place coding is affected by the sound intensity it may not be sufficiently robust to explain auditory frequency discrimination. The neural coding of vowels in the cat's auditory nerve shows a higher degree of robustness of the temporal code compared with the place code.

13. The exact mechanisms of decoding the temporal code of frequency are unknown but similar

neural circuits as those decoding directional information may decode temporal information about frequency.

14. The most important function of cochlea may be that it prepares sounds for temporal coding by dividing the spectra of complex sounds into (narrow) bands before conversion into a temporal code occurs.

15. Auditory nerve fibers and cells in the nuclei of the classical ascending auditory pathways respond poorly to steady state sounds. The discharge rate of most neurons reaches a plateau far below the physiologic range of sound intensities.

16. Changes in intensity or frequency of sounds are coded in the discharge pattern over a larger range of stimulus intensities than constant sounds or sounds with slowly varying frequency or intensity.

17. The response to complex sounds (the frequency or intensity of which changes) cannot be predicted from knowledge about the response to steady sounds or tone bursts.

18. Hearing with two ears improves discrimination of sounds in noise and helps select listening to one speaker in an environment where several people are talking at the same time.

19. Hearing with two ears (binaural hearing) is the basis for directional hearing, which has been of great importance in phylogenic development but it is of less apparent importance for humans than it is in many other species.

20. The physical basis of directional hearing in the horizontal plane is the difference in the arrival time and the difference in the intensity of sounds at the two ears, both factors being a function of the azimuth.

21. The time between the arrival of sounds at the two ears can be detected by neurons that receive input from both ears. The neural processing of interaural intensity differences is more complex and less studied than that of interaural time differences.

22. The physical basis for directional hearing in the vertical plane is the dependence of the elevation on the spectrum of the sounds that reaches the ear canal. This is a result of the outer ears and the shape of the head.

2. INTRODUCTION

All information that is available to the auditory nervous system is contained in the neural discharge pattern of auditory nerve fibers. This information undergoes an extensive transformation in the nuclei of the classical ascending auditory pathways, which performs hierarchical and parallel processing of information. I have shown in the previous chapter that the auditory nervous system is more complex anatomically than that of other sensory system. It is therefore not surprising that also the processing of auditory information that occurs in the ascending auditory pathways is complex and extensive. Recognition of the existence of two parallel ascending pathways, the classical and the non-classical pathways, adds to the complexity of information processing in the auditory system. The interplay between these two systems and the role of the vast descending pathways is not understood. The non-classical auditory system may be analogous to the pain pathways of the somatosensory system [187] and that may explain the similarities between hyperactive disorders of the hearing and central neuropathic pain [192]. It seems reasonable to assume that a better understanding of these aspects of the function of the auditory nervous system is important for understanding many disorders of the auditory system and it is a necessity for developing better treatments of disorders of the auditory system. The introduction of cochlear implants and cochlear nucleus implants (auditory brainstem implants [ABIs]) (see Chapter 11) have made understanding of the anatomy and physiology of the auditory nervous system of clinical importance.

Most studies of the function of the auditory system have aimed at the coding of different kinds of sounds in the auditory nerve and how this code changes as the information travels up the neural axis towards the cerebral cortex in the classical auditory pathways. Peripheral parts of the ascending auditory pathways have been studied more extensively than central portions. The physiology of the auditory nervous system has been studied mostly in experiments in animals such as the rat, guinea pig and cat. Little is known about the difference between the function of the auditory system in small animals and humans.

The information processing that occurs in the non-classical (adjunct or extralemniscal) ascending auditory pathways has not been studied to any great extent and therefore little is known about the coding and transformation of information in these systems. In fact little is known about the activation of the non-classical auditory pathways in humans [220]. The function of the vast descending pathways is practically unknown with the exception of its most peripheral parts. We will therefore in this chapter focus on the processing of auditory information that occurs in the classical ascending auditory nervous system including the auditory cortex.

For humans, speech is the most important sound and it would have been natural to ask the question: How does the auditory nervous system discriminate speech sounds? Nevertheless, that is too complex a question and it is more realistic to ask simpler questions such as how frequency is discriminated. Frequency discrimination is a prominent feature of hearing and its physiologic basis has been studied extensively because it is assumed to play an important role in discrimination of natural sounds. In this section, I will therefore first discuss the representation of frequency in the auditory nervous system as a place code and as a temporal code and thereafter I will discuss the relative importance of these two different ways to code frequency for discrimination of complex sounds.

The neural code of complex sounds undergoes more extensive transformations than that evoked by pure tones. However, much more is known about responses to tones than to complex sounds. The first part of this chapter will be devoted to the neural representation of simple sounds such as tones and clicks and subsequent sections will discuss neural coding of complex sounds such as tones and broad band sounds the frequency or amplitude of which varies at different rates.

Frequency, or spectrum, however, is only one feature of complex sounds. The representation in the nervous system of different other features of natural sounds that are the basis for our ability to discriminate a wide variety of sounds has also been studied. The way sounds change is an important feature of natural sounds and changes in frequency and amplitude of sounds are accentuated in the neural processing of the classical ascending auditory nervous system. Our ability to discriminate changes in the spectrum of complex sounds is also prominent and this ability is assumed to be essential for discrimination of speech.

Changes in frequency (spectrum) and amplitude are prominent features of natural sounds that are important for distinguishing between different sounds. Studies of coding of complex sounds in the auditory nervous system have therefore focused on processing of sounds the frequency and amplitude of which change more or less rapidly. The sounds that are

discussed in this chapter are similar to important natural sounds such as speech sounds but better defined. We will also in this chapter discuss the neurophysiologic basis for directional hearing and the physiological basis for perception of space.

3. REPRESENTATION OF FREQUENCY IN THE AUDITORY NERVOUS SYSTEM

We can discriminate very small changes in the frequency of a tone. In fact even moderately trained individuals can detect the difference between a 1,000-Hz tone and a 1,003-Hz tone (three tenths of 1% difference in frequency). The enormous sensitivity of the human auditory system to changes in frequency has aroused many investigators' curiosity and much effort has been made to determine the mechanism by which the ear and the auditory nervous system discriminate such subtle differences in the frequency of a tone.

3.1. Hypotheses about Discrimination of Frequency

Two hypotheses have been presented to explain the physiologic basis for discrimination of frequency. One hypothesis, the place principle, claims that frequency discrimination is based on the frequency selectivity of the basilar membrane resulting in frequency being represented by a specific place in the cochlea and subsequently, throughout the auditory nervous system. The other hypothesis, the temporal principle, claims that frequency discrimination is based on coding of the waveform (temporal pattern) of sounds in the discharge pattern of auditory neurons, known as phase locking (Fig. 6.1). There is considerable experimental evidence that both the spectrum and the time pattern of a sound are coded in the responses of neurons of the classical ascending auditory nervous system including the auditory cerebral cortices.

The concept that certain features of a sound are coded in the discharge pattern of neurons in the auditory system means that these features can be recovered

---

BOX 6.1

COMPLEX SOUNDS

Complex sounds are sounds that have their energy distributed over a large part of the audible frequency range and the amplitude and the frequency distribution varies more or less rapidly over time. Most natural sounds are complex sounds. Communication sounds such as speech sounds are examples of complex sounds.

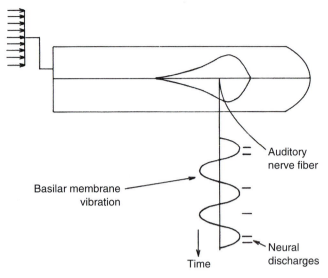

Basilar membrane vibration

Auditory nerve fiber

Neural discharges

Time

FIGURE 6.1 Schematic illustration of the two representations of frequency in the auditory nerve (reprinted from Møller, 1983, with permission from Elsevier).

by analyzing the discharge pattern of neurons in the auditory nervous system. The presence of a certain type of information in the nervous system does not mean that it is utilized for sensory discrimination.

The understanding that place and the temporal representation of frequency can be demonstrated throughout the auditory nervous system, however, does not resolve the question about which one (or both) of these two principles of coding frequency or spectrum is the basis for discrimination of the frequency of sounds. I will discuss the physiological basis for frequency discrimination in more detail in subsequent sections of this chapter.

Studies for the development of the vocoder [39] (see Chapter 11, p. 271) more than half a century ago have shown that speech intelligibility can be achieved using only the (power) spectrum. More recently studies have shown that speech intelligibility can be achieved by either the information about the spectrum of sounds [140] or the temporal pattern [269]. Some modern cochlear implants use only information about the spectrum and achieve good speech intelligibility (see Chapter 11). This indicates that the temporal and the place coding may represent a form of redundancy.

BOX 6.2

FREQUENCY AND SPECTRUM

Frequency and spectrum of sounds are terms that sometimes are used synonymously for describing the physical properties of sounds. While the term frequency of sounds is reserved for pure tones or trains of impulses, the term spectrum is used to describe the properties of sounds that have energy in a certain frequency range. When the spectra of sounds are discussed in Chapter 3 and this chapter, it usually refers to the power spectrum. The power spectrum is a measure of the distribution of the energy of a sound as a function of the frequency. The power spectrum provides an incomplete description of the spectral properties of sounds. The spectrum of a sound can be completely described by a real and an imaginary number for each frequency. The power spectrum is the sum of the squared real and imaginary values of the spectrum. The spectrum of a sound can be obtained from its waveform by a mathematical operation known as the Fourier transformation. Inverse Fourier transformation of a spectrum described by real and imaginary components can reconstruct the waveform. The waveform of a sound cannot be reconstructed from the power spectrum because it is an incomplete description of a sound.

All practical spectral analysis provides measures of the energy in certain frequency bands with finite width and integrated over a certain finite time. An approximation of the spectrum of sounds can be obtained by applying the electrical signal from a microphone to a bank of filters the center frequencies of which are distributed over the range of frequencies of interest. The energy of the output of each filter displayed as a function of the filter's center frequencies is an approximation of the power spectrum. This is similar to the frequency analysis that takes place in the cochlea, with the important difference that the spectrum analysis in the cochlea is non-linear whereas the spectral analysis of sound that is made by equipment or computers is linear.

There is a limitation regarding the relationship between the width of the frequency bands within which the energy is obtained and the time over which the energy is integrated. Thus, obtaining accurate measures of the energy within a narrow frequency band requires a longer observation time than obtaining the energy within a broader band. This means that the product of time and bandwidth is a constant.

3.2. Frequency Selectivity in the Auditory Nervous System

Frequency tuning of single neurons is prominent at all levels of the classical ascending auditory nervous system. Auditory nerve cells of the nuclei of the ascending auditory nervous system and those of the auditory cerebral cortex all show distinct frequency selectivity. This frequency selectivity originates in the frequency selective properties of the cochlea and neural processing in nuclei of the ascending auditory pathways modifies the cochlear frequency selectivity.

Frequency tuning of auditory nerve fibers, cells in nuclei and fiber tracts of the auditory ascending pathways including the cerebral cortex can be demonstrated in animal experiments using several different methods, but it has been studied most extensively in recordings from single nerve cells or nerve fibers using pure tones as stimuli. Frequency threshold curves that map the response areas of neurons with respect to frequency are the most commonly used descriptions of frequency selectivity in the auditory nervous system.

Studies of the response from single auditory nerve fibers lend a window to the function of the cochlea, without having to disturb the function of the cochlea. Such studies can be performed in animals using standard electrophysiological equipment. The discharge pattern of single auditory nerve fibers is controlled by the excitation of inner hair cells and a minimal amount of signal transformation is involved in that process. This is in contrast to the response from cells in the nuclei and fibers of the ascending auditory pathways, and the auditory cerebral cortices where considerable signal processing occurs, thus transforming the response pattern in various ways. The shape of the frequency tuning curves obtained by recordings from cells in the different nuclei are therefore different from those obtained from fibers of the auditory nerve. This is one of the several signs of the transformation of the frequency tuning that occurs in the classical ascending auditory pathways.

The frequency tuning of auditory nerve fibers is a result of the frequency selectivity of the basilar membrane while the coding of the temporal pattern of a sound is a result of the ability of hair cells to modulate the discharge pattern of single auditory nerve fibers with the waveform of the vibration of the basilar membrane (Fig. 6.1). Each point on the basilar membrane can be regarded as a band-pass filter and the vibration amplitude at different points along the basilar membrane provides information about the spectrum of a sound (see Chapter 3).

Each point along the basilar membrane filters the sound that reaches the ear and the hair cells that are located along the basilar membrane convert the vibration into a membrane potential that controls the discharge pattern of individual auditory nerve fibers. The discharge pattern in auditory nerve fibers thereby becomes modulated with a filtered version of the sound rather than the sound itself. This temporal code of sounds in the discharge pattern of auditory nerve fibers thus includes information about the waveform of the vibration at individual points along the basilar membrane. The temporal pattern of the vibration of the basilar membrane contains information about the spectrum of sounds, as does the distribution of vibration amplitudes along the basilar membrane. This means that there is a redundancy of the representation of the spectrum of sounds in the auditory nerve.

Each auditory nerve fiber (type I, see Chapter 5) innervates only one inner hair cell, and the discharges of a single auditory nerve fiber are thus controlled by the vibration of a small segment of the basilar membrane. This is the basis for the frequency selectivity of single auditory nerve fibers. Auditory nerve fibers discharge spontaneously in the absence of external sounds and increase their discharge rates when the vibration of the basilar membrane exceeds the threshold of the hair cell to which the nerve fiber in question connects. The lowest level of sound that produces a noticeable change in a fiber's discharge rate is regarded to be the fiber's threshold. The threshold of a nerve fiber is lowest at a specific frequency and that is the fiber's characteristic frequency (CF). The frequency range of tones to which a single auditory nerve fiber responds widens with increasing sound intensity (Fig. 6.2). This also means that more nerve fibers are activated as the intensity of a tone is increased above its threshold.

A contour of the frequency-intensity range within which an auditory nerve fiber responds with a noticeable increase in its discharge rate (Fig. 6.2) is known as the nerve fiber's frequency threshold curve or frequency tuning curve. Frequency threshold curves have been the most common way of describing the frequency selectivity of single auditory nerve fibers. When such frequency threshold curves are obtained for a sufficiently large number of nerve fibers, the result is a family of tuning curves that covers the entire range of hearing of the particular animal that is studied (Fig. 6.3). The range of hearing of different animal species differs; therefore, the set of tuning curves obtained in different animals will also differ.

The shape of the tuning curves of auditory nerve fibers tuned to low frequencies is different from those tuned to high frequencies but the shape of tuning curves that have similar CF are similar. Nerve fibers that are tuned to high frequencies have asymmetric tuning curves, with the high frequency skirt being

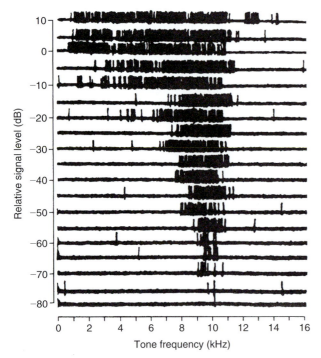

FIGURE 6.2 Illustration of the frequency selectivity of a set of auditory nerve fibers in a guinea pig. The nerve impulses elicited by a tone, the frequency of which is changed from low frequencies to 16 kHz (horizontal scale), are shown. The different rows represent responses to tones of different intensities (given in arbitrary decibel values) (reprinted from Evans, 1972, with permission from The Physiological Society (London)).

very steep and the low frequency skirt much less steep. Nerve fibers that are tuned to low frequencies have tuning curves that are more symmetrical.

The most common ways of studying frequency tuning of auditory nerve fibers has been by obtaining

frequency threshold curves such as those in Fig. 6.3. When different measures of neural activity are used, the frequency tuning of auditory nerve fibers appears differently from threshold tuning curves. Thus, the shape of curves that show a nerve fiber's firing rate as a function of the frequency of a tone stimulus is different from that of frequency tuning curves of auditory nerve fibers (Fig. 6.4A). In a few studies phase-locking of neural discharges has been used to determine the frequency selectivity of auditory nerve fibers in a large range of sound intensities (see p. 99). Yet another method to determine the frequency selectivity of an

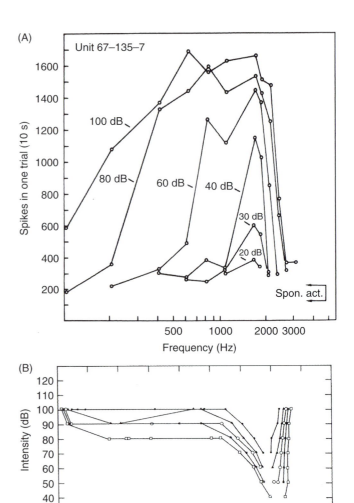

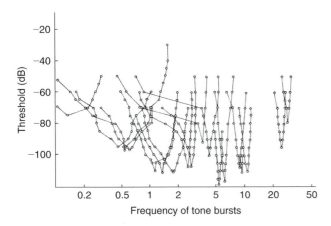

FIGURE 6.3 Typical frequency threshold curves of single auditory nerve fibers in a cat. The different curves show the thresholds of individual nerve fibers. The left-hand scale gives the thresholds in arbitrary decibel values and the horizontal scale is in kHz (reprinted from Kiang et al., 1965, with permission from MIT Press).

FIGURE 6.4 (A) Number of discharges per trial of an auditory nerve fiber of a squirrel monkey stimulated by tones of 10-s duration, shown as a function of the frequency of the tones. The different curves represent sounds of different intensities (in arbitrary decibels) (reprinted from Rose et al., 1971, with permission from The American Physiological Society). (B) Iso-rate curves of the responses from an auditory nerve fiber of a squirrel monkey (reprinted from Geisler et al., 1974, with permission from The American Physiological Society).

auditory nerve fiber determines the sound level required to evoke a certain increase in the firing rate of a single auditory nerve fiber (iso-rate curves). That method also yields tuning curves that are different from frequency threshold curves (Fig. 6.4B).

The non-linear vibration of the basilar membrane (Chapter 3) and the non-linear properties of the neural transduction in hair cells make the conversion of the mechanical stimulation of hair cells into the discharge rate of single auditory nerve fibers to become non-linear. Insufficient understanding of how the firing rate of single auditory nerve fibers are related to the displacement of the basilar membrane complicates interpretation of the results of studies of the frequency selectivity of the auditory system that use different experimental methods.

When two tones are presented at the same time, specific interactions between the two tones may occur. For example, the response elicited by a tone at a fiber's CF can be inhibited (suppressed) by another tone when that tone is within a certain range of frequency and intensity (Fig. 6.5). Inhibitory frequency response areas thus surround the response areas of each auditory nerve fiber. The discharge rate of the response elicited by a tone within the fiber's response area decreases when a second tone with frequency and intensity within one of these inhibitory areas is presented. Such inhibitory areas are usually located on each side of a fiber's (excitatory) response area.

3.3. Cochlear Non-linearity Is Reflected in Frequency Selectivity of Auditory Nerve Fibers

The non-linearity of the basilar membrane motion causes its frequency selectivity to depend on the intensity of sounds that reaches the ear. Cochlear non-linearity that was discussed in Chapter 3 (p. 44), is reflected in the tuning of auditory nerve fibers.

The tuning of auditory nerve fibers broadens at high sound intensities as shown in studies where the frequency selectivity of auditory nerve fibers was determined by analyzing the discharge pattern in response to broad band noise [41, 42, 179, 180]. These studies showed that the frequency selectivity decreased when the intensity of the test sounds was increased above threshold. The reason that the frequency selectivity of single auditory nerve fibers is intensity dependent is the non-linearity of the vibration of the basilar membrane. These studies made use of the fact that the temporal pattern of discharges of single auditory nerve fibers is modulated by the waveform of low frequency sounds and that made it possible to determine the filter function of the basilar membrane over a large range of sound intensities. Analyzing the discharge pattern of single auditory nerve fibers [46, 179, 180] yields measures of the spectral filtering that precedes impulse initiation in auditory nerve fibers.

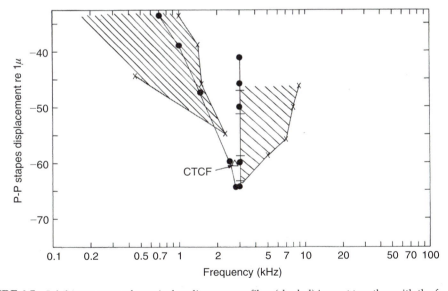

FIGURE 6.5 Inhibitory areas of a typical auditory nerve fiber (shaded) in a cat together with the frequency threshold curve (filled circles). The inhibitory areas were determined by presenting a constant tone at the characteristic frequency of the nerve fiber (marked CTCF) together with a tone, the frequency and intensity of which were varied to determine the threshold of a small decrease in the neural activity evoked by the constant tone (CTCF) (reprinted from Sachs and Kiang, 1968, with permission from the American Institute of Physics).

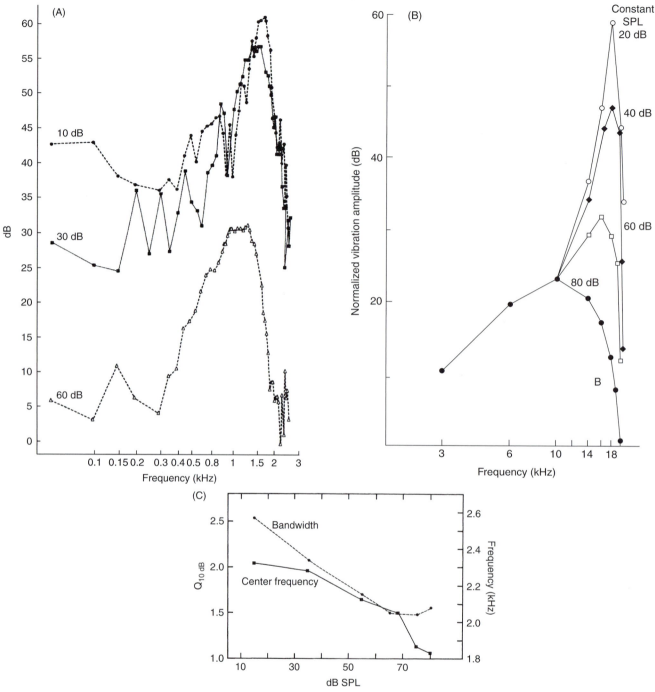

FIGURE 6.6 Comparison between the tuning of a single auditory nerve fiber in a rat (A) and that of the basilar membrane (B) in a guinea pig. (A) Estimates of frequency transfer function of a single auditory nerve fiber in a rat at different stimulus intensities (given in dB SPL), obtained by Fourier transforming cross-correlograms of the responses to low-pass-filtered pseudorandom noise (3.4 kHz cutoff). The amplitude is normalized to show the ratio (in dB) between the Fourier transformed cross-correlograms and the sound pressure and the individual curves would have coincided if the cochlear filtering and neural conduction had been linear (reprinted from Møller, 1999; modified from Møller, 1983, with permission from Elsevier). (B) Vibration amplitude at a single point of the basilar membrane of a guinea pig obtained using pure tones as test sounds at four different intensities. The amplitude scale is normalized, and the individual curves would have coincided if the basilar membrane motion had been linear (reprinted from Johnstone et al., 1986; based on results from Sellick et al., 1982, with permission from the American Institute of Physics). (C) The shift in the center frequency (solid lines) and the width of the tuning of a single auditory nerve fiber (dashed line) in the auditory nerve of a rat as a function of the stimulus intensity. The width is given a "$Q_{10\,dB}$" which is the center frequency divided by the width at 10 dB above the peak (reprinted from Møller, 1977, with permission from the American Institute of Physics).

BOX 6.3

$Q_{10\ dB}$

The width of the frequency tuning of single auditory nerve fibers has been expressed in "$Q_{10\ dB}$" values. The $Q_{10\ dB}$ is the center frequency divided by the width measured 10 dB above the threshold, thus an inverse measure of the broadness of the tuning. The $Q_{10\ dB}$, increases gradually with increasing sound intensity (Fig. 6.6C).

The frequency selectivity of the basilar membrane (Fig. 6.6B) and that of auditory nerve fibers obtained at the same sound intensity (Fig. 6.6A) were remarkably similar and it was evident that these two measures of auditory frequency selectivity change in a similar way when the sound intensity was changed over a large range of sound intensities [179].

It is not only the width of the tuning of the basilar membrane and auditory nerve fibers that change with sound intensity but also the frequency to which the basilar membrane and auditory nerve fibers are tuned shifts when the stimulus intensity is changed (Fig. 6.6C). The shift towards lower frequencies of auditory nerve fibers' CF is also gradual and it occurs over a large range of sound intensities, from near the threshold to above the physiological sound levels (approximately 75 dB above the threshold) [180].

The tuning of the basilar membrane and that of auditory nerve fibers can be displayed in different ways. In Fig. 6.6 the tuning of an auditory nerve fiber [179] and that of the basilar membrane [267] are both shown in a comparable way. If the frequency selectivity were a linear function, the individual curves for the different sound intensities would coincide. They obviously do not do that and this is another indication of the non-linearity of cochlea. The fact that the curves of the response to high sound intensities appear below the curves of the response to sounds of lower intensities is a result of the gain of the cochlear amplifier being lower for high sound intensities than near threshold. The shift of the individual curves can also be interpreted as a sign of the amplitude compression (automatic gain control [AGC]) that occurs in the cochlea (see p. 47).

BOX 6.4

VULNERABILITY OF THE FREQUENCY SELECTIVITY OF SINGLE AUDITORY NERVE FIBERS

Evans' [47] findings were probably the first published evidence that the frequency selectivity of single auditory nerve fibers is physiologically vulnerable. Evans showed that the frequency tuning curves lost their tip when the animal from which they were recorded was exposed to anoxia (Fig. 6.7). Similar changes were seen after poisoning of the cochlea with, for instance, furosemide (a diuretic that is ototoxic). These results, however, were then interpreted as a sign of the presence of a "second filter" that would normally sharpen the tuning of the basilar membrane. As was discussed in Chapter 3 these changes in frequency tuning were caused by loss of the function of the outer hair cells that normally act as "motors" [22].

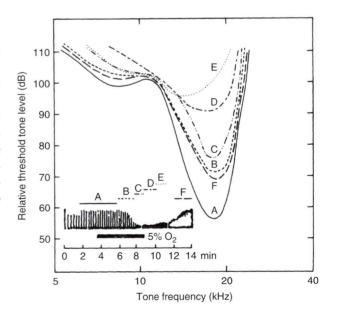

FIGURE 6.7   The effect of anoxia on the frequency threshold curves of an auditory nerve fiber in a guinea pig. The insert shows the amplitude of the CAP recorded from the round window of the cochlea and obtained when the different tuning curves were obtained (reprinted from Evans, 1975).

When frequency threshold tuning curves of auditory nerve fibers were first obtained in the end of the 1950s [105], it was a surprise that the tuning of auditory nerve fibers was very sharp, and thus much sharper than the tuning of the basilar membrane as it was known at that time (using data from von Békésy's studies of human cadaver ears, see [7]). Although it was believed that the tuning of the basilar membrane was the source of the frequency tuning of single auditory nerve fibers, the studies of the responses from single auditory nerve fibers suggested that some kind of sharpening of the basilar membrane tuning occurred before its vibrations were converted into a neural code [49]. Several mechanisms for sharpening of neural tuning were suggested but none were ever supported by results of experimental studies. Eventually, much later, it was shown that the discrepancy between the sharpness of tuning of the basilar membrane and auditory nerve fibers was a result of non-linearity of the basilar membrane [22].

As was discussed in Chapter 3, early measurements of the vibration of the basilar membrane were done at very high sound levels (by von Békésy [7]) while the tuning curves of the auditory nerve fibers were obtained at very low sound levels [105]. When it became possible to measure the vibration of the basilar membrane at low sound levels, its tuning was found to be as sharp as the tuning curves of auditory nerve fibers (Fig. 6.6B) [97]. That frequency threshold curves of single auditory nerve fibers are sharp indicating a high degree of frequency selectivity can therefore be explained by the non-linear behavior of the basilar membrane where outer hair cells are active elements that sharpen basilar membrane tuning (see Chapter 3).

It is not possible to record from single auditory nerve fibers in humans but estimates of the cochlear tuning in humans can be obtained by recording of the ECoG potentials from the ear in connection with masking (two tone masking [32]).

Obtaining psychoacoustic tuning curves has been used to study the effect of injuries to cochlear hair cells in humans and in animals [77] confirming that cochlear tuning becomes broader when hair cells are injured by administration of ototoxic substances (Kanamycin) (see p. 51). Comparison between the results obtained using ECoG methods and psychoacoustic methods shows good agreement, and the obtained tuning curves are similar to those obtained in recordings from single auditory nerve fibers. Interestingly, simultaneous masking and forward masking gave different results in individuals with hearing loss while in individuals with normal hearing the results of the two methods were similar [77].

3.4. Frequency Tuning in Nuclei of the Ascending Auditory Pathways

When studied using conventional methods (frequency threshold tuning curves), practically all cells in all of the nuclei of the ascending auditory pathways including the auditory cerebral cortex show clear frequency selectivity. Frequency selectivity thus seems to be a prominent feature of the responses of single nerve cells in all the nuclei of the classical ascending auditory pathways. Most of the cells in the cochlear nuclei have tuning curves the shapes of which are similar to those of auditory nerve fibers (Fig. 6.8A), but some cells have tuning curves of different shapes (Fig. 6.8B). The shapes of tuning curves of cells of more centrally located auditory nuclei vary more. The difference is greatest in neurons in the auditory cortex but large variations in the shape of frequency tuning curves is also seen in neurons of the superior olivary complex

---

BOX 6.5

COCHLEAR FREQUENCY TUNING DETERMINED USING MASKING

The use of masking to determine the tuning of the cochlea is based on the assumption that a weak tone activates only a few auditory nerve fibers. To obtain a tuning curve the ECoG response (compound action potentials [CAP] from the auditory nerve or the auditory brainstem responses [ABR]) to a weak tone (a few decibels above threshold) is recorded while a masking tone is applied. The intensity of the masking tone is adjusted so that the test tone evokes a reduced response (e.g., two-thirds of the response without a test tone). The test tone and the masker are presented as short tone bursts, and the masker is usually applied immediately before the test tone (forward masking), but it can also be applied at the same time as the test tone (simultaneous masking). This procedure can be used in animals [32] as well as in humans using recordings of ECoG potentials and ABR. Similar measures of frequency selectivity can be obtained in humans using behavioral methods [333] (psychoacoustic tuning curves) [77].

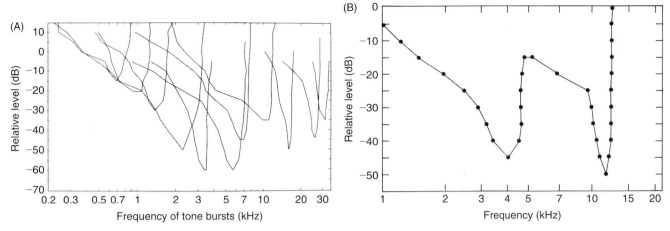

FIGURE 6.8   Frequency threshold tuning curves from cells in the cochlear nucleus of the rat. (A) Frequency threshold curves that are similar to those of auditory nerve fibers (reprinted from Møller, 1969. Unit responses in the cochlear nucleus of the rat to pure tones. *Acta Physiol. Scand.* 75, 530–541, with permission from Blackwell Publishing Ltd). (B) Frequency threshold tuning curves from cells in the cochlear nucleus of the rat with a different shape (reprinted from Møller, 1983, with permission from Elsevier).

(SOC) (Fig. 6.9), the inferior colliculus (IC) (Fig. 6.10) and the medial geniculate body (MGB). Tuning of neurons in the IC is generally much sharper than tuning of auditory nerve fibers (Fig. 6.10) but there are also cells that have much broader tuning than those of auditory nerve fibers.

The diversity of the shapes of the frequency tuning curves from different nuclei can be explained by different degrees of convergence of nerve fibers onto a single nerve cell and the interplay between inhibitory and excitatory influence on a neuron. The convergence of excitatory input may result in broadening of the

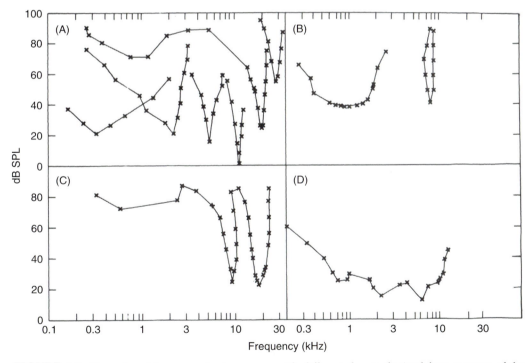

FIGURE 6.9   Examples of frequency tuning curves with different shapes obtained from neurons of the superior olivary complex of the cat (reprinted from Guinan et al., 1972, with permission from the *Journal of Neuroscience*).

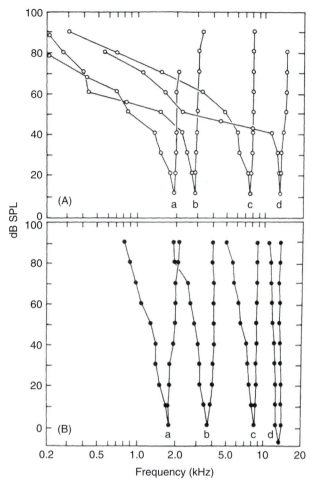

FIGURE 6.10 Frequency tuning in the auditory nerve (A) compared with tuning of some sharply tuned neurons in the inferior colliculus (B) of the cat (reprinted from Suga, 1995 data from Katsuki et al., 1958, with permission from the American Physiological Society).

tuning of a nerve cell that receives its input from many excitatory nerve fibers that are tuned to different frequencies. The interplay between inhibitory input and excitatory input may sharpen the tuning by mechanisms known as lateral inhibition.[1] Sharpening of frequency tuning has been demonstrated in neurons of the MGB by gamma amino butyric acid (GABA[2]) mediated inhibition [219]. The complex pattern of inhibitory, excitatory and facilitatory response areas of neurons in the IC [43] is illustrated in Fig. 6.11.

[1]Lateral inhibition is a term borrowed from vision to explain enhancement of contrast and it is used in connection with the somatosensory system to explain sharpening of sensory response areas on the skin.

[2]GABA is a common inhibitory neurotransmitter in the central nervous system.

3.5. Tonotopic Organization in the Nuclei of the Ascending Auditory Pathways

The different nerve cells of the ascending auditory nervous system are organized anatomically in an orderly fashion according to the frequency to which they are tuned and nerve cells tuned to similar frequencies are located anatomically close to each other. This is known as tonotopic organization. Maps showing the frequency to which neurons are tuned can be drawn on the surface of nuclei as well as in sections of the nuclei of the classical ascending auditory

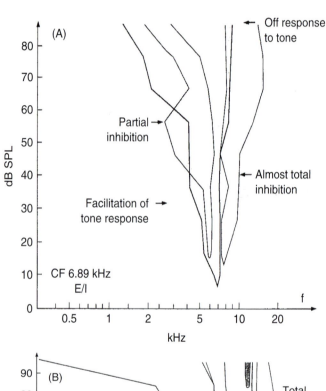

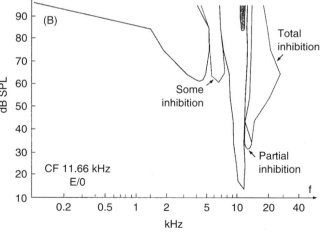

FIGURE 6.11 Four different types of tuning curves found in the inferior colliculus (reprinted from Ehret, G. and Romand, R. 1997. *The Central Auditory Pathway*. New York: Oxford University Press, with permission from Oxford University Press).

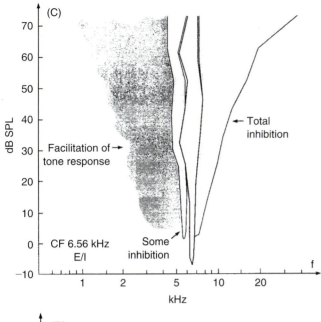

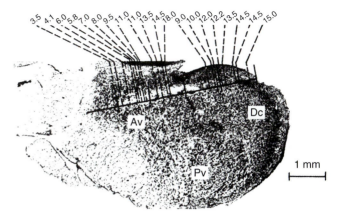

FIGURE 6.12 Anatomical organization of neurons in the cochlear nucleus in the cat according to the frequency to which they are tuned (tonotopic organization). Dc = dorsal cochlear nucleus; Pv = posterior ventral cochlear nucleus; Av = anterior ventral cochlear nucleus (reprinted from Rose et al., 1959, with permission from Johns Hopkins University Press).

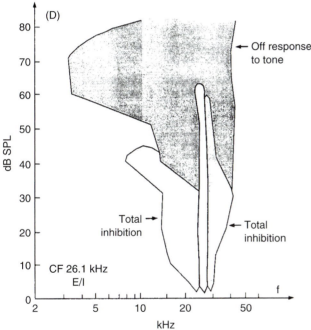

FIGURE 6.11 *(Continued)*

pathways (Fig. 6.12). Also the auditory cortex is anatomically organized according to the frequency to which neurons are tuned (tonotopic organization). These tonotopic maps depend on the separation of sound on the basis of their frequencies that occurs in the cochlea, but they are altered through the processing that occurs in the nuclei of the ascending auditory pathways and the cerebral cortex. The functional importance of the tonotopic organization is unknown

but its prominence and consistency have supported the hypothesis that frequency tuning plays an important role for auditory discrimination (see p. 113).

The nervous system is plastic and its function can change as a result of stimulation or by deprivation from stimulation. An example of that is the change in neural tuning that has been shown to occur in cells of the cerebral auditory cortex. In studies in animals it has been demonstrated that frequency tuning depends on previous exposure to sounds (Fig. 6.13) [111, 12, 319]. Neural tuning along the neural axis of the classical ascending auditory pathways is thus not only different in the different nuclei but expression of neural plasticity is another source of variability in the responses of nerve cells in the ascending auditory pathways, including frequency tuning.

The input to cells in the nuclei of the ascending auditory pathways, and the cerebral cortex, is mediated through synapses. Activation of a cell therefore depends both on the activity in the fibers that impinge on a cell and on the efficacy of the synapses that connect the fibers to the cells. Plastic changes consist of establishment of new connections or elimination of existing connections. Changes in the efficacy of synapses are an important form of neural plasticity. The efficacy of these synapses is subject to change by external and internal processes (expression of neural plasticity), and the response of nerve cells to the same stimuli may therefore differ depending on the degree of expression of neural plasticity.

The maps such as those shown in Fig. 6.12 are not static but can be altered through the expression of

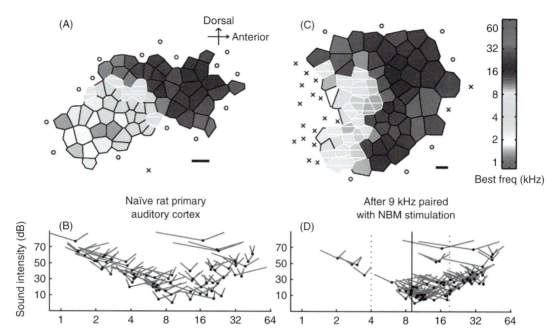

FIGURE 6.13 Illustration of how cortical maps depend on previous sound exposures. The results were obtained in rats in recordings from the primary auditory cortex (A1). (A) and (B) no previous sound exposure, (C) and (D) after exposure to 9 kHz tones simultaneously with electrical stimulation of nucleus basalis (which promote expression of neural plasticity). Penetrations that were either not responsive to tones (O) or did not meet the criteria of A1 responses (X) were used to determine the borders of A1. Each polygon in (A) and (C) represents one electrode penetration. (B) and (D) Tuning curve tips at every A1 penetration indicating the CF, threshold, and receptive field width 10 dB above the threshold for neurons recorded at each penetration. Scale bar, 200 μm (modified from Kilgard and Merzenich, 1998, with permission from the American Association for the Advancement of Science).

neural plasticity. This especially is the case for maps of the cerebral cortex (Fig. 6.13) [44, 111]. The changes in these maps were induced by sound stimulation that was paired with electrical stimulation of the nucleus basalis, which provides arousal and facilitates expression of neural plasticity of the sensory cortex (see Chapter 5, p. 89). The changes that were induced decreased the number of cells that responded best to low frequency sounds and thus shifted the representation of frequency over the surface of the auditory cortex. Similar changes in the spatial representation are seen in other sensory systems [91, 187].

3.6. Extraction of Information from Place Coding of Frequency

The consistency of the anatomical organization of neurons according to the frequency to which they respond (tonotopic organization) suggests that frequency (spectral) tuning is important for extraction of spectral information about a sound. I will discuss this matter in connection with cochlear implants (Chapter 11).

4. CODING OF TEMPORAL FEATURES

There is considerable evidence that the auditory nerve supplies the nervous system with a neural code that is phase locked to the time pattern of the vibration of the basilar membrane, thus band pass filtered versions of the sound that reaches the ear. We know that the temporal pattern (waveform) can be recovered experimentally in recordings from single auditory nerve fibers but little is known about how the nervous system may decode temporal information so that it may be used for discrimination of frequency. This, however, does not prove that information about the frequency of sound is actually extracted from the phase-locked neural responses. (I will discuss the importance of the place and temporal coding of frequency in detail later in this chapter, p. 112).

Temporal coding of frequency has been studied to a lesser degree than frequency tuning in the auditory nervous system and many questions regarding the importance of temporal coding of sounds remain unanswered. It has been questioned whether accurate timing is preserved through synaptic transmission and it is not known how the temporal code of the time pattern of a sound is decoded in the nervous system.

4.1. Coding of Periodic Sounds

The time locking of neural discharges to the waveform of a sound is known as phase locking. Studies of coding of the temporal pattern of sounds have revealed that phase locking is prominent in the auditory nerve. Phase-locking means that more nerve impulses are delivered at a certain phase of the sound than at other phases. Averaging the recorded neural activity to many cycles of a tone is necessary to demonstrate phase locking to a pure tone. Practically, that is done by compiling a period histogram of the responses from single auditory nerve fibers or cells in the nuclei of the ascending auditory pathways. The duration of one period of the sound is divided into a series of bins and the number of nerve impulses that fall into each bin is counted.

The discharges of single nerve fibers are time locked to the waveform of a sound that is within the fiber's response area, at least for frequencies below 5 kHz, but probably even for higher frequencies. Period histograms of the responses to low frequency tones have the shape of half wave-rectified sine waves (Fig. 6.14) [6].

Phase locking of the discharges of single auditory nerve fibers to complex periodic sounds can also be demonstrated. Thus, period histograms of the response to sounds that are the sum of two pure sine waves (tones) of different frequencies have forms similar to the wave shapes of the half wave rectified sound waves (Fig. 6.15) [251]. The two sine waves

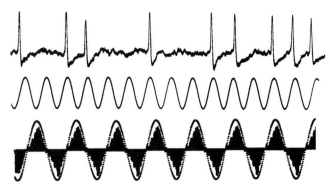

FIGURE 6.14 Phase-locking of discharges in a single guinea pig auditory nerve fiber to a low-frequency tone (0.3 kHz), near threshold (reprinted from Arthur et al., 1971, with permission from the Physiological Society (London)).

must be multiples of each other to get a waveform that repeats itself accurately, and the period histograms are compiled over the period of such a waveform.

Phase locking to pure tones is prominent in many cells of the cochlear nucleus, more so in the ventral cochlear nucleus than the dorsal cochlear nucleus. Time-locking to pure tones and particularly to repetitive clicks can also be observed in many cells of the inferior colliculus and the medial geniculate body. The upper frequency of phase-locking is lower than it is in the auditory nerve. In the MGB it is rarely observed at higher rates than 800 clicks per second [253].

Phase-locking of the discharges of auditory nerve fibers can be demonstrated in response not only to two pure tones [251] but it is also prominent in response to complex sounds. Complex sounds such as speech sounds contain several periodic or quasi-periodic components. The fundamental (vocal cord) frequency is one and other quasi-periodic components are damped

---

BOX 6.6

COMPLEXITY OF EXCITATION OF INNER HAIR CELLS

The histograms in Fig. 6.15 are half wave rectified versions of the basilar membrane vibration because cochlear hair cells are only excited when the basilar membrane is deflected in one direction, namely towards the scala vestibuli (Chapter 3). This, however, is an oversimplification. Some hair cells are in fact excited when the basilar membrane is deflected in the opposite direction and some

are excited when the velocity of the basilar membrane is highest [115]. One of the reasons for the complexity in excitation of the inner hair cells is the active role of outer hair cells [22], the motion of which contributes to excitation of the inner hair cells. Another reason is the viscoelastic coupling between the basilar membrane and the inner hair cells [162, 334] (see Chapter 3).

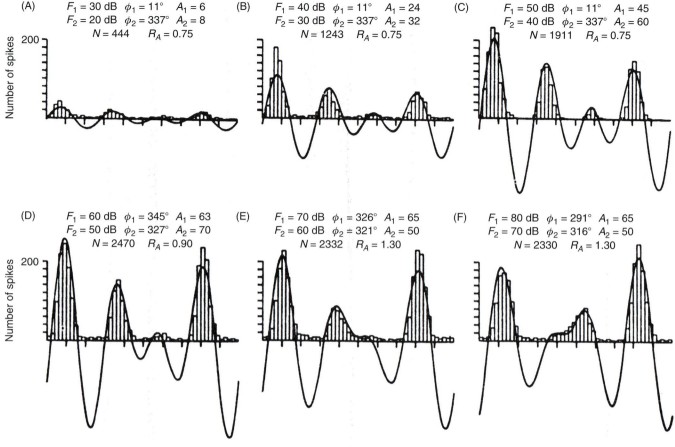

FIGURE 6.15 Period histograms of discharges in a single auditory nerve fiber of a squirrel monkey to stimulation with two tones of different frequencies that were locked together with a frequency ratio of 3:4 and an amplitude ratio of 10 dB. The different histograms represent the responses to this sound when the intensity was varied over a 50-dB range (modified from Rose et al., 1971, with permission from the American Physiological Society).

oscillations the frequency of which is that of the vowel formants. This pattern of damped oscillations is coded in the discharge pattern of auditory nerve fibers [326].

The temporal pattern of a vowel is a mixture of several damped oscillations. In order to determine the formant frequencies on the basis of the temporal pattern it is necessary that each of these damped oscillations are coded independently in a different population of auditory nerve fibers. In the cochlea it is not the sound itself that activates auditory nerve fibers but it is the sound that is filtered by the basilar membrane to which the discharge of auditory nerve fibers phase lock.

BOX 6.7

FORMANTS

The spectrum of a vowel has several peaks, known as formants. Formants are the results of the acoustic properties of the vocal tract and the frequencies of the formants uniquely characterize a vowel. In the time domain, each formant contributes a damped oscillation to the total waveform of a vowel. The frequencies of these damped oscillations are the formant frequencies, and these damped oscillations are repeated with the frequency of the vocal cords, i.e., the fundamental frequency of the vowel in question.

The spectral selectivity of the basilar membrane thus divides the audible spectrum in suitable slices before the waveform is coded in the discharge pattern of auditory nerve fibers [326]. This means that the periodicity of each vowel formant is coded in different populations of auditory nerve fibers. This is known as "synchrony capture" and it enables different populations of auditory nerve fibers to carry the periodicity of different spectral components of a sound. This separation of spectral components may be the most important feature of the frequency selectivity of the basilar membrane (discussed in more detail later in this chapter, p. 118).

The increase in the width of the cochlear filter with increasing stimulus intensity may impair the separation of vowel formants before coding the waveform and thus impair the preservation of phase locking to individual formant frequencies. Such deterioration of frequency acuity of the cochlear filtering may be one

reason why speech discrimination is impaired when the sound intensity is raised above a certain level. Absence of the acoustic middle-ear reflex, which results in the input to the cochlea being greater than normal, has been shown to cause impairment of speech discrimination at high sound intensities (see Chapter 9).

Some nerve cells in the cochlear nucleus fire with great temporal precision in response to transient stimulation such as clicks and tone bursts (Fig. 6.17A) [197] whereas other cells respond with less temporal precision. It is believed that many nerve fibers terminate on the neurons that respond with such great precision and such nerve cells thus act as signal averagers that not only compensate for synaptic jitter but even increase the accuracy of temporal coding of the waveform of the sound stimuli. These neurons respond to transient stimulation with a greater precision than that of their input (from auditory nerve fibers), showing that spatial

---

BOX 6.8

PHASE LOCKING TO BROADBAND SOUNDS IN THE AUDITORY NERVE

Phase locking in auditory nerve fibers can also be demonstrated in response to broad band noise sounds [12, 41, 42]. Since it is necessary to average the responses from a single nerve fiber over a long time, some investigators [179, 181] have used noise that repeats itself many times (pseudorandom noise) in studies of phase locking. Using such noise phase-locking can be readily demonstrated in the discharges from single auditory nerve fibers. The phase-locking does not follow the waveform of the noise sound but it follows the waveform of the noise that has been band-pass filtered by the cochlea. This means that studies of phase locking to noise sounds provide information about the spectral filtering in the cochlea. That fact has been used to determine the properties of the cochlear filters over a large range of sound intensities [179, 180] (see p. 99, and Fig. 6.6A). These studies have demonstrated that phase-locking of auditory nerve impulses occurs over a much larger range of sound intensities than the range over which the average discharge rate increases with increasing sound intensity [46, 180]. The discharge rate of most auditory nerve fibers show a saturation at sound levels as low as 20–30 dB above their threshold. Thus, while the average discharge rate of auditory nerve fibers may be essentially constant

in response to sounds in the entire physiological range of sound intensities, phase locking can be demonstrated over the entire physiological range of sound intensities (Fig. 6.16).

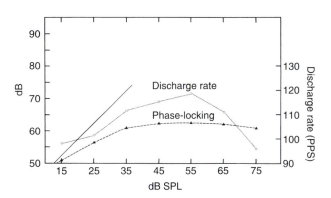

FIGURE 6.16 Average discharge rate as a function of stimulus intensity (dotted line) of an auditory nerve fiber in a rat together with a measure of the fiber's ability to phase lock to the stimulus sound (low-pass filtered noise), shown as a function of sound intensity (reprinted from Møller, 1977, with permission from Elsevier).

integration in the nervous system can improve temporal precision. Specific cells in the anterior ventral cochlear nucleus (AVCN) (bushy cells) have been shown to fire with greater precision in response to low frequency tones than do fibers of the auditory nerve (Fig. 6.17B) [98]. That means that the temporal precision of the discharge of some neurons in the cochlear nucleus has increased rather than decreased as a result of synaptic transmission between auditory nerve fibers and cells in the cochlear nucleus.

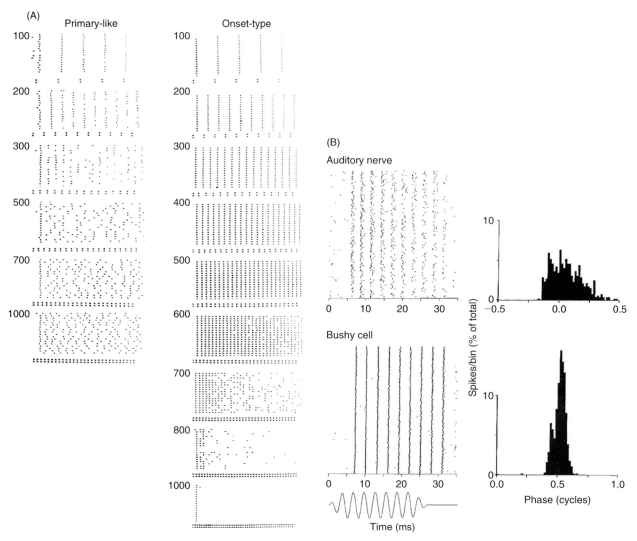

FIGURE 6.17   (A) Discharge pattern of two types of neurons in the cochlear nucleus of a rat in response to clicks of different repetition rates. Each nerve impulse is indicated by a dot and the stimulus clicks are indicated by pairs of dots below the responses. The stimulus clicks were presented in 50-ms long bursts (reprinted from Møller, A.R. 1969. Unit responses in the rat cochlear nucleus to repetitive transient sounds. *Acta Physiol. Scand.* 75, 542–551, with permission from Blackwell Publishing Ltd). (B) Comparison of stimulus synchronization in an auditory nerve fiber (a) and a cell in the AVCN (a bushy cell) (b). Each dot in the rasters (left panels) indicates a spike occurrence to a short tone at the cell's best frequency (0.35 kHz in (A) and 0.34 kHz in (B)); each row of dots is the response to one of the 200 repetitions. The responses are phase locked to the stimulus, as seen in the tendency of spikes to occur at a particular phase angle by graphing the response relative to stimulus phase (right panels). Spikes in the bushy cell are temporally less dispersed than in the auditory nerve and occur in each stimulus cycle, whereas cycles are often skipped in the nerve (adapted from Joris et al., 1994, with permission from the American Physiological Society).

4.2. Extraction of Information from the Temporal Pattern of Neural Discharges

The discussion above regarding coding of the temporal pattern of sounds has focused on periodic sounds but it must be emphasized that the nervous system codes non-periodic sounds in the same manner as periodic sounds. Studies of phase locking in single nerve fibers or nerve cells require averaging of the responses to many stimuli to reduce the variability in the discharges of single nerve fibers. The averaging used in recordings from single nerve cells relies on repeating the same sound many times. The nervous system uses another method to reduce the statistical variability in the response pattern of single auditory neurons, namely averaging of the response of many neurons. This yields a result from a single presentation thus unlike studies of neural discharges, which require averaging of the responses to many (identical) stimuli.

This difference between the normal function of the nervous system in extracting temporal information and the methods used in studies of phase-locking has to be taken into account when interpreting the results. It may have caused an underestimation of the ability of the nervous system to process very small time intervals (see p. 117).

In order to use that temporal code to determine the frequency of a sound, the nervous system must determine the interval between nerve impulses that are time locked to individual waves of a sound. Models have been proposed to explain the ability of the nervous system to determine the time between the arrival of a sound at the two ears that is the basis for directional hearing (see p. 112) [89, 100, 300, 322].

Determining the time interval between two waves of a sound wave is more complex than determining the time difference between neural activity that originates in the two ears. However, similar principles and similar neural circuitry could be used for decoding temporal information in sounds. Licklider [137] proposed a variation of Jeffress's model for detecting the time intervals between individual waves of sound. This model is based on auto correlation analysis to determine the intervals between sound waves.

The auto correlation model for decoding time intervals requires a set of delays and multiplies (or coincidence detectors), thus similar to models of directional hearing. For decoding temporal information in sounds the frequency of which is higher than 1 kHz, axons of different lengths may serve as the required variable delay lines. Each axon is assumed to be connected to a nerve cell that also receives input from another axon of a different length, the target nerve cell acting as a coincidence detector. Such an array of axons of different length may be found in the cochlear nucleus, while the neurons that act as coincidence detectors may be located in the medial superior olivary nuclei (MSO) [69]. Other investigators [127] have found evidence from studies in cats that some neurons in the inferior colliculus have different delays thus providing the delay lines needed for determining the frequency of sounds of relatively high frequency.

Determining the frequency of low frequency sounds requires delays that are much longer than what can be accomplished by axons of different length. The delays necessary to explain echolocation in bats are also much longer than the interaural delay that can be generated by axons. The delays associated with echo localization in bats are between 0.4–18 ms. Delays of that length have been assumed to be accomplished by an "inhibitory gate" that has a variable time [292]. Similar mechanisms may be used for determining the frequency of low frequency sound such as the fundamental frequency of vowel sounds.

BOX 6.9

CORRELATION ANALYSIS

Auto correlation analysis is similar to cross correlation assumed to be used for determination of the delay between the sounds that reach the two ears. When auto correlation analysis of a signal is done it is common to sample the signal, multiplying the amplitude of the signal and a replica of the signal, sample by sample, and then adding the values. That process is repeated after the signal and its replica is shifted one sample relative to each other. This is then continued for as many delays as is required.

The resulting auto correlation function appears as series of values that are the function of the delay. This is the way auto correlation analysis is done using digital computers. However, a neural system that performs auto correlation analysis does not obtain the correlation point by point as done by a computer. Instead, the nervous system is assumed to use an array of multipliers, one for each delay. The output of each multiplier neuron provides a continuous signal that is the correlation at one delay as a function of time.

The human ear can discriminate changes of approximately 3 Hz in a 1,000 Hz tone. This means that tones of 1,000 Hz and 1,003 Hz can be differentiated. The difference in the length of one period of 1,000 Hz and a 1,003 Hz tone is three thousandth of the period length of a 1,000 Hz tone, thus 3 µS. This is about the same time difference as can be discriminated in the arrival time of sounds at the two ears (see p. 117).

It has been suggested that neurons in the superior olivary complex may have the ability to detect time intervals. Neurons in the superior olivary complex acting as coincidence detectors can discriminate time differences between the arrival of sounds at the two ears.

The neurons in the ventral nucleus of the lateral lemniscus (VNLL) in the mammalian auditory system also seem to specialize in coding temporal information, as shown in the echolocating bat and dolphin [31]. These cells have no or little spontaneous activity and are broadly tuned. Langner and Schreiner [127] presented evidence that variable delays needed for determining the frequency of a sound from the temporal pattern of nerve impulses may exist in the inferior colliculus (IC). Many neurons in the IC have an intrinsic periodicity of firing (like an oscillator) that might be used as a time base and a variable delay. Suga and his co-workers found that the auditory cortex of the flying bat makes a map from the measured difference between time of the emission of a sound and the arrival of the echo at the bat's ear [290].

The neurons in the cochlear nucleus that respond to transient sounds with a single discharge require a certain duration of silence before they can fire again [197]. This may be an example of inhibition to determine the time interval between sounds, and it may be similar to the inhibition based variable delays suggested by Suga [286] for explaining the bat's discrimination of delays between transmittal and receipt of an echo.

5. IS TEMPORAL OR PLACE CODE THE BASIS FOR DISCRIMINATION OF FREQUENCY?

Discrimination of both tones and complex sounds rely either on the frequency analysis of the basilar membrane (place principle) or on neural analysis of the code of the sounds' temporal pattern (temporal principle) or on a combination of both. While the question about the neural basis for discrimination of frequency was earlier mostly of academic interest, the advent of cochlear and auditory brainstem implants has made understanding of the physiological basis for frequency discrimination of great practical importance (see Chapter 11).

Several criteria must be met to make a specific type of coding of frequency (place or temporal) a candidate for providing the basis for discrimination of frequency by the auditory system. These basic criteria are:

1. The code must be present and accurately preserved in the nervous system.
2. The code must be robust (independent of sound intensity).
3. The code must be interpreted (decoded) in the nervous system.

The reason that the code of frequency must be robust and independent of the sound intensity is related to the general finding that frequency discrimination of both tones and complex sounds (speech sounds and music) is largely independent on the intensity of the sounds. If frequency discrimination would depend on sound intensity it would interfere with discrimination of complex sounds such as speech sounds and music.

While there is ample evidence that both these two representations of frequency are coded in the auditory nervous system, it is not known which one of these two principles is used by the auditory system in the discrimination of natural sounds or for the discrimination of unnatural sounds, such as pure tones in experiments done in the laboratory under more or less natural conditions. It may be that the place and the temporal principle of frequency discrimination may be used in parallel by the auditory system for discrimination of sounds of different kinds.

Designing experiments that can determine whether it is the place principle or the temporal principle that is the basis for frequency discrimination is difficult because the spectral and temporal properties of sounds are closely linked together and the temporal pattern of a sound cannot be manipulated experimentally without also altering its spectrum.

5.1. Temporal Hypothesis for Frequency Discrimination of Complex Sounds

Coding of the temporal pattern of sounds in the discharge pattern of auditory nerve fibers is the basis for the temporal hypothesis of frequency discrimination [128, 137]. The temporal hypothesis has earlier been regarded less important than the place principle for discrimination of frequency. Two reasons have been given. It was assumed that the temporal code could not be preserved in synaptic transmission and it was not known how the temporal code could be decoded. While many studies in animals have shown that nerve impulses in the auditory nerve are phase-locked to the waveform of the basilar membrane vibration, at least up to 5 kHz and probably higher [6, 179, 180], it was

---

BOX 6.10

TONOTOPIC ORGANIZATION IN NEONATAL DEAF ANIMALS

Studies [131, 275] showed that tonotopic (cochleotropic) organization in the inferior colliculus exists in neonatal deaf mice but it can be modified by electrical stimulation of the cochlea. Electrical stimulation of a single location of the cochlea could expand the response areas and degrade the cochleotropic organization.

Competing stimulation of two locations on the basilar membrane, however, maintained the frequency representation of each sector of the inferior colliculus without expanding the response areas. Synaptic activity in the auditory cerebral cortex is abnormal in congenitally deaf cats [116].

---

generally assumed that phase locking deteriorated after the first synapse in the cochlear nucleus. This assumption was supported by the fact that it was necessary to average discharges over considerable time in order to recover the time pattern of sounds. However, these assumptions did not take into account the spatial integration that occurs in neurons that receive many inputs, which in fact improve the accuracy of temporal coding compared to what is present in individual auditory nerve fibers.

5.2. Place Hypothesis for Frequency Discrimination of Complex Sounds

The cochlea normally separates sounds into narrow frequency bands that activate different populations of hair cells and thereby different populations of auditory nerve fibers. This is the basis for the place hypothesis. Individual nerve fibers are tuned to different frequencies, and frequency tuning is a characteristic feature of nerve cells throughout the ascending classical auditory nervous system. Nerve cells in the ascending auditory pathways are anatomically organized according to the frequency to which they are tuned (tonotopic organization).

This is assumed to be the result of the frequency tuning of the basilar membrane.

The fact that neurons in all parts of the classical ascending auditory pathways respond best to sounds of a certain frequency and that neurons are anatomically organized according to the frequency to which they respond best (tonotopic organization) indicates that the place representation of sounds in the cochlea is maintained throughout the ascending classical auditory nervous system. This has been taken as an indication that the nervous system uses place information as a basis for discrimination of frequency.

However, the frequency threshold tuning curves of single nerve cells that have been used to establish the tonotopic organization may not reflect the function of the auditory system under normal conditions because frequency threshold tuning curves are obtained by determining the threshold to pure tones presented in a quiet background thus at very low sound intensities. Tuning curves of auditory nerve fibers obtained by using noise stimuli at intensities within the physiologic range of hearing (Fig. 6.6A) are more representative for the function of the auditory system under normal conditions than frequency threshold curves.

BOX 6.11

REPRESENTATION OF VOWELS IN AUDITORY NERVE FIBERS

Studies of the neural representation of vowels in the responses from single auditory nerve fibers [259, 260, 326] have shown that the discharge rates of large populations of auditory nerve fibers reproduce the formant frequencies of vowels only at low sound intensities. The discharge rates of many auditory nerve fibers collected in the same animal and plotted as a function of the CF have distinct

peaks that correspond to the formants of the vowels only when the vowels were presented at low intensity (Fig. 6.18). These peaks became less distinct as the sound intensity was increased and at physiologic sound levels these peaks were poorly defined. Spectral separation of vowel formants based on the (average) discharge rates of single auditory nerve fibers thus becomes poor at sound intensities

BOX 6.11   (cont'd)

in the range of conversational speech. This means that the place representation of vowel formants cannot be regarded as sufficient to discriminate formant frequencies, which is necessary for discrimination of vowels.

Several studies have shown lack of robustness of the place code of frequency. Studies of the tuning of the basilar membrane using recordings of the cochlear microphonic potential [85] showed a considerable shift in the location of the maximal response. Studies of the tuning of auditory nerve fibers have shown that the frequency to which nerve fibers were tuned shifted and became broader when the sound intensity was increased from threshold levels to the physiologic range of sound intensities (Fig. 6.6A&C) [179, 180, 335].

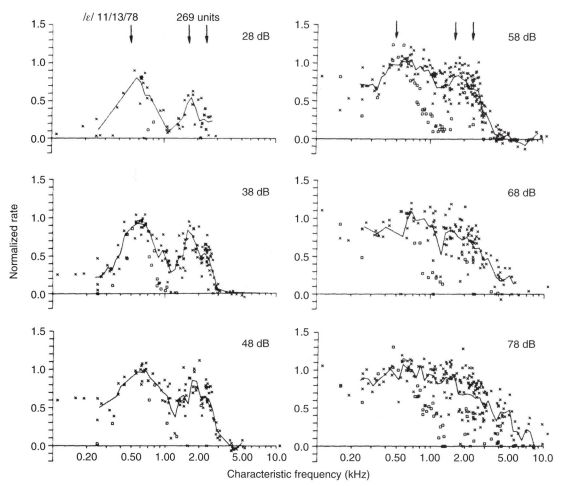

FIGURE 6.18 Normalized discharge rates of auditory nerve fibers in a cat in response to a synthetic vowel /ɛ/ presented at different sound intensities (in dB SPL). The arrows mark the frequency of the three vowel formants (reprinted from Sachs and Young, 1979, with permission from the American Institute of Physics).

BOX 6.12

WHAT PROPERTIES OF BASILAR MEMBRANE TUNING ARE IMPORTANT FOR FREQUENCY DISCRIMINATION?

It has been suggested that it might not be the peak of the envelope of vibration of the basilar membrane that is important for frequency discrimination, but instead the entire envelope or the edges (slopes) of the envelope. It has also been suggested that frequency discrimination may rely on the steep high frequency slope of tuning curves of single auditory nerve fibers corresponding to the slopes (skirts) of the frequency tuning of a single point on the basilar membrane. It has been claimed that the location of the skirts of the frequency tuning curves of the basilar membrane might vary less than the location of the peak of the basilar membrane motion when the sound intensity is changed. The high frequency skirts of frequency

threshold tuning curves of cells in the cochlear nucleus are extremely steep [196] and that might therefore be used by the auditory system to detect differences in the spectrum of sounds and thus be the basis for frequency discrimination according to the place principle.

However, the slope of the high frequency skirts of such functions changes with sound intensity to a similar extent as the shift in the frequency of the peak of frequency threshold tuning curves. The same is the case for the mechanical tuning curves of the cochlea (Fig. 6.6B), thus making the slope of cochlea frequency tuning equally dependent on the sound intensity as the center frequency of the cochlear and auditory nerve tuning.

Since the pitch of sounds changes little with sound intensity [285], the findings that the tuning of the basilar membrane depends on the sound intensity placed serious doubt on the place coding of frequency as a basis for frequency discrimination. Also, frequency discrimination of speech and musical sounds is known to change very little with sound intensity. (For a discussion of the use of the spectrum of sounds for the discrimination of speech sounds see page 118 and chapter 11, page 273.)

5.3. Preservation of the Temporal Code of Frequency

It has been believed that phase locking deteriorates as the information travels along the ascending auditory pathways. The reason is that synaptic transmission implies a certain amount of jitter,[3] and that "blurs" the time pattern of the neural code. However, that assumes that only few nerve fibers terminate on a nerve cell. Nerve cells on which many fibers terminate act as spatial integrators and such cells thereby increase the temporal precision of phase locking. The assumption that phase locking of neural discharges deteriorates in synaptic transmission may be incorrect and the need

to convert the temporal code into a spike rate code or a spatial code is not as urgent as earlier assumed.

Studies have confirmed that synaptic transmission can enhance temporal precision. Thus, neurons with many synapses perform a spatial integration of input and that increases the temporal procession in a similar way as signal averaging enhances signals that are buried in noise.

When many nerve fibers converge on one nerve cell such as occurs in the cochlear nucleus, the result is spatial integration of neural activity and that can enhance temporal precision and thus counteract its deterioration by synaptic jitter (see Fig. 6.17).

Evidence that the temporal coding of frequency is preserved over a large range of sound intensities comes from studies of the temporal code of synthetic vowels by Young and Sachs [326] who found that the temporal structure of such sounds was coded in the discharge pattern of auditory nerve fibers over a large range of sound intensities.

In experimental studies where recordings are made from single auditory nerve fibers the probabilistic nature of the discharges of auditory nerve fibers that "blur" the temporal pattern of the neural code of sounds makes it necessary to average the discharges in the response from a single nerve fiber to many repeated presentations of the same stimuli. The nervous system, however, does not use this method to reduce statistical variability by integrating the discharge patterns of many nerve fibers to obtain a stable response. Thus this (natural) spatial integration in a target neuron can reduce the statistical variability of neural discharges in response to a single stimulus presentation while recordings from only one nerve fiber require the responses of many stimuli to be added to obtain a similar reduction in the statistical

---

[3]Jitter means that the time at which nerve impulses occur vary randomly, thus a lack of temporal precision. Thus, phase-locking to the waveform of low frequency sounds does not occur precisely to a certain phase of the sound. Only the average number of nerve impulses is higher at a certain phase of a sound than at other phases. The random variation around that mean value is a result of the probabilistic nature of synaptic transmission, which occurs in hair cells and cells in the various nuclei of the ascending auditory pathways.

BOX 6.13

TEMPORAL PRECISION CODING OF WAVEFORM SOUNDS

Studies have shown that improvement of temporal precision of coding of the waveform of sounds occurs in certain cells of the cochlear nucleus. These nerve cells fire more precisely than auditory nerve fibers in response to transient sounds (Fig. 6.17) [197]. Some neurons in the cochlear nucleus (bushy cells in the AVCN) have been shown to fire with great precision in response to low frequency tones [98]. As a comparison, the response that the same stimulus evokes in fibers of the auditory nerve show great variations in their firing to that kind of stimulus (right panels in Fig. 6.17B). The discharges of the bushy cell are temporally less dispersed than in the auditory nerve and occur in exactly the same way in each cycle of the stimulus tone, whereas auditory nerve fibers show a great variability in relation to the waveform of the stimulus sound (tone).

These examples show that precision of timing is not only preserved, but also even improved through synaptic transmission. While phase locking of neural discharges in auditory nerve fibers decreases gradually above a certain frequency, spatial integration in neurons of the nervous system may enhance the temporal coding to an extent that compensates for the decrease of phase locking in the auditory nerve.

BOX 6.14

WAVEFORM CODING OF VOWEL SOUNDS IN THE AUDITORY NERVE

The waveform of vowels can be regarded as composed of a series of damped oscillations where the frequency of the oscillations is that of the formants. Histograms of auditory nerve fiber responses to synthetic vowels that show the distribution of discharges over one period of the fundamental frequency of the vowel reveal the periodicity of the damped oscillation that correspond to each formant. The histograms from neurons with CF near the frequency of a formant will show a periodic pattern with the frequency of the formant. The spectra of these histograms show harmonics of the fundamental frequency of the vowel and the formants appear as peaks in the envelope of these spectra (Fig. 6.19). When information such as that displayed in Fig. 6.19 was compiled the formants were coded in the time pattern of the discharges of single auditory nerve fibers over a large range of stimulus intensities for three different vowels.

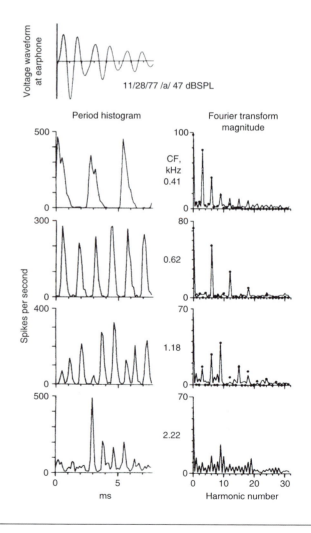

FIGURE 6.19 Period histograms (left column) of the responses from four different auditory nerve fibers of a cat in response to stimulation with a synthetic vowel /a/. Right column shows Fourier transforms of these histograms. The electrical signal applied to the earphone is shown on top (reprinted from Sachs and Young, 1979, with permission from the American Institute of Physics).

variability of the discharges. This difference in the way experimental data are processed and the way the central nervous system extracts information must be considered when experimental data are evaluated.

5.4. Preservation of the Place Code

Ample experimental evidence shows that frequency tuning is preserved throughout the ascending auditory pathways and preserved at least to the different divisions of the cerebral cortex. The frequency representation as demonstrated by tuning curves of cells and fibers changes as information ascends up the neural axis of the ascending auditory pathways. The diversity in width and shapes of tuning curves increases as the information ascends in the auditory pathways. The importance of that is unknown.

5.5. Robustness of the Temporal Code

Temporal coding is also depending on the time it takes for the EPSP to reach the threshold of target neurons [186] and that is expected to make timing of nerve impulses depend on the stimulus intensity. However, studies of directional hearing have revealed that temporal information is transmitted through several synapses without introducing uncertainties. The results of studies of binaural hearing show convincingly that the auditory nervous system can detect very small time intervals (or rather very small differences in time intervals). Time differences in the order of 5 μS between the arrival of sounds at the two ears can be detected (see p. 145). The reason for that is probably the spatial integration that occurs when many nerve fibers impinge on a target neuron.

5.6. Robustness of the Place Code of Frequency

The fact that frequency tuning changes with the intensity of a sound means that the place principle lacks the robustness that is assumed to be necessary to explain psycho-acoustic findings regarding discrimination of frequency (see p. 99) and this makes the place principle an unlikely candidate for the basis of auditory frequency discrimination. Zwislocki [335] has stated "Therefore, if the intensity-dependent shift in the cochlear excitation maximum found in gerbils has a counterpart in human cochleae, as appears likely, the excitation maximum cannot constitute an adequate physiological code for pitch."

5.7. Coding of Speech Sounds

We have already discussed some physiological aspects on processing of speech sounds by the auditory system showing that the temporal properties of vowels are accurately represented in the firing pattern of single fibers of the auditory nerve. Observations regarding the effect of pathologies of the auditory nerve support the importance of such temporal coding.

It is well known that speech discrimination is more impaired in patients with hearing loss from injury of the auditory nerve than it is in individuals with the same threshold elevation from cochlea injury. Studies of patients with vestibular Schwannoma (see Chapter 9) and in patients in whom the auditory nerve has been injured by surgical manipulations (Fig. 9.29) show that such patients have a varying degree of hearing loss but their speech discrimination is always decreased more than expected from their pure tone audiograms.

If frequency discrimination is based on temporal coding of sounds in the auditory nerve, information from different locations along the basilar membrane must appear temporally coherent when it enters the central nervous system so that the interval between individual sound waves can be determined accurately. Injury to the auditory nerve is associated with a decrease in neural conduction velocity that usually differs among the fibers of the auditory nerve as indicated by the broadening of the compound action potentials (CAP) recorded directly from the exposed intracranial portion of the auditory nerve in response to click sounds after surgical manipulation (injury) of the auditory nerve [185]. The uneven neural conduction time in auditory nerve fibers causes an increased temporal dispersion and thus impairs the temporal coherence of nerve impulses that arrive at the neurons of the cochlear nucleus. Impairment of temporal coherence of neural activity in the auditory nerve is believed to be responsible for the impairment of speech discrimination in patients with injuries to the auditory nerve such as from surgical manipulation [185] or from disease processes such as vestibular Schwannoma. Morphological studies have shown that there is only a small variation in the diameters of different auditory nerve fibers in young individuals [281], indicating that the conduction velocity among different auditory nerve fibers varies very little. The variation in axon diameters increases with age and that may explain why some elderly individuals have poor speech discrimination (Fig. 5.3). The poor speech discrimination associated with auditory nerve injuries and aging may thus be a result of impaired coherence of auditory nerve impulses indicating that temporal coding is important for speech discrimination (see Chapter 9).

Vestibular Schwannoma and surgical manipulation of the auditory nerve could also injure the efferent fibers, which would cause a change in the function of the outer hair cells. However, the effect on hearing function from severance of the efferent bundle (done in connection with vestibular neurectomy to treat Ménière's disease) has been found to be minimal [261].

5.8. A Duplex Hypothesis of Frequency Discrimination

Contemporary research indicates that both place and temporal coding are important for frequency discrimination in the auditory nervous system. This was recognized already in 1949 by Wever who presented the volley theory (described in his book [307]) that suggested that both place and temporal coding were used for frequency discrimination. Wever suggested that the temporal coding was most important at low frequencies and that the place coding of frequency was most important at high frequencies. In the mid frequency range both principles would work side by side for frequency discrimination.

5.9. Cochlear Spectral Filtering May Be Important in Other Ways than Frequency Discrimination

The spectral filtering in the cochlea divides the audible spectrum into narrow portions before coding the sound into a pattern of nerve impulses in the auditory nerve. This means that the temporal pattern within limited parts of the spectrum become coded in different populations of nerve fibers. The frequency of the sound within a narrow frequency band can be determined by measuring the time interval between individual waves of the filtered waveform that is coded in the pattern of neural discharges of individual auditory nerve fibers. The cochlea divides the spectrum of sounds into frequency bands of suitable width before the sound is coded into the discharge pattern of individual auditory nerve fibers. This division reduces the requirements regarding coding of details of the waveform of complex sounds. Without such spectral division of the spectrum of natural sounds into narrow frequency bands, coding of the temporal pattern would require coding of fine details of the time pattern of sounds in the discharge pattern of auditory nerve fibers and that likely would exceed the limits for neural coding. The spectral separation also reduces the demand on the neural circuitry that decodes the temporal information. Decoding of the "raw" sound wave would likely overwhelm any neural discriminator of temporal information.

As an example, analysis of the temporal pattern of vowel sounds is more likely to provide accurate information about formant frequencies if the temporal analysis is performed separately in narrow bands, each of which contain no more than one formant. This phenomenon is known as "synchrony capture."

When only one formant is contained in such frequency bands, the output of such a filter is a damped oscillation, the frequency of which is the formant frequency. The formant frequency can thus be determined accurately by measuring the interval between two waves of the output of such filters, which simplifies decoding of temporal information about frequency (see p. 116).

Spectral selectivity in the cochlea deteriorates at high sound intensities and in individuals with injured cochleae. The decreased spectral separation may impair "synchrony capture" and the resulting deterioration of speech discrimination may be caused by impairment of temporal coding of frequency because of the widening of cochlear tuning. Absence of the acoustic middle-ear reflex, which results in less amplitude compression before sounds reach the cochlea, is associated with impaired speech discrimination at high sound intensities, probably because of widening of the cochlear filters [18].

That division of the spectrum of complex sounds into bands prior to analysis by the auditory nervous system is important for speech discrimination is supported by the observation that different bands of the speech spectrum contributed independently to speech discrimination (articulation score, defined as a listener's correct perception of non-sense syllables from a standardized list). This observation was made by Harvey Fletcher at Bell Telephone during the Second World War and not published until some time after the war (in Fletcher's 1953 book, which has been reprinted 1995 by the Acoustical Society of America, with Jont B. Allen as editor).

5.10. Speech Discrimination on Spectral Information Only

Experience from development of the channel vocoder[4] [39, 265] and more recently, from cochlear implants [139, 141] (see Chapter 11) shows that very coarse power spectral information using only a few broad frequency bands can provide good speech discrimination. The experience

---

[4]The channel vocoder (Voice Operated reCorDER) was developed in the 1950s–1960s for transmitting telephone signals over long lines such as transoceanic cables. Its principle is to divide the spectrum of speech into a few bands at the transmitting end of such lines and converting the energy in these bands into electrical signals that can be transmitted to the receiving end of a long cable using less bandwidth than speech sounds (known as analysis-synthesis telephony). These signals are then used to synthesize the speech at the receiving end. Channel vocoders never came into practical use for transmitting telephone signals because better and less expensive broad band communication systems that could transmit many telephone lines (satellites and later fiber optic cables) became available before vocoder systems were fully developed.

from cochlear implants showing that satisfactory speech discrimination can be achieved by devices that have only a few channels supports the hypothesis that neither fine spectral resolution nor the coding of temporal information are necessary for obtaining good speech discrimination. The fact that little improvement is achieved by increasing the number channels of cochlear implants above eight supports the hypothesis that fine spectral resolution is not necessary for discrimination of speech sounds [54, 140]. These matters are further discussed in connection with cochlear implant processors in Chapter 11.

5.11. Conclusion

The fact that studies indicate that temporal information plays a greater role than place coding in discrimination of complex sounds such as speech sounds does not mean that spectral analysis (the place principle) cannot provide the basis for speech intelligibility. This observation underlines that the auditory system possesses a considerable redundancy with regard to the role of frequency discrimination as a basis for speech discrimination.

6. CODING OF COMPLEX SOUNDS

Most studies of coding of sounds in the classical ascending auditory pathways have employed simple sounds such as pure tones and clicks. The recorded responses have been analyzed by determining the threshold of firing (frequency threshold curves) and the distribution of nerve impulses during and after the presentation of tone bursts (post-stimulus time [PST] histograms).

Natural sounds have broad spectra that change more or less rapidly and changes in frequency and amplitude are prominent features of natural sounds. Changes in the amplitude and spectrum carry important information in such sounds as speech sounds. Many studies have demonstrated that changes in the frequency and amplitude of sounds are enhanced in the discharge pattern of the nuclei of the classical ascending auditory pathways indicating that the nervous system transforms sounds and enhances aspects of sounds that are rich in information while suppressing features of sounds that carry little or no information. A steady sound such as a pure tone does not provide any information after it has been switched on, except perhaps what information its duration might provide. Many nerve cells in nuclei of the auditory system only respond to tones when turned on or off and that is one indication that the auditory nervous system "filters" sound with regard to their information contents.

In the following the response from single auditory nerve fibers and cells of the nuclei of the classical ascending auditory pathways to steady tones and tone bursts will be described first because that has been the traditional way to study the function of the auditory system. We will then proceed to describe how various

---

BOX 6.15

CHOICE OF STIMULI FOR STUDIES OF THE AUDITORY SYSTEM

The choice of stimuli for studies of the function of the auditory nervous system has been affected more by technical possibilities of generation and description of the sounds than by how well they represent natural sounds. Tone bursts have been the most common stimuli in studies of the auditory system but these stimuli are specialized sounds that are different from natural communication sounds. Pure tones are easy to generate and to describe, whereas it was more difficult to generate complex sounds before the development of computer systems had progressed. Currently it is possible to synthesize nearly any kind of sound on inexpensive laboratory computers.

Complex sounds are also more difficult to describe than pure tones, another factor that has detracted investigators from using complex sounds. Amplitude modulated (AM) sounds and tones the frequency of which change at different rates resemble natural sounds such as speech but are easier to generate and describe. Although they are less complex than natural sounds, they are more appropriate for studying the transformation of information in the auditory system than pure tones. Natural sounds normally appear together with a background of other sounds but stimuli used in studies of the auditory system are usually presented in a background of silence and this is another example of the un-naturalness of stimuli used in many studies of the auditory nervous system.

parts of the classical auditory nervous system respond to complex sounds such as amplitude modulated (AM) sounds and tones with rapidly varying frequencies.

6.1. Response to Tone Bursts

Neurons of the auditory system are commonly characterized and classified by their response to tone bursts. The classification based on the use of tone bursts as stimuli, however, provides little insight in how the auditory system responds to natural sounds, but it is reviewed here because of its extensive use in earlier studies.

Post-stimulus time (PST) histograms of the response of single auditory nerve fibers to short bursts of broad band noise reveals an initial high rate of firing followed by a (exponentially) decrease in firing rate (Fig. 6.20). When the sound is switched off (end of the tone burst), the discharge rate falls below the fiber's firing rate in silence (the spontaneous rate). The firing gradually returns to the rate it had before the tone was switched on. PST histograms such as those seen in Fig. 6.20 are compiled by adding the number of discharges from a single auditory nerve fiber to many presentations of the same sound. PST histograms therefore represent the average firing pattern of a nerve fiber. When no sound is presented, only the spontaneous activity is seen (Fig. 6.20). The discharge rate measured during the time that the stimulus tone is on increases with increasing sound level as seen from the histograms in Fig. 6.20. Note that the histograms in Fig. 6.20 only cover the intensity range from a fiber's threshold up to approximately 40 dB above threshold, thus only covering stimulus intensities below what is regarded as the physiological range of sound intensities, which is 50–75 dB above the threshold of hearing.

Practically all auditory nerve fibers have spontaneous activity similar to the example shown in Fig. 6.20. The spontaneous discharge rate may be a result of mechanical stimulation of hair cells by vibrations of the cochlear fluid that is not induced by outside sounds or as a result of random release of quanta of neurochemicals at the hair cell–nerve fiber synapse.

In response to continuous tones the discharge rate of auditory nerve fibers increases as the sound intensity is raised above the fiber's threshold but the discharge rates for most nerve fibers reaches a plateau at sound intensities well below physiological levels (Fig. 6.21A). Thus, most auditory nerve fibers have a small dynamic range in response to continuous tones.

The dynamic range of an auditory nerve fiber is related to its spontaneous activity (Fig. 6.21B) [135, 136, 221]. A few fibers that have low spontaneous rate

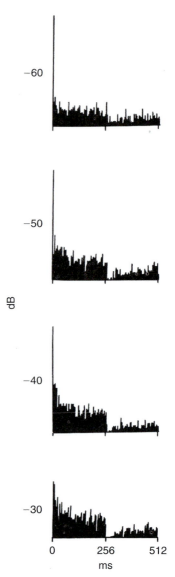

FIGURE 6.20 Post stimulus time histograms of the responses to 50-ms long bursts of broad band noise from a typical auditory nerve fiber in a cat. The stimulus level is given in arbitrary dB values. Obviously, the threshold of this fiber is slightly lower than -70 dB (reprinted from Kiang et al., 1965, with permission from MIT Press).

tend to have a larger dynamic range than fibers with high spontaneous activity (Fig. 6.21B). These nerve fibers may thus communicate information about the intensity of a sound over most of the audible intensity range (from threshold to 80–90 dB above). Fibers with high spontaneous activity and low threshold have a small dynamic range and saturate 20–30 dB above threshold.

While the shapes of the PST histograms of the discharges of different auditory nerve fibers in response to tone bursts are similar, the shape of the PST histograms of

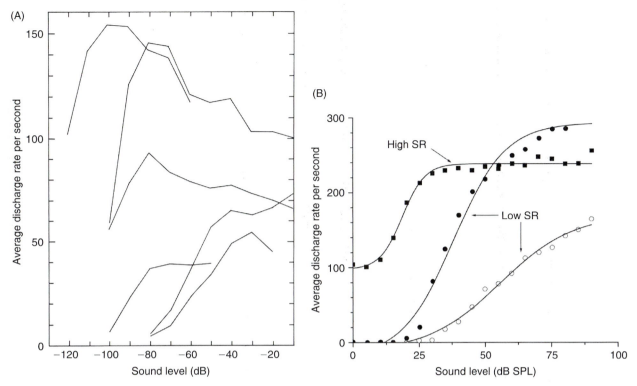

FIGURE 6.21 (A) Stimulus response curves of single auditory nerve fibers in the cat. The discharge rate is shown as a function of the stimulus intensity for continuous tones at the CF of the fiber from which recordings were made. The sound level is given in arbitrary dB values. The threshold is slightly below 100 dB for the fibers studied, except the top left curve where it is approximately 120 dB (modified from Kiang et al., 1965, with permission from the MIT Press). (B) Stimulus response curves for three different auditory nerve fibers in a guinea pig to tones at the fiber's CF. Squares show the response from a nerve fiber with a low threshold (below the threshold of compound action potentials, CAP) and a high spontaneous activity (84.4 spikes/s). Filled circles show the response from a nerve fiber with a threshold near the CAP threshold and low spontaneous activity (0.2 spikes/s). The open circles represent the response of a fiber with threshold near that of the CAP and no spontaneous firings (reprinted from Müller et al., 1991, with permission from Elsevier).

the response from nerve cells in the cochlear nucleus (CN) to tone bursts varies between cells. The shape of such PST histograms has been used to classify the response pattern of CN neurons and the best known and widely used classification was presented by Pfeiffer [230] who divided the neurons in the CN into four different groups according to their response to tone bursts (Fig. 6.22). While this classification separates nerve cells according to their response to tone bursts it is doubtful whether this classification also separates neurons with regard to their responses to complex sounds.

The response to tone bursts of neurons in nuclei that are located more centrally vary within wide limits. Many cells respond best to the onset of a tone burst indicating a preference for transient sounds. Since many cells respond poorly to continuous sounds, it is difficult to obtain records of their discharge rate as a function of sound intensity.

6.2. Coding of Small Changes in Amplitude

Tone bursts represent changes in stimulus intensity that are far greater than that of natural sounds and the responses to tone bursts do not provide information about how small changes in sound intensity are coded in the auditory nerve and nuclei of the auditory nervous system. The response to a small rapid increase or decrease in the intensity of a tone can be illustrated by observing the discharges of a neuron in response to a continuous sound the intensity of which is increased and decreased stepwise (Fig. 6.23). It is seen that a small increase in the intensity of the sound results in a large but brief increase in the discharge rate of the nerve cell from the cochlear nucleus illustrated in Fig. 6.23. When the sound intensity is decreased, the discharge rate decreases briefly below its steady state discharge rate then gradually returns to its steady state rate.

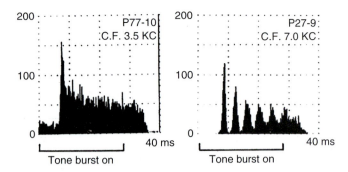

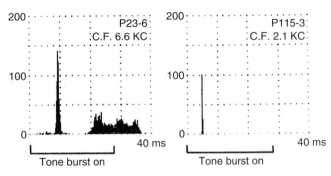

FIGURE 6.22 Post stimulus time histograms of the responses of cells in the cochlear nucleus of cats to tone bursts. Each histogram represents one class of units: A = primary-like; B = chopper; C = pause; and D = onset (reprinted from Pfeiffer, 1966, with permission from Elsevier).

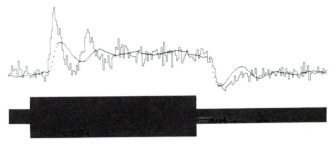

FIGURE 6.23 Period histogram of the response from a cell in the cochlear nucleus of a rat to tones the intensity of which was changed up and down in a stepwise fashion. The dots show calculated response obtained from the response to a tone that was amplitude modulated by pseudorandom noise (reprinted from Møller, 1979, with permission from John Wiley).

The auditory nervous system has mechanisms that normally compress the range of sound intensities and such "automatic gain control" enhances the reproduction of small changes in amplitude of sounds. The cochlea contributes to this automatic gain control which also gets contributions from the interplay between inhibitory and excitatory response areas of nerve cells in the ascending auditory pathways, similar to what has been studied extensively in the visual system. The abundant descending pathways in the auditory nervous system may also contribute to such automatic gain control.

While the response to changes in a sound's intensity (Fig. 6.23) illustrates how neural discharges respond to small stepwise changes in a sound's intensity it does not provide information about the effect of the rate with which the intensity of a sound is varied. That can be studied in experiments where tones or noise that are amplitude-modulated with a sinusoidal waveform are used as stimuli (AM tones). Such AM sounds are similar to many natural sounds such as speech sounds and communication and warning sounds of various animals. Recordings of the responses from single auditory nerve fibers and cells of the nuclei of the classical ascending auditory pathways to AM sounds yield results that reflect processing of natural sounds more

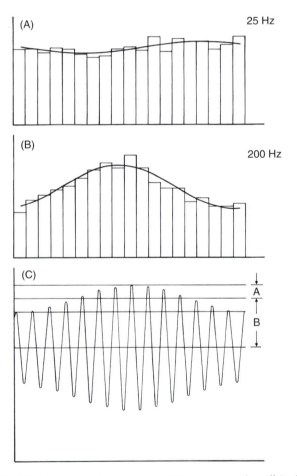

FIGURE 6.24 Period histograms of the response of a cell in the cochlear nucleus of a rat to amplitude modulated tones. The frequency of the tone was 15 kHz, equal to the cell's CF. Histograms of the response to two different modulation frequencies. (A) Modulation frequency 25 Hz. (B) Modulation frequency 200 Hz. (C) One period of the modulated sound. A is the modulation; B is the mean amplitude of the stimulus. A/B is the modulation depth (reprinted from Møller, 1974, with permission from Elsevier).

BOX 6.16

DIFFERENT WAYS TO DISPLAY MODULATION OF NEURAL RESPONSES

Several measures of the modulation of the neural discharges have been used to describe the coding of AM sounds in the discharge pattern of auditory nerve fibers and cells of auditory nuclei. One such measure is the ratio between the amplitude of the modulation of the histograms and the modulation of the stimulus sounds. An example of that is shown in Fig. 6.25 where the solid line in the graph shows the relative modulation, expressed in decibels (the scale to the left of the graph) together with the phase angle between the modulation of the sound and that of the histograms. Such graphs (known as Bode plots) show modulation transfer functions. A modulation of 100% of the

histograms corresponds to 0 dB on that scale. Zero degrees means that a sine wave that fits the histogram falls exactly over the sine wave of the modulation of the stimulus tone. A negative phase angle means that the modulation of the histogram lags the modulation of the sound [171].

Another frequently used measure of modulation of neural discharges introduced by Goldberg and Brown [69] as a measure of the ability of a nerve fiber to follow the waveform of low frequency tones (phase-locking) is called the synchronization index. It is defined as the fraction of the nerve impulses that occurs phase locked to the sound (or the envelope of an AM sound).

---

closely than the responses to tone bursts or sounds the amplitude of which changes stepwise. (The sounds used to obtain the results in Fig. 6.23 may be regarded as a tone that is modulated by a rectangular waveform.)

Many systematic studies have been published of the responses of single auditory nerve fibers to amplitude-modulated tones and noise [57, 99, 175], and from cells in the CN [57, 58, 176, 244] and the IC [242, 243] (for a review see [99]). A few studies of the responses from neurons of the auditory cortex have also been published [8, 40, 83, 138, 264]. We will begin by discussing results of studies in the CN because they are the most extensive.

The discharge pattern of single nerve cells in the CN in response to sinusoidal AM tones is modulated by the waveform of the modulation of the stimulus sound as seen from modulation period histograms[5] of the discharges in response to AM tones (Fig. 6.24). Neurons in the CN reproduce the (sinusoidal) modulation waveform of AM tones and noises faithfully in their discharge pattern over a large range of modulation frequencies (Fig. 6.25) [171].

The shape of the modulation transfer function and the maximal modulation gain varies from cell to cell (Fig. 6.26). For most cells, the modulation transfer function changes from a low-pass type at low sound intensities to a band-pass type at higher sound intensities (Fig. 6.27). The transfer functions of many units are narrow and such cells may be regarded as "tuned" to a certain modulation frequency.

[5]Modulation period histograms show the distribution of nerve impulses over one period of the modulation in a similar way as period histograms used to study coding of the waveform of a sound in the discharge pattern of single auditory nerve fibers (see Fig. 6.14).

The modulation waveform is reproduced with a high degree of fidelity over a range of 60–80 dB thus covering the physiologic range of the hearing sense. This is a much larger range than the range of stimulus intensities over which the discharge rate increases with increasing stimulus intensity for pure tones. This means that nerve cells in the CN can follow the envelope of AM sounds at stimulus intensities above which the steady discharge rate has reached its saturation level.

While coding of steady state sounds in neurons in the CN only covers a small range of sound intensities, the envelope of amplitude modulation of a sound is reproduced with a high degree of fidelity over a large range of sound intensities in the discharge pattern of neurons in the CN. The period histograms of the response from cells in the CN to sinusoidal amplitude modulation of a tone or noise have very little distortion [171] indicating that these cells reproduce the waveform of the envelope of a sound with great fidelity in the temporal waveform of their discharges.

Some investigators have studied the responses from cells in the CN to amplitude modulated tones by using bursts of tones where part of the bursts were sinusoidal amplitude modulated [57, 58]. The responses obtained in these studies were illustrated by PST histograms of the response covering the entire duration of the sound (150 ms [57]) (Fig. 6.28). These investigators confirmed that the modulation transfer function changed from a low pass shape to a band-pass shape when the sound level was increased (58).

Frisina and co-workers [57] compared the modulation gain of different types of CN cells based on their response to tone bursts using the classification described by Pfeiffer [230] (Fig. 6.22). The modulation gain at 0.15 kHz was highest for "on" type units where it was

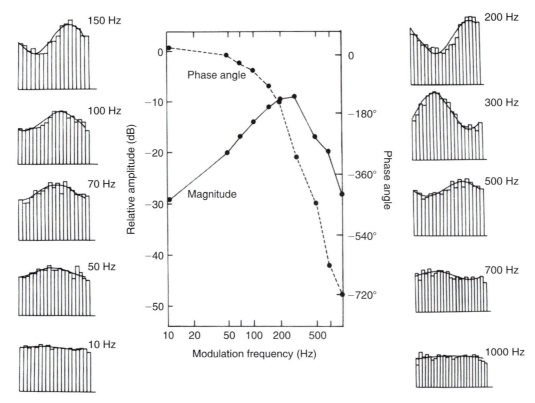

FIGURE 6.25   Period histograms of the responses from a cell in the cochlear nucleus of a rat to amplitude modulated tones for modulation frequencies between 0.010 and 1 kHz. The middle graph is the modulation transfer function (Bode plot) showing the gain defined as the ratio between the relative modulation of the histograms divided by the modulation of the carrier tones. The frequency of the carrier tones was equal to the individual cell's CF. The solid line shows the magnitude (in decibels) and the dashed lines show the phase angle of the modulation transfer function (reprinted from Møller, A.R. 1972. Coding of amplitude and frequency modulated sounds in the cochlear nucleus of the rat. *Acta Physiol. Scand.* 86, 223–238, with permission from Blackwell Publishing Ltd).

approximately 2.7 dB at 50 dB [57] and less for "chopper" type cells (−7.5 dB). Primary-like cells had modulation gains of approximately −9.4 dB. These results may be interpreted to show that the further CN cells are from being primary-like the better they encode amplitude modulation. Primary-like neurons receive fewer auditory nerve fibers than the other types of CN nerve cells. The ability to encode the modulation waveform of an AM sound may thus be related to the number of auditory nerve fibers that terminate on a cell (the degree of convergence of auditory nerve fibers onto a cell in the CN).

Auditory nerve fibers reproduce the modulation waveform in their discharge pattern to a greater extent for sounds of low intensity than sounds of high intensity (Fig. 6.30) [30]. The modulation transfer functions of auditory nerve fibers are low-pass functions with cut-off frequencies around 1 kHz (Fig. 6.31) [102]. The cut-off frequency for fibers with a high characteristic frequency (CF) is higher than for fibers with a low CF. Fibers that have a low spontaneous rate have the strongest coding of the modulation waveform of AM

tones at CF but the influence of the sound intensity on the reproduction of the modulation was different for different nerve fibers [30, 56]. Fibers with medium high spontaneous rate and fibers with high spontaneous rate reproduced the envelope of AM tones to a lesser degree.

While most studies of the neural coding of amplitude modulation have been done in quiet, studies by Rhode and Greenberg [244] showed that the phase locking of both auditory nerve fibers and CN cells is relatively resistant to background noise. Frisina et al. [56] showed that background noise could increase or decrease the reproduction of the modulation waveforms in the discharge pattern of auditory nerve fibers depending on the spontaneous activity of the fibers.

Cells in the CN reproduce the modulation waveform in their discharge pattern to a greater extent than auditory nerve fibers (Fig. 6.32) [102, 175, 227]. This difference is greatest for modulation frequencies between 0.1 and 0.3 kHz. That cells in the CN respond better to small changes in the amplitude of a sound (amplitude modulation) than auditory nerve fibers

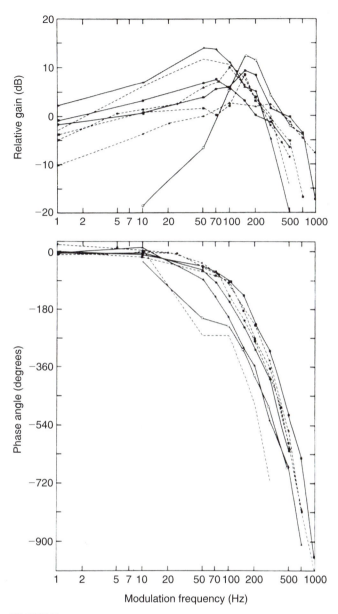

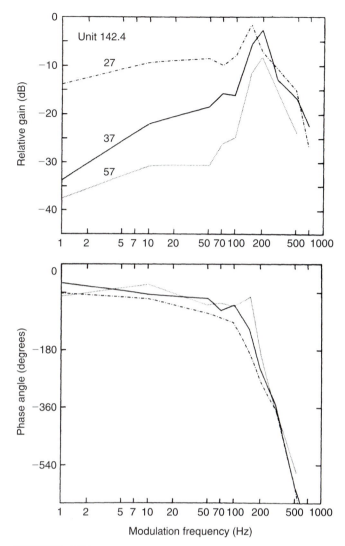

FIGURE 6.26 Modulation transfer functions (gain functions) of eight typical cells in the cochlear nucleus of a rat. The frequency of the carrier tones was equal to the CF of the units (ranging from 0.95 to 30 kHz), and the sound intensity was 20 dB above the threshold of the cells (reprinted from Møller, A.R. 1972. Coding of amplitude and frequency modulated sounds in the cochlear nucleus of the rat. *Acta Physiol. Scand.* 86, 223–238, with permission from Blackwell Publishing Ltd).

FIGURE 6.27 Modulation transfer functions of a cell in the cochlear nucleus of a rat where the different curves represent different sound intensities given in dB SPL (approximately the same as dB above threshold). The modulation depth was 20% and the frequency of the carrier tone was equal to the cell's CF (15.2 kHz) (reprinted from Møller, A.R. 1972. Coding of amplitude and frequency modulated sounds in the cochlear nucleus of the rat. *Acta Physiol. Scand.* 86, 223–238, with permission from Blackwell Publishing Ltd).

may seem paradoxical but it is probably a result of the fact that cells of the CN receive input from many auditory nerve fibers. Such convergence of primary nerve fibers onto CN cells causes averaging of inputs from many auditory nerve fibers by CN cells and that enhances small changes in the discharge rate of auditory nerve fibers in a similar way as the signal averaging techniques used to recover evoked potentials from

a background of other signals (noise). This hypothesis is supported by the finding that the modulation gain is related to the number of auditory nerve fibers that converge onto a CN cell [57, 58]. The larger dynamic range of the response to amplitude modulation of CN cells compared with auditory nerve fibers may also, partly, be a result of the fact that CN cells receive both excitatory and inhibitory input from the auditory nerve.

The reproduction of amplitude modulation in neurons in the superior olivary complex is similar to that of neurons in the cochlear nucleus providing a precise temporal coding of the envelope of sounds as

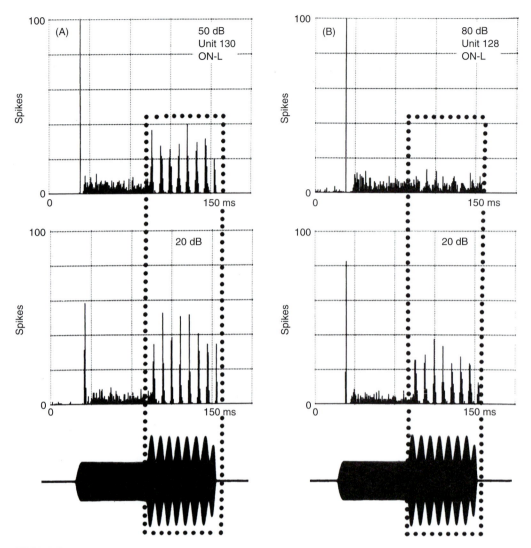

FIGURE 6.28   Post stimulus time histograms of the response of a cell in the cochlear nucleus of a cat to tone bursts the last half of which were amplitude modulated. The response to two different sound intensities is shown (reprinted from Frisina et al., 1990, with permission from Elsevier).

evidenced from studies of the response to amplitude modulated sounds of neurons in the superior olivary nuclei [101, 120].

In an early study Erulkar et al. [45] showed that some cells in the IC preferentially responded to AM tones. The cell depicted in Fig. 6.33 only responded (with a single discharge) at a certain phase of the modulation. Later systematic studies confirmed that coding of AM sounds in neurons in the IC is prominent [243]. The modulation transfer functions of cells in the IC change from low-pass functions to band-pass functions when the stimulus intensity is increased (Fig. 6.34), thus similar to cells in the CN, but the peak in the modulation transfer function occurs at a lower modulation frequency than in the CN. The variations in the response pattern between different cells in the

IC are larger than in the CN. The responses to the modulation waveform is robust and adding masking noise affects the shape of the modulation transfer functions only to a small extent even when the noise level exceeds the sound level of the AM tones from which the response is derived (Fig. 6.35).

Unlike auditory nerve fibers and cells in the CN, the average discharge rate of cells in the IC is affected by the amplitude modulation. The discharge rate increases with the modulation frequency and reaches a maximum around 0.5 kHz for sinusoidal AM tones [127], a modulation frequency that is higher than that at which the modulation waveform is reproduced to the greatest extent.

In one study [264], it was shown that the modulation transfer functions of cortical neurons were band pass

BOX 6.17

RESPONSES TO TONES: INHIBITORY AND EXCITATORY

It was mentioned above (p. 99) that the response from auditory nerve fibers to a tone could be inhibited by another tone ("two tone inhibition"). Thus, the discharge rate evoked by a tone at CF decreases when a second tone with a slightly higher or slightly lower frequency is added to the tone at CF. Neurons in the CN have similar excitatory and inhibitory response areas. If two tones are presented simultaneously, one excitatory and one inhibitory, a cell in the CN will respond to a decrease in the amplitude of the excitatory tone by a decrease in its discharge rate while an increase in the intensity of the inhibitory tone will cause a decrease in the discharge rate. If the inhibitory tone is sinusoidal amplitude modulated, the period histogram will be modulated in a similar way as modulation of the excitatory tone but the modulation of the histograms will be of opposite phase (180° difference) reflecting that an increase in the amplitude of a tone that inhibits the response will cause the discharge rate to decrease while an increase of the intensity of an excitatory tone will cause an increase in the discharge rate (Fig. 6.29).

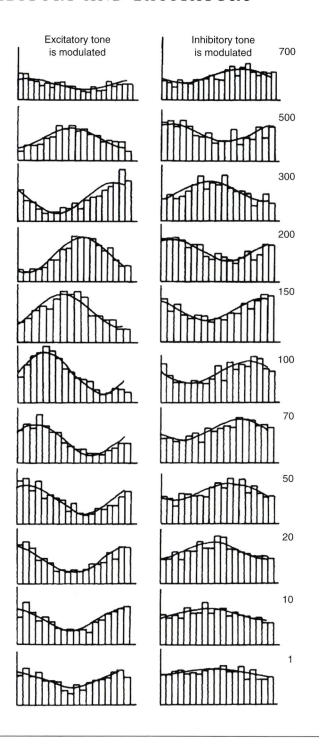

FIGURE 6.29 Period histograms of the response to amplitude modulated tones of a cell in the cochlear nucleus of a rat. Two tones were presented simultaneously, one excitatory tone (at CF = 4.5 kHz) and the other tone (5.5 kHz) located in the cell's inhibitory response area. The histograms in the left-hand column were obtained when the excitatory tone was modulated and the inhibitor was un-modulated; right-hand column histograms were obtained when the inhibitory tone was modulated and the excitatory tone was un-modulated. The modulation frequency is given on the histograms (reprinted from Møller, A.R. 1975. Dynamic properties of excitation and inhibition in the cochlear nucleus. *Acta Physiol. Scand.* 93, 442-454, with permission from Blackwell Publishing Ltd).

BOX 6.18

RESPONSES TO SOUNDS THAT ARE MODULATED BY NOISE

Using non-periodic sounds for modulation of tones or noise has several advantages in studies of the coding of the amplitude modulation of sounds. The response to sinusoidal modulated tones or noise only represents the response to one modulation frequency at a time and neural discharges may phase-lock to the modulation waveform. To avoid these problems and get a more representative stimulus, some investigators have used broad band signals to modulate tones or noise [193]. Many natural sounds are similar to sounds that are modulated with non-sinusoidal waveforms.

When low-pass filtered noise is used to modulate a sound such as a tone, the entire modulation transfer function can be obtained from a single recording. Such studies used pseudorandom noise that is (random) noise that repeats itself periodically [193, 215]. Analysis of the response from single cells to sounds that are amplitude modulated with pseudorandom noise can be done by compiling a period histogram of the responses over one period of the pseudorandom noise. These histograms are then cross-correlated with the waveform of the modulation (the pseudorandom noise) and the resulting cross-correlograms are estimates of the impulse response of the system under test. Fourier transforms of such correlograms yield an estimate of the frequency transfer function [199]. When the pseudorandom noise is used to modulate a sound (tone or noise) the resulting cross spectra are estimates of the modulation transfer function and thus similar to the modulation transfer functions that are obtained by compiling the results from using sinusoidal modulations with different frequencies of the modulation. Studies of the response from cells in the CN have shown that the modulation transfer functions that are obtained using these two different methods are similar [193].

type (Fig. 6.36). The peak in the modulation transfer function occurs at a much lower frequency than what is the case for cells in the IC and the CN. These investigators used both sinusoidal modulation and modulation with a rectangular waveform. It is interesting that the response is approximately the same for ipsilateral and contralateral stimulation.

Efferent fibers of the auditory nerve also code the modulation waveform of AM sounds in a way that depends on the frequency of the modulation [74]. Efferent fibers respond best to the modulation between modulation frequencies of 0.1 to 0.14 kHz. A tone that is amplitude modulated 30% produced approximately 100% modulation of the discharges in efferent fibers, thus a modulation gain of approximately 3. This is about twice the modulation gain of afferent auditory nerve fibers [74]. The reason that the modulation is reproduced to a greater extent in efferent fibers than in afferent auditory nerve fibers is probably due to the greater degree of spatial averaging of the information that occurs in the efferent fibers.

6.3. Response to Tones with Changing Frequency

It was mentioned earlier in this chapter that fibers of the auditory nerve only respond to tones within a certain range of frequencies and sound intensities (Fig. 6.2)

and that the frequency selectivity of auditory nerve fibers depends on the intensity of the stimulus sound (Fig. 6.6). Cells in the nuclei of the classical ascending auditory pathways show similar frequency selectivity although the size and shape of the response areas of such cells varies as a result of the transformation that occurs in the ascending auditory pathways. When the frequency and intensity of a stimulus tone is within the response area of an auditory nerve fiber, or a cell, its discharge rate becomes a function of the frequency of the tone (Fig. 6.4A). A histogram of the distribution of discharges as a function of the frequency of the tone (Fig. 6.37) is a description of the area of an auditory fiber or of a nerve cell in an auditory nucleus. When the frequency of the stimulus tone is varied slowly, such histograms show a similar frequency range of response as is obtained using steady tones (as shown in Fig. 6.4A). Such histograms are broader and their tips are less sharp.

While the response area of auditory nerve fibers is little affected by the rate of change of the frequency of a stimulus tone (Fig. 6.39), the response area of single nerve cells in the CN and other nuclei of the classical ascending auditory system change radically when the rate with which the frequency of a sound is changed.

The responses of single auditory nerve fibers of the cat to tones the frequency of which was changed at

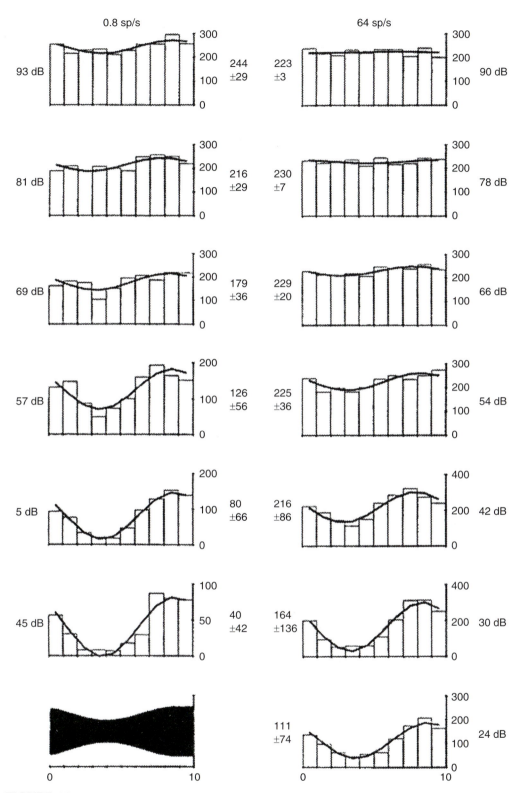

FIGURE 6.30   Period histograms of the response from auditory nerve fibers in guinea pigs to amplitude modulated sounds at different sound intensities. The two rows of histograms are from two nerve fibers with different spontaneous activity (left column: 0.8 spikes per second and right column: 64 spikes/s (reprinted from Cooper et al., 1993, with permission from the American Physiological Society).

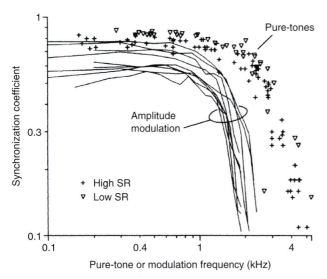

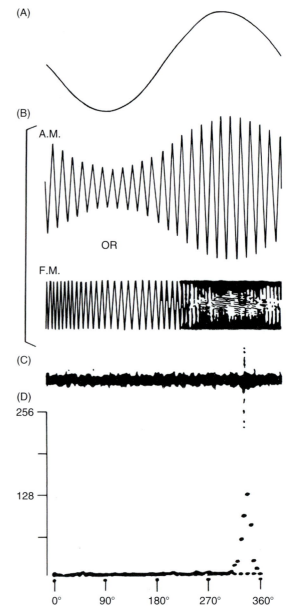

FIGURE 6.31 Modulation transfer functions of the response to amplitude modulated sound for auditory nerve fibers together with the degree of phase locking (synchronization index) as a function of frequency for auditory nerve fibers (reprinted from Joris and Yin, 1992, with permission from the American Institute of Physics).

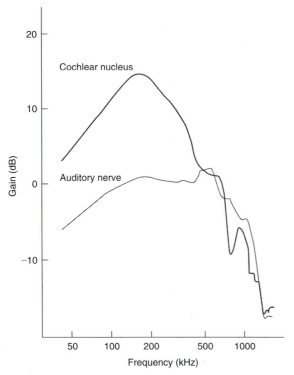

FIGURE 6.33 Response of a cell in the inferior colliculus to amplitude modulated tones. The modulation was sinusoidal and the tone was modulated 50% (reprinted from Erulkar et al., 1968, with permission from Elsevier).

FIGURE 6.32 Modulation transfer functions of a cell in the cochlear nucleus of a rat and of an auditory nerve fiber. The vertical scale shows gain (in decibels). The modulation transfer functions were obtained from analysis of the response to tones that were amplitude modulated with pseudorandom noise. The bottom curves are coherence functions which show the degree of significance of the gain functions (reprinted from Møller, A.R. 1976. Dynamic properties of primary auditory fibers compared with cells in the cochlear nucleus. *Acta Physiol. Scand.* 98, 157–167, with permission from Blackwell Publishing Ltd).

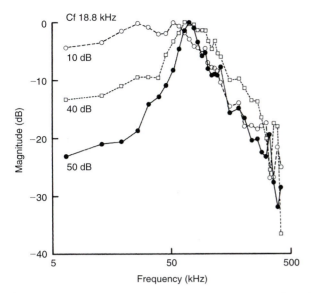

FIGURE 6.34 Modulation transfer functions of a cell in the inferior colliculus of a rat obtained at three different sound intensities (10, 40 and 50 dB above the cell's threshold). The transfer function changes from low-pass to band-pass with increasing sound intensity (reprinted from Rees and Møller, 1987, with permission from Elsevier).

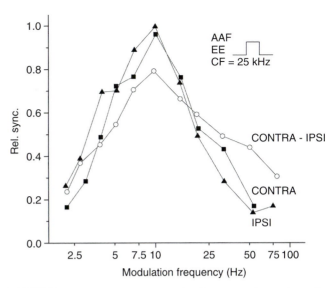

FIGURE 6.36 Modulation transfer functions obtained in the anterior auditory field (AAF) of the auditory cerebral cortex of a cat to contralateral (c), ipsilateral (I) and bilateral (chi) stimulation with amplitude modulated tones (reprinted from Schreiner and Urbas, 1986, with permission from Elsevier).

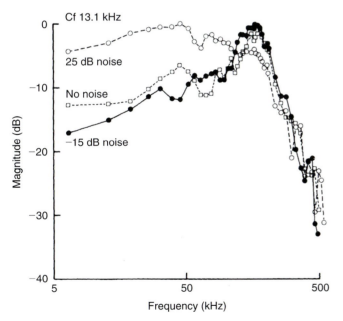

FIGURE 6.35 Similar graphs as in Fig. 6.34 with different levels of background noise. (reprinted from Rees and Møller, 1987, with permission from Elsevier).

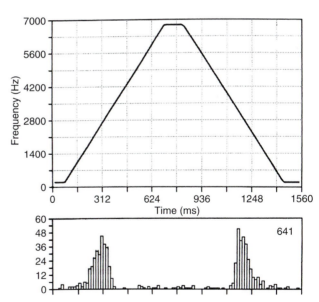

FIGURE 6.37 Period histograms of the response of an auditory nerve fiber in a cat (lower graph) to tones the frequency of which was varied up and down (upper graph) in a range that extended over the range that the fiber responded (reprinted from Sinex and Geisler, 1981, with permission from Elsevier).

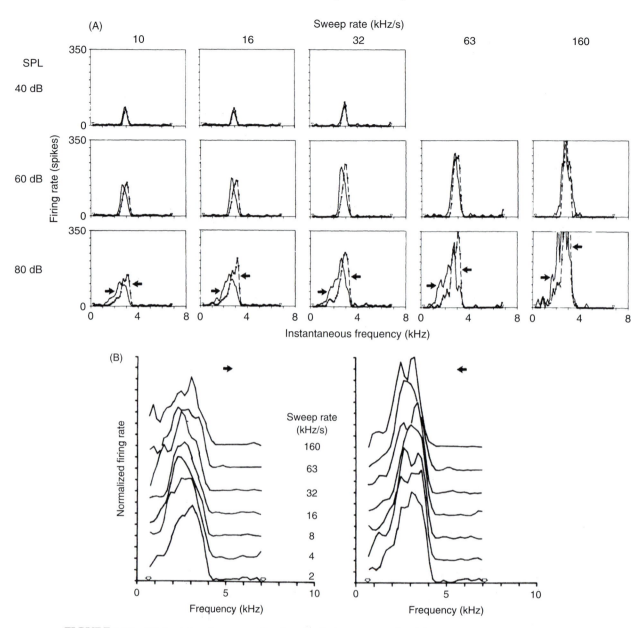

FIGURE 6.38   (A) Iso-intensity curves showing the discharge rates of auditory nerve fibers as a function of the frequency of the tone stimulus at different intensities (given by legend numbers). Solid lines: steady tones; dashed lines: response to tones the frequency of which was changed at different rates (reprinted from Sinex and Geisler, 1981, with permission from Elsevier). (B) Similar data as in (A) but normalized to the same height of the histograms. Left column: the change in the frequency of the tones was from low to high; right column: frequency change from high to low (reprinted from Sinex and Geisler, 1981, with permission from Elsevier).

different rates show a slight increase in the height of the histograms with little change in the shape of the histograms as the rate of change was increased (Fig. 6.38A) [274]. This means that the tuning of single auditory nerve fibers is little affected by the rate of change in the frequency of a continuous tone. The fact that the shape of the tuning curves is not affected by the rate of change in the frequency of the stimulus tone can be demonstrated

when the histograms are displayed on a normalized scale (Fig. 6.38B).

The response of neurons in the CN to tones the frequency of which changes rapidly is different from that of auditory nerve fibers and histograms of the responses become narrower and higher when the rate of change in frequency of the tone is increased from a low rate [173]. This indicates that the response area becomes

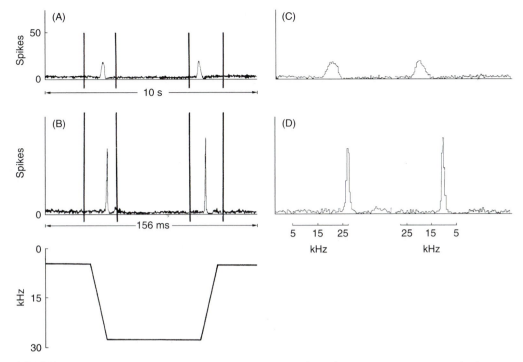

FIGURE 6.39 Period histograms of a cell in the cochlear nucleus of a rat in response to tones the frequency of which was varied between 5 and 25 kHz at different rates. (A) and (C): slow rate; (B) and (D): fast rate. The top histograms ((A) and (C), slow rate) show the responses obtained when the duration of a full cycle was 10 s and the lower histograms ((B) and (D), fast rates) show the responses obtained when the duration of a complete cycle was 156 ms. The change in the frequency of the stimulus tone was accomplished by having a trapezoidal waveform control the frequency of the sound generator (E). The two left-hand graphs ((A) and (B)) are histograms of a full cycle of the modulation and the right-hand graphs ((C) and (D)) show the details between the vertical lines in the left-hand graphs (reprinted from Møller, 1974, with permission from the American Institute of Physics).

narrower when the frequency of a stimulus tone changes rapidly and more nerve impulses are delivered closer to the cell's characteristic frequency (Fig. 6.39). The total number of nerve impulses, however, is only slightly dependent on the rate of change in frequency of the stimulus tone, which means that it is mostly a redistribution of nerve impulses that occurs when the rate of change in the frequency of the stimulus tone is increased.

The sharpening of the responses areas of cells in the CN is yet another demonstration of the complexity of the auditory nervous system and the extension of the information processing that occurs in a nucleus of the classical auditory nervous system.

The response from neurons in other nuclei of the classical ascending auditory pathways to sounds the frequency of which changes fast has not been

---

BOX 6.19

RESPONSE TO TONES WITH RAPIDLY CHANGING FREQUENCY

The height of the histograms does not continue to grow as the rate of change in frequency is increased but reaches a maximum height when the frequency of the stimulus tone is changed at a certain rate (Fig. 6.40). The height of the peaks in the histograms of the response to tones increases more for tones between 45 and 65 dB above threshold than for tones of lower intensity (such as

25 dB above threshold). The rate of change at which the histograms reach their maximal height is different when the frequency of the stimulus tone is rising compared with falling. The effect of the direction of the change in frequency is most pronounced when the intensity of the stimulus tone is high (Fig. 6.40).

BOX 6.19 (*cont'd*)

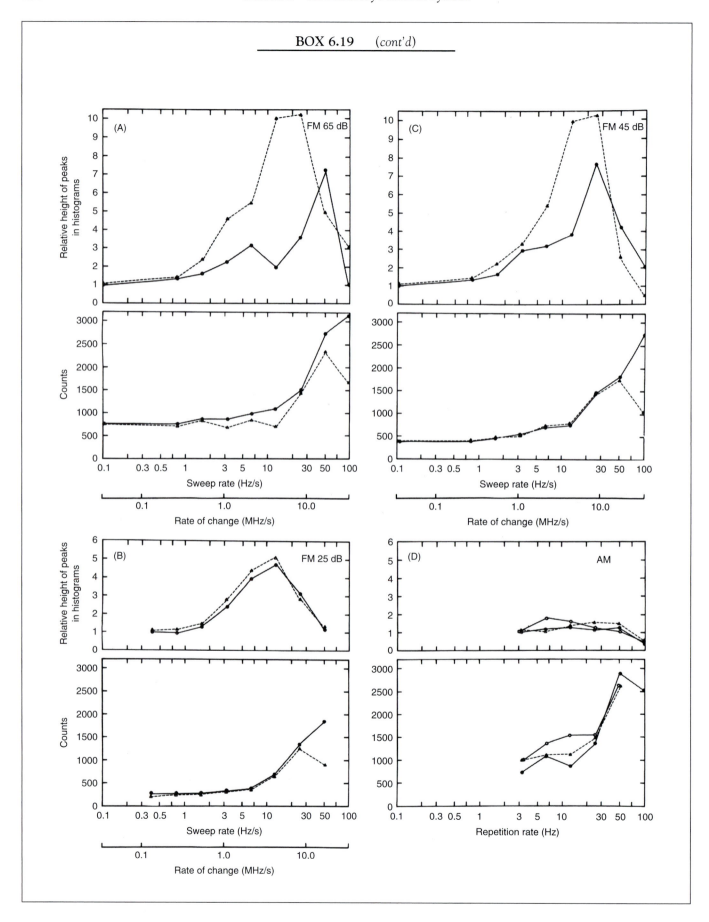

BOX 6.19   (cont'd)

FIGURE 6.40 The relative height of the histograms of the response of a nerve cell in the cochlear nucleus of a rat (CF of 22 kHz) in relation to the rate of change of the frequency of the tones used as stimuli. The height of the histogram peaks of the response obtained when the frequency is changed at a slow rate was set to the value of 1.0. The height of the histograms of the response to tones of increasing frequency (solid lines) is different from those to tones of decreasing frequency (dashed lines). The results depicted in the three graphs marked FM ((A), (B), and (C)) were obtained at three different stimulus intensities (65, 45, and 25 dB above the threshold for that cell). The number of spikes in the peaks of the histograms is shown in the graphs below. The graphs labeled (D) shows the response to amplitude modulated sounds (modified from Møller, A.R. 1971. Unit responses in the rat cochlear nucleus to tones of rapidly varying frequency and amplitude. *Acta Physiol. Scand.* 81, 540–556, with permission from Blackwell Publishing Ltd).

BOX 6.20

DIFFERENCES IN RESPONSE TO TONES WITH RAPIDLY VARYING FREQUENCY

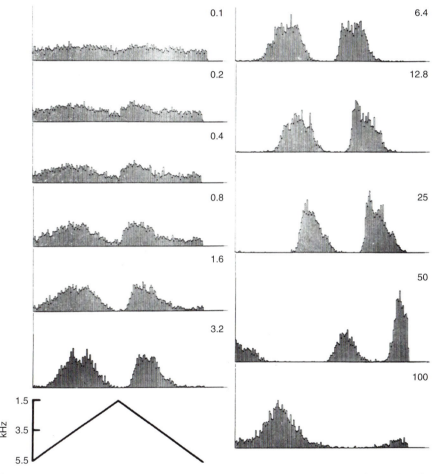

FIGURE 6.41 Period histograms of the responses of a cell in the cochlear nucleus of a rat that show little frequency selectivity to tones of slowly varying frequency but a pronounced frequency selectivity when the frequency of the tones are varied rapidly. The rate of frequency change was varied by changing the rate by which the cycle of change in frequency of the stimulus tones are repeated (given by legend numbers). The cell's CF was approximately 3.5 kHz and the stimulus intensity was 47 dB SPL (reprinted from Møller, A.R. Unit responses in the cochlear nucleus of the rat to sweep tones. *Acta Physiol Scand* 76: 503–512, 1969, with permission from Blackwell Publishing Ltd).

(Continued)

BOX 6.20   *(cont'd)*

Some nerve cells in the CN have a high rate of spontaneous activity and such cells may not respond to tones with constant or slowly varying frequency. Such neurons may, however, respond vigorously to tones with rapidly varying frequency and show pronounced frequency selectivity (Fig. 6.41) [197]. When a tone with constant intensity and a frequency equal to the CF of a nerve cell in the CN is superimposed on a tone the frequency of which is varied up and down, some nerve cells show little frequency selectivity when the frequency of the variable tone is changed slowly but they show clear and pronounced frequency selectivity when the frequency is changed at a high rate (Fig. 6.42).

Some nerve cells in the CN show mainly inhibition of their spontaneous firing when stimulated with pure tones of constant or slowly varying frequency. When the frequency of the tone is changed at a high rate, such units change their response pattern to become excitatory with similar enhancement of the responses as seen in other cells (Fig. 6.43).

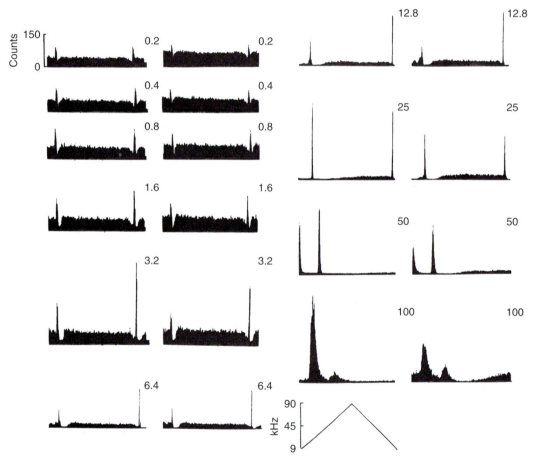

FIGURE 6.42   Period histograms similar to those in Fig. 6.41. A tone with changing frequency, 25 dB above threshold, was presented in a quiet background (left and third column), and when presented together with a constant tone (at CF, and 32 dB above threshold) (second and fourth columns). The frequency was changed between 9 and 90 kHz (insert below). The rate of frequency change was varied by changing the rate by which the cycle of change in frequency of the stimulus tones are repeated (given by legend numbers) (reprinted from Møller, A.R. 1971. Unit responses in the rat cochlear nucleus to tones of rapidly varying frequency and amplitude. *Acta Physiol. Scand.* 81, 540–556, with permission from Blackwell Publishing Ltd).

BOX 6.20   *(cont'd)*

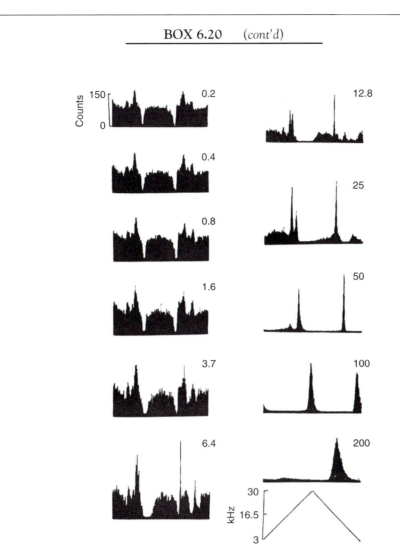

FIGURE 6.43 Period histograms of the response from cells in the cochlear nucleus of rats to tones the frequency of which was varied at different rates (similar to Figs 6.41 and 6.42). This cell had a high spontaneous rate and responded to tones with slowly varying frequency (mainly with inhibition of its spontaneous firing). The rate of frequency change was varied by changing the rate by which the cycle of change in frequency of the stimulus tones are repeated (given by legend numbers) (reprinted from Møller, A.R. 1971. Unit responses in the rat cochlear nucleus to tones of rapidly varying frequency and amplitude. *Acta Physiol. Scand.* 81, 540–556, with permission from Blackwell Publishing Ltd).

studied as systematically as it has in the CN. It is therefore not known if this sharpening of the response to tones with rapidly changing frequency is preserved or further enhanced as the information travels along the neural axis towards the auditory cortex.

One of the first studies of the response from nerve cells in the auditory system to tones with rapidly changing frequency showed that such neurons often did not respond to steady tones but only responded to rapid changes in frequency (Fig. 6.44) [308]. Neurons in the auditory cortex responded preferentially to

a specific direction of change in frequency of a tone as illustrated in the response to sinusoidal frequency modulated tones (Fig. 6.44) [308]. The response adapted rapidly to steady tones. These results were obtained from cells in the AI in unanaesthetized cats [308].

More recently, extensive studies of the responses from nerve cells in the AI of the barbiturate anesthetized cat [83] and posterior auditory field [82] showed a preference to sounds the frequency of which changed at a high rate. Most neurons responded preferentially to tones the frequency of which changed at rates of

(A)

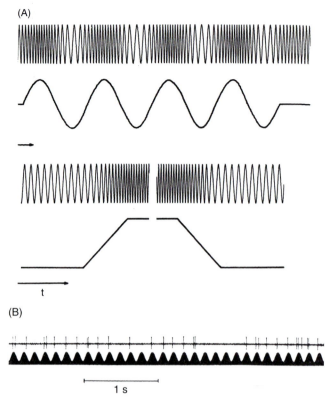

(B)

1 s

FIGURE 6.44 (A) Illustration of a sinusoidal frequency modulated tone and a ramp modulated tone (reprinted from Whitfield and Evans, 1965, with permission from the American Physiological Society). (B) Response from a cell in the auditory cerebral cortex in response to a tone at 11.6 kHz and 75dB SPL that was within the cell's response area. The first part of the stimulation was a constant tone and after a brief period the tone was frequency modulated as indicated by the modulation shown below the discharges. Note that the cell adapted rapidly to a continuous tone (11.6 kHz) but when that tone was frequency modulated with a sinusoidal waveform (the frequency changed +/- 4.75%) the nerve cell began to respond again. The time bar is 1 s, the amplitude calibration is 2 mV (reprinted from Whitfield and Evans, 1965, with permission from the American Physiological Society).

1 MHz/s or higher. These neurons thus seem to prefer rates of change in frequency that are in the same range as the most preferred range of neurons in the CN (cf. Fig. 6.40). The results of these studies [83] indicate that preference for rapid change in frequency may occur throughout the classical ascending auditory pathways. However, these and other studies were performed under barbiturate anesthesia that suppresses neural activity in the auditory cortex and therefore, these results must be interpreted with caution.

6.4. Selectivity to Other Temporal Patterns of Sounds

Most studies of coding of sounds in the auditory nervous system have focused on the frequency or spectrum of sounds. However, features that are not directly related to the spectrum of sounds such as the duration of sounds

and intervals between sounds are also coded in the response of neurons in the auditory nervous system.

Most neurons in the IC respond only to the beginning of long tone bursts of sounds but some neurons in the IC respond in accordance with the duration of a sound [24]. These neurons fire more nerve impulses when sounds of a specific duration were presented (Fig. 6.45). The duration to which the neurons

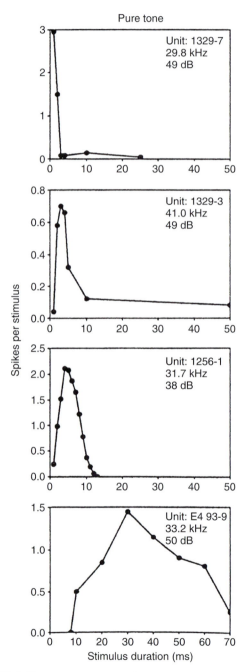

FIGURE 6.45 Duration tuning of neurons in the inferior colliculus of a bat (the big brown bat). The graphs show the responses of four different nerve cells to tone bursts of different duration (horizontal axis) (reprinted from Casseday et al., 1994, with permission from the American Association for the Advancement of Science).

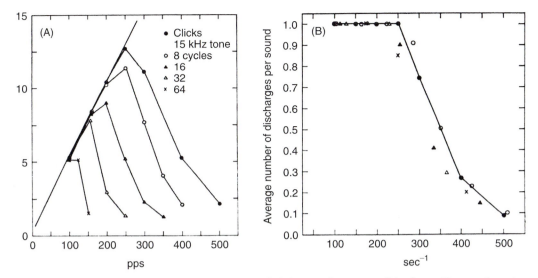

FIGURE 6.46 (A) Discharge rate as a function of click rate of a nerve cell in the cochlear nucleus that responds to each transient sound. Response to tone bursts of different duration (given in number of cycles of the 15 kHz stimulus tone) are shown. (B) Same data as in the upper graph, reported as a function of the inverse of the silent period (reprinted from Møller, A.R. 1969. Unit responses in the rat cochlear nucleus to repetitive transient sounds. *Acta Physiol. Scand.* 75, 542–551, with permission from Blackwell Publishing Ltd).

responded with the largest number of discharges (best duration) varied from 1 to 30 ms. Application of bicuculline, a GABA$_A$ antagonist, eliminated the duration tuning, thus indicating that GABAergic inhibition is involved in the duration selectivity of these neurons.

Some neurons in the CN respond to transient sounds with a single discharge, and the discharges follow repetitive stimuli up to a certain rate, above which they cease to fire (Fig. 6.17A) (Fig. 6.46) [197]. The tuning to the time pattern of sounds was found to be unrelated to the spectrum of the sounds as shown be reversing the polarity of every other click (from

condensation to rarefaction), which changes the spectrum but did not change the way these neurons responded [197]. These CN cells were thus not "tuned" to the repetition rate of sounds but rather to the silent interval between two sounds. This specificity to the silent period between two sounds may be a result of inhibition that is released a certain time after a cell is activated. Such duration tuning may be important for echolocating bats and for discrimination of communication sounds including speech sounds, which have silent periods, the duration of which carry information.

BOX 6.21

CELLS TUNED TO DURATION OF SILENCE ARE AFFECTED BY DESCENDING ACTIVITY

The selectivity to duration of silence may be similar to the inhibition based variable delays which Suga and co-workers [55] suggested could explain the bat's ability to discriminate delays between transmittal of a sound and receiving of an echo. Yan and Suga [320] showed that the response of such "delay tuned" neurons is enhanced by electrical stimulation of neurons in the auditory cortex. That means that a descending tract (corticofugal system, see Chapter 5 and Fig. 5.10) can affect the tuning to time intervals of neurons in the IC. In other studies the same

authors [321] showed that that neurons in the medial geniculate body are more sharply tuned to delays than the delay tuned neurons in the IC. The corticofugal descending system, together with the classical ascending auditory system, constitute a closed feedback loop that may be the ingredients for self-adjustment of some forms of sound analysis. These results were obtained in the bat but similar analysis may occur in the auditory system of other mammals including that of humans.

In addition to tonotopic organization, neurons in the nuclei of the classical ascending auditory pathways are also anatomically organized with regards to other features of sounds [245, 254]. Suga and his co-workers have demonstrated spatial organization of the cerebral cortex to a variety of features of complex sounds [288, 290]. Extensive studies of the coding of complex sounds in flying bats have revealed, among other properties, that sounds are not only organized on the surface of the cortex in accordance with their frequency (tonotopic organization) but also by other features of sounds, which have their own areas on the cortex [288, 290]. Changes in frequency of sounds are specifically represented in the auditory cerebral cortex of the flying bat (Fig. 6.47) [286]. Neurons in certain anatomical areas of the cerebral cortex respond only to tones the frequency of which changes rapidly and neurons that respond to tones of constant frequency are separated anatomically from those neurons.

The navigational sounds of the bat are rich in harmonics and the different harmonics are represented in anatomically different areas of the auditory cortex. The bat's cortex also has topographical maps of change in frequency that corresponds to the velocity of the bat in relation to the object that reflects its navigational sound. The delay between the emission and receiving of the reflected sound corresponds to the distance to an object. The amplitude of the echo is a measure of the angle to the reflecting object this is a measure of the relative size of the object and this is also represented on the surface of the auditory cortex together with the distance to the object. This means that different areas of the auditory cortex are devoted to processing of different features of a sound, of which frequency (tonotopicity) is only one feature.

The features of the sounds described above are important for the bat but it is not difficult to recognize similarities between bats' navigational sounds and the communication sounds of many mammals. The auditory system of other mammals including ourselves may therefore perform similar kinds of analysis of sounds and it may be expected that the change in frequency and amplitude of sounds and intervals between sounds are represented on the surface of the cortex of other mammals including humans, as specific topographical maps.

6.5. Coding of Sound Intensity

Coding of sound intensity is much less obvious than coding of frequency and it has been much less studied despite the fact that it is a prominent feature of sound perception. Therefore, little is known about how sound intensity is coded in the auditory nervous system and it is not known how the threshold of hearing is established.

Coding of the intensity of a sound has been assumed to be related on the discharge rate in auditory nerve fibers. However, the discharge rate of most auditory nerve fibers in response to continuous tones reaches asymptotic rates (plateau) at stimulus levels that vary from fiber to fiber. In most fibers, such a plateau is reached only 20–40 dB above the threshold of the nerve fibers (see Fig. 6.21). Only a few auditory nerve fibers have discharge rates that increase monotonically for increasing stimulus intensity over a large range of stimulus intensities (Fig. 6.21B).

6.6. Conclusion

Processing of sound in the auditory nervous system is complex and the results obtained using simple sounds such as pure tones do not describe the function of the system to other kinds of sounds. Studies done at threshold sound levels cannot be used as a basis for understanding the function of the auditory system at physiologic sound levels.

Enhancement of the response to sound with rapidly changing frequency or spectrum, as demonstrated in the response of cells in the nuclei of the ascending

---

BOX 6.22

FREQUENCY-MODULATED SOUNDS FOR NAVIGATION AND PREY LOCALIZATION

The flying bat emits short bursts of high frequency tones the frequency of which changes and it uses these FM sounds for navigation as well as for localizing prey (echolocation). They determine the distance to an object that reflects their sound by determining the time it takes for the echo to arrive at their ears. The frequency of the reflected sound is slightly different from the emitted sound and that difference is directly related to the velocity of the bat relative to the reflecting object (Doppler shift).

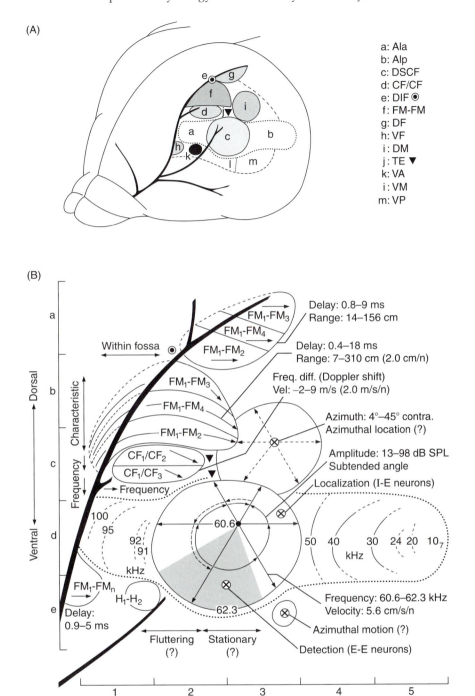

(A)

a: Ala
b: Alp
c: DSCF
d: CF/CF
e: DIF ◉
f: FM-FM
g: DF
h: VF
i: DM
j: TE ▼
k: VA
i: VM
m: VP

(B)

Dorsal

Ventral

Within fossa ◉

Characteristic

Frequency

a

b

c

d

e

FM$_1$-FM$_3$
FM$_1$-FM$_4$
FM$_1$-FM$_2$

FM$_1$-FM$_3$
FM$_1$-FM$_4$
FM$_1$-FM$_2$

CF$_1$/CF$_2$
CF$_1$/CF$_3$

Frequency

100
95
92
91
kHz

FM$_1$-FM$_n$
H$_1$-H$_2$

Delay:
0.9–5 ms

Delay: 0.8–9 ms
Range: 14–156 cm

Delay: 0.4–18 ms
Range: 7–310 cm (2.0 cm/n)

Freq. diff. (Doppler shift)
Vel: −2–9 m/s (2.0 m/s/n)

Azimuth: 4°–45° contra.
Azimuthal location (?)

Amplitude: 13–98 dB SPL
Subtended angle

Localization (I-E neurons)

60.6

50   40   30   24 20   10$_7$
kHz

62.3

Frequency: 60.6–62.3 kHz
Velocity: 5.6 cm/s/n

Azimuthal motion (?)

Detection (E-E neurons)

Fluttering
(?)

Stationary
(?)

1   2   3   4   5

Anterior ◄────► Posterior

◄─1.0 mm─►
P.p.r.

FIGURE 6.47   Cortical representation of sounds that are important to the flying bat (reprinted from Suga, 1994, with permission from Springer-Verlag).

auditory pathways, may be an example of a more general principle of signal processing in the auditory nervous system that emphasizes the important information while discarding information about sounds that contain little or no information. Discarding unimportant information is important for achieving optimal usage of the capabilities of the nervous system.

The descending auditory systems may adjust the analysis that occurs in the nuclei of the classical ascending auditory pathways.

7. DIRECTIONAL HEARING

Hearing with two ears is the basis for directional hearing, which is an ability to determine the direction to a sound source in the horizontal plane (azimuth). This has had enormous importance for many species of vertebrate animals during evolution. Many animals depend totally on the ability to determine the direction of sounds emitted by other animals, who may be foes or prey. The owl is one example of an animal that relies on sound in identifying its food.

Hearing with two ears is not only important because it makes it possible to determine the direction to a sound source. Many tasks are improved by listening with two ears and hearing with two ears is better overall than hearing with one ear. The discrimination of sounds in a noisy background is better with two ears than with one. The "unmasking" from hearing with two ears benefits from both a time (phase) difference between the sounds and also from intensity differences [9]. Studies have shown that the advantage of hearing with two ears is greater if the masking noise in the two ears is different such as shifted by 180°. Hearing with two ears makes it possible to listen to one speaker in an environment where many people talk at the same time ("cocktail-party effect") as described by Cherry [81]. Hearing with two ears gives an impression of space such as occurs naturally when listening to music in nature. Few people have the opportunity to do that and stereophonic reproduction of music is a good substitute because it makes use of the basic principles of intensity and timing difference in binaural hearing [9]. If hearing is impaired more in one ear than in the other ear, the advantages of hearing with two ears diminishes. People often become aware that they have an asymmetric hearing loss because they have difficulties

in understanding speech where many other people are talking.

Making lesions in specific parts of the brain in animals has been important in finding which parts of the brain are involved in specific tasks such as determining the direction to a sound source [223] and more recently by cooling specific parts of the cerebral cortex [145].

7.1. Physical Basis for Directional Hearing

The direction to a sound source in the horizontal plane is signaled by the difference in the time of arrival (interaural time difference [ITD]) and the intensity of the sound at the ears (interaural level differences [ILD]). Low-frequency sounds are localized mainly on the basis of the difference in arrival time at the ears while high-frequency sounds are mainly localized by the difference in intensity of the sound at the ears. This suggests that sound localization in the horizontal plane is based on dual mechanisms. The differences in the sound that reaches the two ears are also the basis for improved speech discrimination in noise and the perception of "space," which is important when listening to sounds such as music under normal conditions [10].

Differences in the spectrum of the sound that reaches the two ears also play a role in binaural hearing and it is essential for directional hearing in the vertical plane.

In nature, a sound travels directly to the observer whereas, in rooms, reflections from the walls make sound reach the ears from many different directions. That makes directional hearing become more complex because a sound emitted at one place in a room through reflections from walls, ceiling and floor will seem to come from many directions. It is mainly the first sound that reaches the ears that the auditory system uses to determine the direction to a sound and

---

BOX 6.23

INSECTS' USE OF DIRECTIONAL HEARING TO AVOID BATS

The importance of directional hearing is not limited to birds and mammals. Thus certain insects (e.g. night moths) that are prey for flying bats use directional hearing to avoid being eaten. These moths have only four cells that are sensitive to the echo locating sounds of bats [246, 247]. These four cells are located on the thorax of the moth, two on each side. One pair has a low sound threshold and one pair has a high sound threshold. When only the most sensitive pair of these cells is activated, by a bat that is far

away, the moth turns away from the bat. That means that the moth must know the direction to the sound source (the bat) and it obtains that information from the difference in the sound that reaches the two receptors. When the moth's high threshold sound receptors are activated (together with the most sensitive pair of receptors), it is an indication that the bat is close to the moth, and it closes its wings and falls to the ground; the fastest possible escape route.

BOX 6.24

PHYSICAL BASIS OF DIRECTIONAL HEARING

Both the differences in the arrival time (ITD) and the difference in the intensity (ILD) of the sound at the two ears are determined by the physical shape (acoustic properties) of the head and the outer ears, together with the direction to the sound source (Chapter 2, Figs 2.3–2.6). The sound arrives at the same time at the two ears when the head is facing the sound source (azimuth = 0°) and directly away from the sound source (azimuth = 180°). At any other azimuth, sounds reach the two ears with a time differences and the sound intensity at the entrance of the ear canals of the two ears is different. The difference in the sound intensity has a more complex relationship to the azimuth than the interaural time difference (Chapter 2, Fig. 2.6).

---

that reduces the effect of reflected waves that might cause confusion about the direction of a sound. This is known as the precedence effect [63, 302, 332].

It is incompletely understood how we determine the direction to a sound in the vertical plane. Because the head is symmetric with regard to the medial plane, movements of the head in the vertical plane (elevation) do not create any time or intensity differences in the sound that reaches the two ears. The sounds that reach the two ears are therefore the same, independent of the elevation of the sound source. However, the spectrum of a broad band sound that reaches the ear canal changes systematically with the elevation (Chapter 2, Fig. 2.7) and that is assumed to be the physical basis for discrimination of the elevation of a sound source [236]. The reason that we do so well in determining the elevation to a sound source overall is that most natural sounds are broad band sounds. Changes in the spectrum of a sound may also be responsible for our ability to determine the azimuth to a sound using only one ear. Turning the head plays an important role in directional hearing under difficult circumstances such as with one ear and in noisy environments.

7.2. Neurophysiologic Basis for Sound Localization

Neurons that receive input from both ears are the neural substrate for detecting interaural time differences. This type of neuron can be found at several levels of the ascending auditory pathways. The most peripheral level at which this type of neuron has been found is in the superior olivary complex, especially the middle and lateral superior olivary nucleus (MSO, LSO; see Chapter 5).

Jeffress [89] in 1948 presented a hypothesis that describes how detection of interaural time differences may occur. Jeffress's model consists of an array of neurons that receive input from both ears. These neurons only fire when the inputs from the two ears arrive at the same time. If the axons that lead to these neurons had the same length the neurons would fire only for sound arriving at the ears at the same time, thus sounds that came from a source that was located directly in front of the observer or directly behind. The delay in the axons that lead input to these coincidence neurons is different for different neurons. The difference in the neural delays to the two inputs makes a neuron fire when sounds arrive at the two ears with that delay. If, e.g., the input to such a coincidence neuron from the left ear is delayed 50 µS relative to the right ear, the two inputs will coincide when sounds arrive 50 µS earlier at the left ear than at the right ear. This occurs when a sound source is located approximately 9° to the left of the midline (90° corresponds to 650 µS [see Chapter 2, Fig. 2.3]. Therefore, sounds that arrive at the two ears at different intervals will activate different ones of these coincidence detectors. With a sufficient range of delays in these delay lines the entire range of azimuths can be covered.

It is remarkable that Jeffress's model has maintained its validity in view of how little was known about the function and the anatomy of the auditory nervous system at the time it was proposed – now more than 50 years ago. Jeffress's model consisted of two delay lines, one from each ear. Later studies have found support for a slightly altered model with only a single delay line (Fig. 6.48) [100]. This is the only major disagreement between Jeffress's model and the recent modified model by Joris which has a single delay line whereas Jeffress assumed dual (bilateral) delay lines (Fig. 6.48). This modified model is based on anatomical studies of the MSO [100]. The model of Jeffress, however, works equally well with a single delay line as the originally proposed bilateral delay lines.

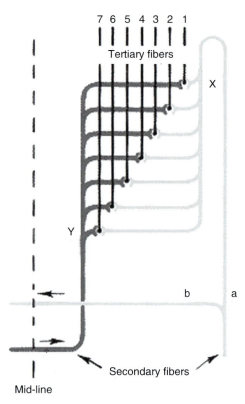

7 6 5 4 3 2 1
Tertiary fibers
X
Y
b   a
Secondary fibers
Mid-line

FIGURE 6.48 Anatomical features of the Jeffress model. The original delay line/coincidence model is shown for the right side of the brainstem (modified from Joris et al., 1998, with permission from Elsevier).

The anatomical basis for this model is neurons in the MSO that consists of a long sheet of cells that are oriented in one plane. The input to these MSO neurons is from bushy cells in the anteroventral cochlear nucleus (AVCN) and ipsilateral afferent fibers terminate on the dendrites of the MSO cells [100].

The MSO neurons have input from both ears. Since the input is delayed various amounts of time by the

different delay lines different neurons in the MSO become assigned to different delays of sounds arriving at the two ears and thus different azimuths. A specific neuron will therefore fire only in response to sounds that come from a certain direction. Which neuron fires tells higher nervous centers what the delay was between the arrival of a sound at the left and the right ear and thus the azimuth of the sound source. The range of delays needed to cover the entire horizontal plane is 650 µS and that range of delays can be realized by axons of different lengths as shown in Fig. 6.48.

This model of directional hearing requires that the discharges that are delivered to these MSO neurons have the temporal information in the acoustic signal preserved. There are neurons in the cochlear nucleus that respond very precisely to the time pattern of the stimulation [98, 197]. These neurons can thus be the basis for this model

Some neurons in the central nucleus of the inferior colliculus (ICC) exhibit similar response patterns as the neurons in the MSO nucleus. Rose et al. [250] found neurons that responded best to specific interaural delays and different neurons had different preferred delays. Later, extensive studies by Yin and Kuwada [324] confirmed these studies and showed that many neurons in the ICC responded best to binaural sounds when the sound in one ear was delayed a certain time relative to the sound in the other ear. The connections from the IC to the SC are important in the ability to move the head towards a sound source.

The role of the auditory cortex in directional hearing has been studied in animal experiments using ablations (for a review, see [223]). These studies in animals and humans showed only a small or moderate deficit from total or partial ablation of the auditory cortex. A more recent study in cats using inactivation of specific parts of the cortex by cooling [145] showed that inactivation of A1 and PAF resulted in increased

---

BOX 6.25

MSO NEURONS ACT AS COINCIDENCE DETECTORS

Several investigators have found experimental evidence that neurons in the superior olivary MSO are selective with regard to interaural time intervals [323] and neurons in the LSO are sensitive to interaural intensity differences [100, 300, 322]. That neurons in the MSO can act as coincidence detectors has also been shown experimentally [2] and it has been found that these neurons (as required by the Jeffress model) will fire only when two inputs arrive exactly at the same time. The basis for the function of neurons that act as coincidence detectors is their bipolar dendrites; one dendrite receives input from only one ear [2]. The neuron performs a nonlinear summation of the excitatory input to each of these two dendrites and this particular morphology enriches the computational power of these neurons compared with neurons that do not have such dendrites [2].

errors in determining the direction to a sound source in the horizontal plane in the opposite hemifield. Inactivation of other auditory cortical areas has no noticeable effect. Perhaps the connections in the midbrain from the IC to the SC provide the ability to move the head towards a sound source in the absence of the function of the auditory cortex.

The corpus callosum that connects the two sides' auditory cortex is important for binaural hearing, for sound localization and for fusing a sound image on the basis of sounds that reach the two ears [134]. Severance of the corpus callosum impairs sound localization and it is even more impaired when one hemisphere (temporal lobe) is removed. There are some indications that elderly people have difficulties in fusing auditory images and that has been related to impairment of the connections between the auditory cortices. This has implications for selecting hearing aids.

In order to be the basis for discrimination of direction to a sound source, the neural code of the interaural time difference must not be affected by other factors than the direction to a sound source. Differences in the intensity of sounds that reaches the two ears could affect the neural coding of time differences in the arrival of sounds at the two ears. The assumption that the latency of the response decreases when the intensity of the stimulus is increased is, however, only valid for large changes in the intensity of sounds such as

tone bursts, the intensity of which increases rapidly from below hearing threshold to a high intensity. The latency of small variations in the intensity of a sound such as the envelope of amplitude-modulated sounds does not change noticeably when the sound intensity is changed [186]. This means that the neural discharges evoked by a relatively small amplitude changes will not change noticeably with the intensity of the sound. The intensity dependence on the latency of neural responses observed in experiments using tone bursts may thus be regarded mostly as an experimental artifact due to the use of sounds that do not resemble natural sounds.

The use of the interaural time difference requires that the coding of the temporal pattern of the sound be preserved until the neural activity from the two ears can be compared. The demand on maintaining the temporal code is high because the differences in time of arrival of sounds at the two ears that can be detected are very small (5–10 μS) as shown in psycho acoustic experiments [299]. This corresponds to an angle of 1–2° (the interaural time difference between arrival of sounds at the two ears for a 90° turn of the head is 650 μS) (Chapter 2, Fig. 2.3) [51]. This means that neurophysiologic studies should be able to demonstrate a similar sensitivity to interaural time difference of neurons that are to be regarded as candidates for explaining the results of psychoacoustic studies.

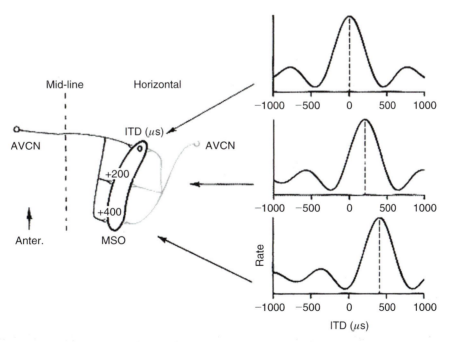

FIGURE 6.49   Current view of the Jeffress model in the cat. Input from one ear enters a binaural processor consisting of a bank of cross-correlators each of which processes the signal at a different ITD. The cells for which the internal delay exactly offsets the acoustic ITD are maximally active (reprinted from Joris et al., 1998, with permission from Elsevier).

It has been assumed that neurons, which compare the time of arrival of sounds at the two ears, must be located peripherally in the ascending auditory nervous system because it is believed that temporal information is degraded in synaptic transmission. As mentioned above, precision of temporal information may in fact improve through synaptic transmission because of spatial integration (see p. 110).

Detection of inter-aural intensity differences contributes to our ability to localize a sound source in the horizontal plane. The interaural intensity difference is a much more complex function of azimuth (see Figs 2.5 and 2.6 in Chapter 2) and the difference in the intensity of the sounds that reach the two ears also depends on the frequency of the sound and that complicates studies of the neural mechanisms for detecting interaural intensity differences.

Neurons that receive excitatory input from one ear and inhibitory input from the other ear (EI neurons) are likely to be involved in detecting interaural intensity differences. Such neurons have been found in the SOC, mainly in the LSO, in the DNLL and the ICC [86] (Fig. 6.50).

Jeffress has suggested that intensity differences may be converted to time differences because the latency of the response of most neurons decreases with increasing sound intensity [89]. This would make it possible to use the same neural circuitry that detects interaural time difference for detecting interaural intensity differences.

Some sound localization is possible in persons with only one hearing ear. This ability depends on the change in the sound that reaches the ear when the head is turned. Interaction between the intensity of sounds that reach the ear and proprioceptive input from neck muscles are presumably used in monaural sound localization. Recent studies indicate that neural circuitry in the cerebellum and the dorsal cochlear nucleus are involved [224].

7.3. Localization in the Vertical Plane

Humans are able to localize sounds in the vertical plans (elevation) but so far little is known about the physical basis for discrimination of elevation. It is not possible to obtain information about the elevation of a sound source based on the difference between the sound that reaches the two ears because that is not affected by the elevation of a sound source. This is because the ears and the head are symmetrical around the midline in most animal species including humans. Despite that, psychophysical experiments show that we can discriminate 4–6° in the vertical plane (elevation). The ability to localize the direction to a sound

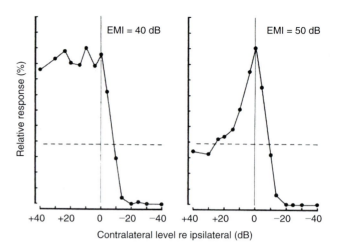

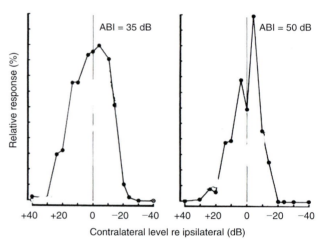

FIGURE 6.50 Illustration of the sensitivity of neurons in the central nucleus of the inferior colliculus in the cat to interaural intensity differences. ABI = average binaural intensity; and EMI = excitatory monaural intensity (reprinted from Irvine, 1987, with permission from Elsevier).

source in the vertical plane is assumed that it is based on changes in the spectrum of a sound as a function of the elevation of the sound source (Fig. 2.7).

7.4. Representation of Auditory Space (Maps)

The ability of the auditory system to detect interaural time differences and interaural intensity differences is the basis for perception of space from sound. While mapping of a neuron's receptive field with regard to frequency and intensity is common (frequency tuning curves) (see p. 98) it is also possible to map a neuron's receptive field regarding the location of a sound source in space. Maps that relate auditory space are three-dimensional and therefore require information about the elevation besides the azimuth. While the tonotopic maps are a projection of the receptor epithelium on the basilar membrane, space maps are the result of processing of neural information with regards to differences in the sounds that reach the two ears. Knudsen and his co-workers [114] have called such maps "computational maps" to distinguish them from tonotopic maps that are projections of an anatomical structure (such as the basilar membrane).

The regions in an auditory nucleus of the barn owl (the MLD nucleus) where maps of auditory space have been studied most extensively differ from those that contain maps of the cochlea (tonotopic organization). Yet, the analysis of the spectrum of a sound is one factor that seems to be important for perception of space, particularly regarding elevation. How these two regions are related remains to be elucidated.

Many neurons in the SC receive auditory input from the IC and may neurons in the SC respond to sound. Some neurons in the mammalian SC are sensitive to interaural intensity differences (Fig. 6.51A) [318]. (These neurons are located in the deep layers of the SC while neurons in the superficial layers respond to visual stimuli.) Visual and auditory perception of

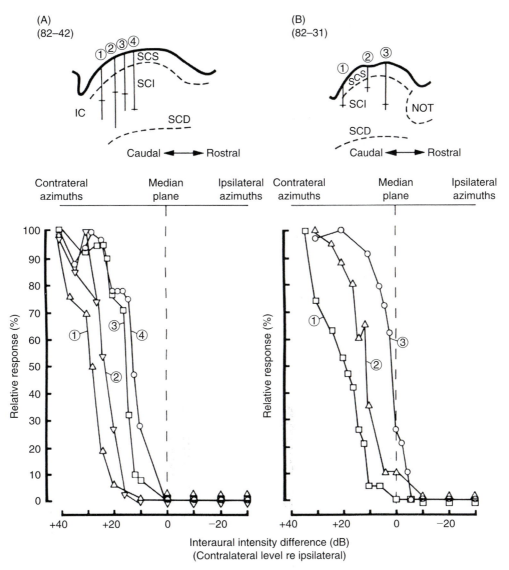

FIGURE 6.51   Sensitivity to interaural intensity difference of neurons in the superior colliculus of the cat. (A) Normalized interaural intensity difference sensitivity of cells at different locations in the deep layer of the superior colliculus. The neurons were excited from one ear and inhibited from the other ear. The stimuli were noise bursts at 60 or 70 dB SPL. (B) Locations of the neurons depicted in (A). NOT = nucleus of the optic tract; and SCD = deep layer of the superior colliculus (reprinted from Wise and Irvine, 1985, with permission from the American Physiological Society).

BOX 6.26

SPATIAL SELECTIVITY OF NEURONS IN THE SUPERIOR COLLICULUS

Some neurons in the superior colliculus respond only to sound from a limited region of space. Such neurons have sharp peaked response curves regarding interaural time and interaural intensity (Fig. 6.52) [153]. These neurons are silent in response to binaural sounds outside these regions and they are silent in response to monaural sounds. That means that some neurons only respond when the sound comes from a certain place in space (three-dimensional direction). Such neurons may have been overlooked in conventional neurophysiologic experiments because it is common to search for neurons using monaural "search" sounds while a microelectrode is advanced through a nucleus. The neurons that showed the greatest spatial selectivity (Fig. 6.52 A & C) did not respond to monaural stimulation at all.

The similarities between the responses of the neurons from the superior colliculus (Fig. 6.52) and those from the MLD in the barn owl are striking. The optic tectum in the barn owl (corresponding to the SC in mammalians) has maps of auditory space. As in mammals, most neurons in these nuclei receive input both from the visual and the auditory systems as mentioned above.

While the SC in the cat is dominated by cells that respond selectively to the direction of a moving visual stimulus there are cells in the deep layer of SC that also respond both to auditory and somatosensory stimuli [238]. Visual and auditory cells respond not only to motion but such cells also have pronounced directional selectivity. However, in the auditory domain similar cells are rare. The mechanism of direction selectivity in these cells seems to be suppression of the response in the 'non-preferred' direction rather than facilitation in the "preferred" direction.

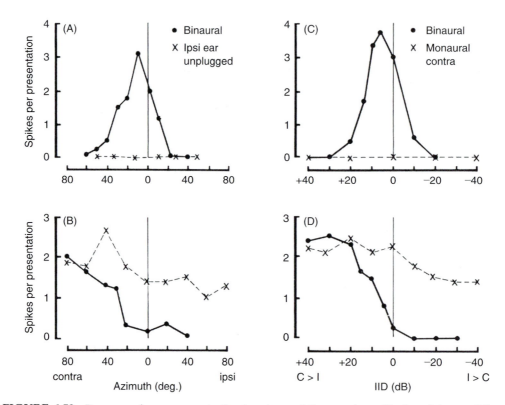

FIGURE 6.52 Responses from neurons in the deep layer of the superior colliculus of the cat. (A) and (B): Responses of two different cells to binaural sound and monaural (dashed lines, one ear plugged). (C) and (D): Response as a function of interaural intensity difference in two other neurons from the superior colliculus (reprinted from Middlebrooks, 1987, with permission from the American Physiological Society, and Irvine, 1992).

space is integrated in the SC as the SC is involved in righting, movements of the eyes and of the head. The SC also coordinates other sensory inputs such as somatosensory input and input from the vestibular organ. The fact that many sensory systems converge onto the SC may explain why other senses can take over to create perception of space if one sense is impaired.

The neurons in the SC that respond to sound are organized topographically according to their sensitivity to interaural intensity difference (Fig. 6.51B).

8. EFFERENT SYSTEM

Little is known about the function of the efferent system. The most peripheral parts, the olivocochlear system, have been studied to a greater extent than other parts of the efferent system. Efforts to find influence of the olivocochlear efferents on hearing functions including speech discrimination have been mainly negative. Thus the effect of severance of the efferent bundle (done in connection with vestibular neurectomy) on hearing function is minimal [261].

It has been suggested that the efferent system may be involved in the "toughening" of the ear regarding noise induced hearing loss (see Chapter 9) and evidence has been presented that both the medial and the lateral olivocochlear efferent systems provide adjustment of

the set point of activity in the outer hair cells and afferent processes that are the respective postsynaptic targets of these fibers [130].

There have been few studies of the function of more central parts of the abundant efferent system. Experiments in the flying bat have shown that the frequency selectivity of cortical neurons is modified by the corticofugal system that consists of descending connections from the auditory cortex to the thalamus. Thus, Suga and his coworkers [331] found that inactivation of cortical neurons affected the tuning of neurons in the IC and the MGB (Fig. 6.53). Since the tuning of cortical neurons depends on the tuning of neurons in the medial geniculate body, these descending connections complete a closed feedback loop that may adjust the frequency selectivity based on the sounds that reach the ear.

9. NON-CLASSICAL PATHWAYS

Coding of sounds in the non-classical pathways is poorly known. A few studies have shown evidence that neurons in the nuclei of the non-classical pathways (mostly the IC) [4, 294] respond less distinct to sounds than neurons in the classical pathways, showing broad frequency tuning and slow response to transient sounds. A few studies have concerned the response

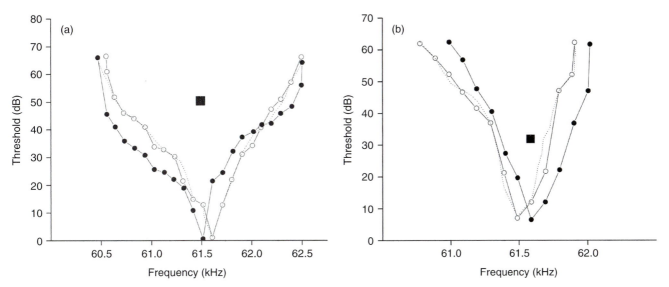

FIGURE 6.53   Changes in tuning curves of a neuron in the ventral division of the medial geniculate body (B) a neuron in the dorso-posterior division of the inferior colliculus of mustached bats (A) after focal inactivation of cortical neurons, the CFs of which are indicated by squares. The tuning curves obtained before inactivation (open circles) are shown together with tuning curves obtained during inactivation (filled circles) and after recovery from inactivation (dashed lines). The inactivation was done by injecting a local anesthetic (Lidocaine) in a location of the auditory cortex (data from Zhang et al., 1997, reprinted with permission from *Nature*).

from neurons in the medial and dorsal portions of the MGB [121]. Neurons in the ICX and DC for the IC also respond to other sensory modalities [4].

10. EFFECT OF ANESTHESIA

Most of the published studies on the physiology of the auditory nervous system have been done in anesthetized animals. Anesthesia affects the responses to a degree that depends on the type of anesthesia and the anatomical location of the structure from which recordings are made. Results of recordings from anesthetized animals may therefore not truly reflect the normal condition, and the tuning and the variability in tuning in the auditory nervous system may be different in awake animals. Generally, the effect of anesthesia is greater in the central parts of the auditory pathways than it is in more peripheral structures. While the effect of commonly used surgical anesthesia regimen on the responses from auditory nerve fibers is small, anesthesia can have a noticeable effect on the response from cells in the CN [50] and the effect increases along the neural axis of the classical ascending auditory pathways. The responses of neurons of the auditory cortex are more affected by anesthesia than those of nuclei of the ascending pathways. The results obtained in experiments in animals under anesthesia may therefore not be a valid description of the response in awake animals.

7

Evoked Potentials from the Nervous System

1. ABSTRACT

1. All neural structures of the ascending auditory pathways can generate sound evoked electrical potentials that can be recorded by an electrode placed on the respective structure.

2. Compound action potentials (CAP) recorded directly from the intracranial portion of the auditory nerve in small animals are different from those recorded in humans because the eighth cranial nerve is longer in humans than in small animals (2.5 cm in humans and approximately 0.8 cm in the cat).

3. In humans, the latency of the main negative peak of the CAP recorded with a monopolar electrode from the intracranial portion of the human auditory nerve is approximately one millisecond longer than that of the N_1 component of the action potential (AP) recorded from the ear.

4. Evoked potentials recorded with a bipolar electrode from a long nerve such as the human auditory nerve represent propagated neural activity.

5. The responses recorded from the auditory nerve to continuous, low frequency sounds is the frequency following response (FFR).

6. The response recorded from the surface of a nucleus (such as the cochlear nucleus and the inferior colliculus) in response to transient sounds has an initial positive–negative deflection, which is generated by the termination of the nerve that serves as the input to the nucleus. The slow deflection that follows is generated by dendrites and the fast components riding on the slow wave are somaspikes generated by firings of nerve cells.

7. Far-field evoked potentials are the potentials that can be recorded from locations that lie far from the anatomical location of their generators, such as the surface of the scalp.

8. Neural activity in many of the structures of the classical ascending auditory pathways, but not all, give rise to far-field evoked potentials that can be recorded from electrodes placed on the scalp.

9. Auditory brainstem responses (ABR) and the middle latency responses (MLR) are far-field responses that are used in diagnosis and research.

10. Propagated neural activity in a nerve or a fiber tract in the brain may generate stationary peaks in the far-field potentials when the propagation is halted, or when the electrical conductivity of the medium surrounding the nerve changes or when the nerve or fiber tract bends.

11. The far-field potentials from nuclei depend on their internal organization.

12. The normal ABR consists of five prominent and constant vertex positive peaks that occur during the first 10 ms after presentation of a transient sound. These peaks are labeled by Roman numerals, I–V. Most studies of the neural generators of the ABR have concentrated on the generators of these vertex positive peaks.

13. Peak I and II of the human ABR are generated exclusively by the auditory nerve (distal respective proximal portion), while peaks III, IV, V have contributions from more than one anatomical structure. Other anatomical structures of the ascending auditory pathways, contribute to more than one peak.

14. Peak III is mainly generated by the cochlear nucleus.
15. The sharp tip of peak V is generated by the lateral lemniscus, where it terminates in the inferior colliculus on the side contralateral to the ear from which the response is elicited.
16. The individually variable slow negative potential following peak V (SN_{10}) is generated by (dendritic) potentials in the contralateral inferior colliculus.
17. The middle latency response (MLR) is composed of the potentials that occur during the interval of 10–80 ms or 10–100 ms after presentation of a stimulus sound.
18. The neural generators of the MLR are less well understood than those of the ABR. Potentials generated in the cerebral cortex contribute to the MLR and muscle (myogenic) responses may also contribute to the MLR.
19. The "40 Hz response" is a far field response that results from summation of components of the evoked potentials that repeat every 25 ms.
20. The frequency following response (FFR) may be recorded from electrodes on the scalp in response to low frequency tones.

2. INTRODUCTION

Evoked potentials can be divided into near-field and far-field potentials, where near-field potentials are the evoked potentials that can be recorded from electrodes placed on the cochlea or directly on specific structures of the auditory nervous system. Auditory evoked potentials are important tools for diagnosis of disorders of the ear and the auditory system. Auditory brainstem responses (ABR) are the most used auditory potentials in the clinic but middle latency responses (MLR) are used in special situations. Studies of evoked potentials have contributed to understanding of the function of the ear and the auditory nervous system. In this chapter, I will discuss the near-field and far-field potentials from the auditory nervous system. The neural generators of the ABR will also be discussed.

3. NEAR-FIELD POTENTIALS FROM THE AUDITORY NERVOUS SYSTEM

Evoked potentials recorded directly from a nerve or a nucleus are known as near-field potentials whereas far-field potentials are the evoked potentials that can be recorded at a (large) distance from the active neural structures. The near-field potentials have large amplitudes and usually represent the neural activity in only one structure whereas far-field potentials, such as the ABR, have small amplitudes and often have contributions from many neural structures as well as muscles. Studies of electrical potentials recorded directly from exposed structures of the ascending auditory pathways have helped to understand how far field auditory evoked potentials, such as the ABR, are generated (see p. 167). Recordings of evoked potentials generated by different parts of the auditory nervous system are important in intraoperative neurophysiologic monitoring that is done for the purpose of reducing the risks of surgically induced injuries.

Below, I will discuss the electrical potentials that can be recorded directly from structures of the classical ascending auditory pathways in response to sound stimulation. I will first discuss evoked potentials recorded directly from the auditory nerve and then discuss responses recorded from nuclei of the ascending auditory pathways.

3.1. Recordings from the Auditory Nerve

Recordings of the response from the exposed auditory nerve have been done extensively in animals [23, 284] and more recently in humans who underwent operations where the central portion of the eighth cranial nerve was exposed [80, 205]. Recordings in animals have provided important information about the function of the ear and recordings in humans have won clinical use in monitoring of the neural conduction in the auditory nerve when the nerve has been at risk of being injuring because of surgical manipulations [185].

The waveform of the compound action potentials (CAP[1]) in response to click stimulation recorded from the intracranial portion of the eighth cranial nerve using a monopolar recording electrode typically has two negative peaks (N_1, N_2) (Fig. 7.1) thus similar to the AP recorded from the round window of the cochlea as described in Chapter 4.

In the cat the latency of the N_1 in the response recorded from the auditory nerve in the internal auditory meatus is approximately 0.2 ms longer than that of the AP recorded from the round window (Fig. 7.2 [109]).

---

[1]In the following, we will use the term compound action potentials (CAP) for the potentials recorded from the exposed auditory nerve, although they are similar to the potentials that are recorded from the cochlea, and which are called action potentials (AP) (p. 57).

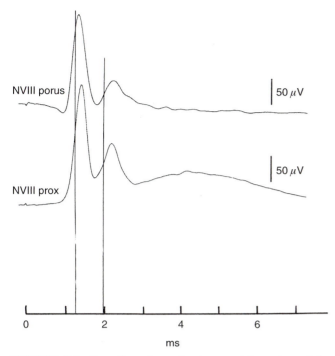

FIGURE 7.1 Recordings from the intracranial portion of the auditory nerve in a rhesus monkey, at two different positions, near the porus acousticus and near the brainstem. The stimuli were clicks presented at 107 dB PeSPL (peak equivalent sound pressure level) and at a rate of 10 pps (modified from Møller and Burgess, 1986, with permission from Elsevier).

The auditory nerve in a small animal, such as the cat, is approximately 0.8 cm long [59]. The difference between the latency of the N_1 of the AP and that of the response from the intracranial portion of the auditory nerve is the travel time in the auditory nerve from the ear to the recording site.

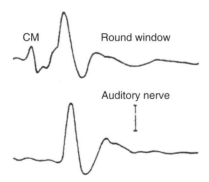

FIGURE 7.2 Comparison between recording from the round window of the cochlea and from the intracranial portion of the auditory nerve in a cat using a concentric electrode. The stimulation was clicks. M is the cochlear microphonic potential (modified from Peake et al., 1962, with permission from the American Institute of Physics).

Because the auditory nerve in small animals is very short, any recording site on the auditory nerve will be close to the cochlea and the cochlear nucleus and potentials that originate in the cochlea and the cochlear nucleus are conducted to the recording site by passive conduction in the eighth cranial nerve and the surrounding fluid. Intracranial recordings from the auditory nerve using a monopolar recording electrode will therefore not only yield potentials generated in the auditory nerve but also potentials that originate in the cochlea (mostly cochlear microphonics [CM]) and in the cochlear nucleus. These passively conducted potentials thus do not depend on the nerve being able to conduct propagated neural activity through depolarization of nerve fibers. (Passive conduction is also the reason that recordings from the cochlea in small animals contain potentials that originate in the cochlear nucleus as was discussed in Chapter 4.)

The contributions of evoked potentials from the ear and the cochlear nucleus to the responses recorded from the auditory nerve can be reduced by using bipolar recording techniques [201]. Some investigators [228] have used a concentric electrode for recording from the intracranial portion of the auditory nerve to reduce the contamination of the neural response by the CM. However, a concentric electrode consisting of a sleeve with an insulated wire inside does not provide true bipolar recording because the two electrodes (the center core and the sleeve) do not have identical electrical properties. A concentric recording electrode is anyhow much more spatially selective than a monopolar electrode and the response recorded from the internal auditory meatus using a concentric electrode has no visible CM component (Fig. 7.2).

The most commonly used stimuli in connection with recordings of the CAP from the intracranial portion of the auditory nerve have been clicks or short bursts of tones or noise. Several studies have shown that the amplitude of the CAP response increases with increasing stimulus level in a similar way as the AP recorded from the round window of the cochlea. The main reason for that is that more nerve fibers fire as the stimulus intensity is increased. The latency of the response decreases with increasing stimulus intensity, mainly because the generator potentials in the cochlear hair cells rise more rapidly at high stimulus intensities than at low stimulus intensities [188]. Cochlear non-linearities also affect the latency differently at different stimulus intensities (see Chapter 3) and that contributes to the dependence of the latency on the stimulus intensity [179]. The conduction velocity of nerve fibers and the synaptic delays are independent of the level of excitation and thus do not contribute to the intensity dependence of the latency of the CAP recorded from the auditory nerve.

The amplitude of the CAP elicited by transient stimuli decreases when the stimulus rate is increased above a certain rate. Above a certain stimulus rate the responses elicited by the individual stimuli overlap, and the amplitude of one of the two peaks may increase because the N_1 peak of one response coincides with the N_2 peak of the previous response. When the rate of the stimulus presentation is increased beyond approximately 700 pps the amplitude of the response decreases rapidly. The latency of the response increases slightly when the stimulus rate is increased.

Recordings of auditory evoked potentials from the exposed auditory nerve in humans have helped in the understanding of some of the differences between the human auditory nervous system and that of small animals often used in studies of the auditory system. Several investigators [80, 205, 280] reported at about the same time that the latency of the CAP recorded from the exposed intracranial portion of the auditory nerve in humans is longer than it is in animals when recorded in a similar way. The reason for that is that the eighth cranial nerve in humans is 2.5 cm [125], thus much longer than it is in the animals such as the cat (approximately 0.8 cm [59]). The latency of the main negative peak of the CAP recorded from the intracranial portion of the auditory nerve in response to loud clicks is approximately 2.7 ms [205, 211] thus approximately 1 ms longer than the AP component of the electrocochlear graphic (ECoG) potentials recorded from the ear. Compare that to a difference of approximately 0.2 ms in the cat (Fig. 7.2 [228]).

In individuals with normal hearing a monopolar electrode placed on the exposed intracranial portion of the eighth nerve records a triphasic potential in response to click stimulation (Fig. 7.4A) as is typical for recordings with a monopolar electrode from a long nerve. The latency of the response decreases with

---

BOX 7.1

HISTORICAL BACKGROUND

It was probably Ruben and Walker [255] who first reported on recordings from the exposed intracranial portion of the eighth cranial nerve. These investigators recorded click evoked CAPs from the auditory nerve during an operation for sectioning of the eighth nerve for Ménière's disease, using a retromastoid approach to the cerebello-pontine angle. The waveform of the recorded potentials was complex and it had several peaks and valleys (Fig. 7.3). Ruben and his coauthor suggested that the responses had contributions from cells of the cochlear nucleus. Examination of their recordings (Fig. 7.3) indicates that the intracranially recorded CAP had a longer latency in humans than in the cat but the authors did not speculate on the reason for the longer latency. (Accurate assessment of the latency of the potentials from their published recordings is not possible because the record does not show the time the stimulus was applied.)

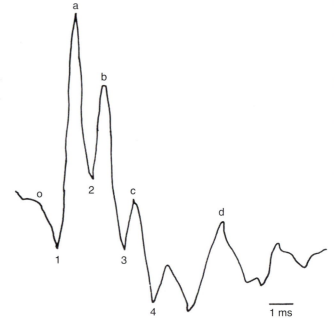

FIGURE 7.3 Recordings from the intracranial portion of the eighth nerve in a patient undergoing an operation for Ménière's disease (reprinted from Ruben and Walker, 1963, with permission from Lippincott Williams and Wilkins).

increasing stimulus intensity (Fig. 7.4B) and the amplitude of the main peak of the CAP increases with increasing stimulus intensity (Fig. 7.4A) similar to what is seen in studies in animals. The response from the exposed intracranial portion of the auditory nerve to short tone bursts has a similar waveform as the responses to click sounds but the latencies are slightly longer (Fig. 7.5A) [205].

A monopolar recording electrode placed on a long nerve along which an area of depolarization propagates will record a characteristic triphasic potential (Fig. 7.6). The initial positive deflection is generated as the area of depolarization approaches the recording electrode. The large negative deflection is generated when the area of depolarization passes directly under the recording electrode. The following small positivity is generated when the area of depolarization is leaving the location of the recording electrode. If the propagation of neural activity in such a nerve is brought to a halt, for instance by injury to the nerve, a monopolar electrode placed near that location would record a single positive potential. Such a potential is known as the "cut end" potential and described by Gasser and Erlangen (1922) and Lorente de No [143].

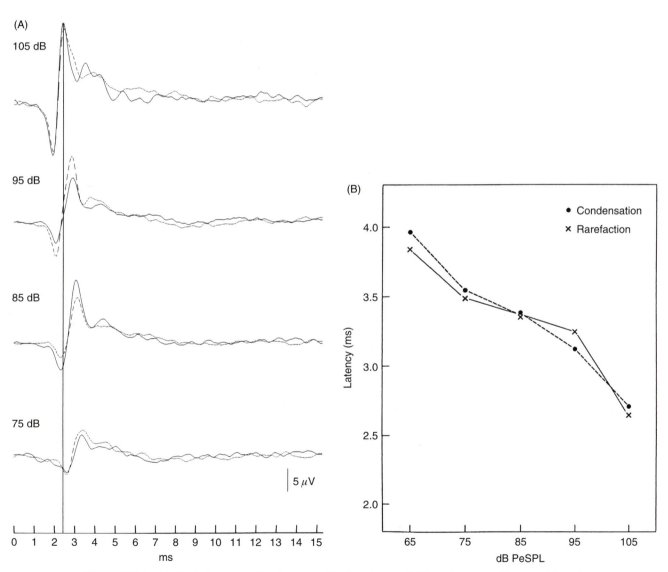

FIGURE 7.4   (A) Typical compound action potentials directly recorded from the exposed intracranial portion of the eighth nerve in a patient with normal hearing. Responses to condensation (dashed lines) and rarefaction (solid lines) clicks are shown for different stimulus intensities (given in dB PeSPL). (B) Latency of the negative peak in the CAP shown in (A) (reprinted from Møller and Jho, 1990, with permission from Elsevier).

BOX 7.2

INTRAOPERATIVE NEUROPHYSIOLOGIC MONITORING

Recording from the intracranial portion of the auditory nerve requires that the eighth cranial nerve be exposed in its course in the cerebellopontine angle. That occurs in some operations such as those to treat vascular compression of cranial nerves. Whenever such recordings are done, it must be assured that the auditory nerve is not injured by the surgical dissection necessary to expose the nerve. Therefore, ABR must be recorded during such dissections to monitor the conduction velocity in the auditory nerve (for details about monitoring neural conduction in the auditory nerve, see Møller [185]).

If the recording electrode is placed on the auditory nerve near the porus acousticus it will be approximately 1.5 cm from the cochlea and it will therefore not record any noticeable potentials from the cochlea (CM or SP). (The total length of the auditory nerve in humans is approximately 2.5 cm and the length of the nerve between the point where it enters into the skull cavity from the porous acousticus to its entrance into the brainstem is approximately 1 cm.) A recording electrode that is placed near the porus acousticus will be approximately 1 cm from the cochlear nucleus and the potentials generated in the cochlear nucleus will be attenuated before they reach the recording electrode provided that the eighth nerve in its intracranial course is submerged in fluid. The amplitude of the evoked potentials generated in the cochlear nucleus will be greater when recording from a location on the auditory nerve that is close to the brainstem and thus near the cochlear nucleus. If the eighth nerve is free of fluid in its intracranial course, it will act as an extension of the recording electrode that is placed anywhere on the nerve and it may record potentials from the cochlear nucleus of noticeable amplitude.

A bipolar recording electrode placed on a nerve with one of its two tips located more peripherally than the other will under ideal circumstances only record propagated neural activity. The waveform of the compound action potential recorded from a nerve with a bipolar electrode is different from that recorded by a monopolar electrode and is more difficult to sinterpret.

BOX 7.3

DISTINGUISHING BETWEEN PROPAGATED AND ELECTRONICALLY CONDUCTED POTENTIALS

The fact that the latency of the response from the auditory nerve to click sounds increases when the recording electrode is moved from a location near the porus acousticus toward the brainstem (Fig. 7.5) is an indication that at least the main portion of the recorded potentials are generated by the propagated neural activity in the auditory nerve [205]. The latency of passively conducted potentials would not change when the recording electrode is moved along the auditory nerve but their amplitude would decrease when the recording electrode is moved away from their source. The response from the exposed intracranial portion of the eighth nerve to low intensity click sounds often yields a slow deflection of a relatively large amplitude. That component is probably generated in the cochlear nucleus and conducted passively in the auditory nerve to the site of recording. This slow component of the response is more pronounced at low stimulus intensities because the amplitude of the evoked response from the cochlear nucleus decreases at a slower rate with decreasing stimulus intensity than that generated by propagated neural activity in the auditory nerve.

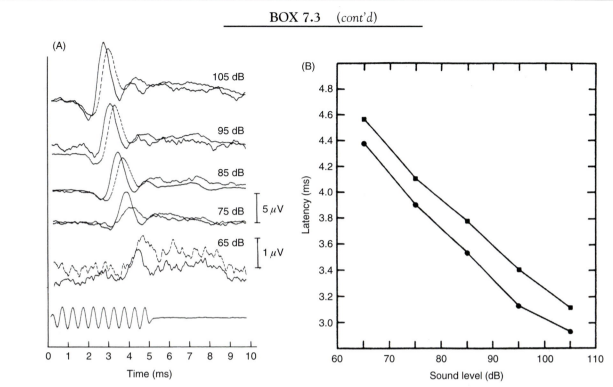

BOX 7.3 (cont'd)

FIGURE 7.5 (A) Similar recordings as in Fig. 7.4 but showing the response to tone bursts recorded at two locations along the intracranial portion of the exposed auditory nerve. The solid lines are recordings close to the porous acousticus and the dashed lines are recordings from a location approximately 3 mm more central. The stimuli were short 2 kHz tone bursts. The sound pressure give is in dB PeSPL. (B) The latency of the main negative peak of the CAP recorded from two different locations as shown in (A) (approximately 3 mm apart) on the exposed eighth nerve as a function of the stimulus intensity (reprinted from Møller and Jannetta, 1983, with permission from Taylor & Francis).

Comparison between bipolar and monopolar recordings from the exposed intracranial portion of the auditory cranial nerve [201] further supports the assumption that click evoked potentials recorded from the auditory nerve with a monopolar recording electrode, at least at high stimulus intensities, is mainly the result of propagated neural activity.

More space is required for placing a bipolar recording electrode on a nerve compared with using a monopolar recording electrode, but the intracranial portion of the auditory nerve in the human is sufficiently long to allow the use of bipolar recording electrodes.

The conduction velocity of the auditory nerve in humans has been determined from bipolar recordings from the exposed intracranial portion of the auditory nerve. The difference in the latency of the CAP recorded at two different locations on the exposed intracranial portion of the auditory nerve has been used to determine the conduction velocity [202]. The value arrived at, approximately 20 m/s, is similar to what has been estimated on the basis of the fiber diameter of the auditory nerve fibers [129].

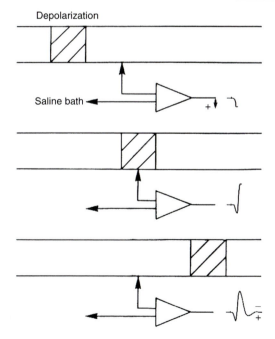

FIGURE 7.6 Illustration of recordings from a long nerve in which an area of depolarization travels from left to right, using a monopolar electrode.

BOX 7.4

INTERPRETATION OF POTENTIALS RECORDED BY BIPOLAR ELECTRODES

The potentials that are recorded by a bipolar recording electrode placed on the intracranial portion of the auditory nerve can be understood by assuming that the bipolar electrode consists of two monopolar electrodes, each one recording the potentials at two adjacent locations along the nerve and that the amplifier to which the electrodes are connected senses the difference between the electrical potentials that the two electrodes are recording (Fig. 7.7). The electrical potentials generated in a nerve by propagated neural activity appear with a slight time difference at the two tips of such a bipolar recording electrode, the time difference being the time it takes the neural activity to travel the distance between the two tips. Under ideal circumstances, passively conducted potentials will appear equal at the two electrodes and thus not result in any output from the differential amplifier to which the electrodes are connected. To achieve such ideal performance of a bipolar recording electrode, the two tips of the electrode must have identical recording properties and be placed so that they both record from the same population of nerve fibers. While that is rarely achieved in practice, a bipolar electrode is less sensitive to potentials generated by passively conducted potentials than a monopolar recording electrode. If the two tips of the bipolar recording electrode have different recording characteristics or are not placed exactly symmetrical on the nerve, passively conducted potentials may appear differently at the two tips and thus appear as an output from the amplifier to which the bipolar electrode is connected [201].

If no passively conducted potentials reach the recording electrodes the response recorded by a bipolar recording electrode will be the same as the potentials recorded by a monopolar electrode from which is subtracted a delayed version of the same response (Fig. 7.8). The difference between such a simulated bipolar recording and a real bipolar recording is a measure of the amount of passively conducted potentials that are recorded by monopolar recording electrode.

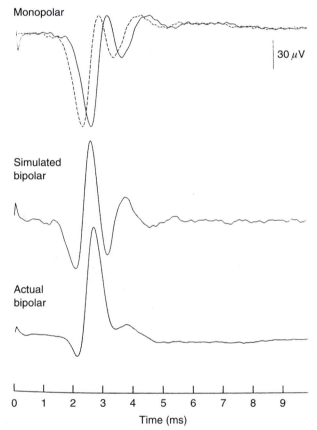

FIGURE 7.8   Recordings from the intracranial portion of the auditory nerve in a patient whose vestibular nerve was just cut. Rarefaction clicks presented at 98 dB PeSPL. Top curves: monopolar recordings by the two tips of a bipolar electrode. Middle curves: computed difference between the response recorded by one tip (monopolar recording) and the same response shifted in time with an amount that corresponds to the distance between the two tips of the bipolar electrode. Lower curves are the actual bipolar recording (reprinted from Møller et al., 1994, with permission from Elsevier).

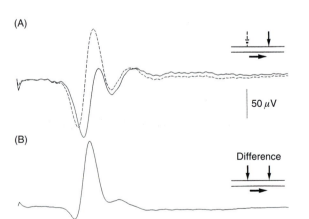

FIGURE 7.7   (A) Separate recordings from the exposed intracranial portion of the eighth cranial nerves with two electrodes placed approximately 1 mm apart. (B) The difference between the recordings by the two electrodes in (A) (reprinted from Møller et al., 1994, with permission from Elsevier).

Direct recording of responses from the eighth nerve is now in general use in monitoring neural conduction in the auditory nerve in patients undergoing operations in the cerebellopontine angle. Such potentials can be interpreted nearly instantaneously [184, 208] because of their large amplitudes. Changes in the function of the nerve from stretching or from slight surgical trauma that may occur during surgical manipulations can thereby be detected almost instantaneously because only few responses need to be added (averaged) in order to obtain an interpretable record. Similar monitoring of neural conduction in the auditory nerve can be achieved by recording the ABR but it takes much longer to obtain an interpretable record because of the small amplitude of the ABR (see p. 163).

Click evoked compound action potentials recorded from the intracranial portion of the eighth nerve changes in a systematic fashion when the auditory nerve is injured such as from surgical manipulations or by heat from electrocoagulation [178]. Recorded centrally to the location of the lesion, the latency of the main negative peak increases and its amplitude decreases. The main negative peak also becomes broader because the prolongation of the conduction time in different nerve fibers is different. More severe injury causes the amplitude of the initial positive deflection to increase and that is a sign that neural block has occurred in some nerve fibers (Fig. 7.9).

The frequency following response (FFR), as the name indicates, is a response that follows the waveform of the stimulating sound. FFR can be demonstrated in the response from the auditory nerve to low frequency tones and tones that are amplitude modulated at low frequencies. The source of the FFR is phase locked discharges in nerve fibers. Some investigators have named these potentials the neurophonic response. FFR has been recorded from the auditory nerve in animals [276, 277] and from the exposed intracranial portion of the auditory nerve in humans [214, 215]. The FFR recorded from the human auditory nerve is similar to that in the cat recorded by bipolar electrodes [277]. When recorded directly from the exposed intracranial portion of the auditory nerve (Fig. 7.10) in humans, the FFR is prominent in the frequency range from 0.5 to 1.5 kHz [214].

Recordings of the FFR from the auditory nerve in animals and in humans have contributed to understanding of the function of the cochlea. At high stimulus intensities the frequency following responses are the results of excitation of the basilar membrane at a location that is more basal than the location tuned to the frequency of the stimulation [276]. This is a sign of

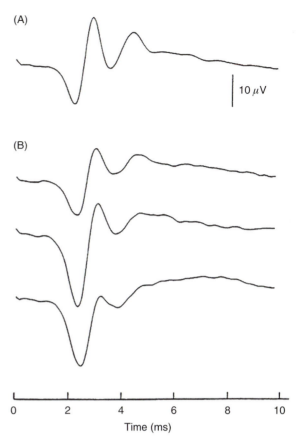

FIGURE 7.9   Change in the CAP as a result of injury to the intracranial portion of the auditory nerve in a patient undergoing an operation where the auditory nerve was heated by electrocoagulation (reprinted from Møller, 1988).

non-linearity of the basilar membrane vibration (see Chapter 3).

The waveform of the recorded responses to stimulation with a 0.5 kHz tone is a distorted sinewave (Fig. 7.13). As a first approximation, the waveform of the responses indicates that auditory nerve fibers are excited by the half wave rectified stimulus sound, thus a deflection of the basilar membrane in one direction. The waveform of the response to high sound intensity tones (104 dB SPL) is more complex than the response to tones of lower intensities and has a high content of second harmonics, similar to a full-wave rectified sinewave. That indicates that hair cells respond to deflection of the basilar membrane in both directions at high stimulus intensities, thus supporting the findings in animal experiments that some inner hair cells respond to the condensation phase of a sound while other inner hair cells respond to the rarefaction phase [278, 336].

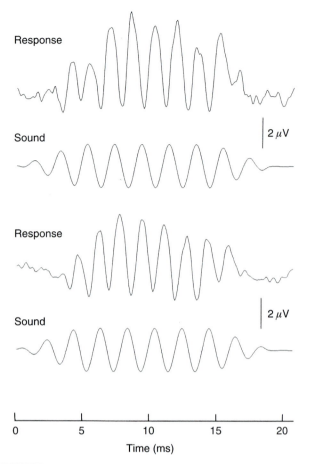

Time (ms)

FIGURE 7.10 Responses recorded from the exposed intracranial portion of the auditory nerve to stimulation with 0.5 kHz tones at 113 dB SPL. Rarefaction of the sound is shown as an upward deflection (reprinted from Møller and Jho, 1989, with permission from Elsevier).

The distortion of the response to low frequency pure tones could also be a result of what has been known as "peak splitting" [256, 268]. The distortion of the waveform of the responses from the human auditory nerve seems to be less than it is in the cat at the same sound pressure level. In the studies of the responses from the exposed eighth nerve in humans, the ABR was monitored during the surgical exposure to ensure that the surgical manipulations of the auditory nerve did not cause noticeable change in the neural conduction in the auditory nerve.

3.2. Recordings from the Cochlear Nucleus

Recordings of the responses from the exposed cochlear nucleus to various kinds of sound stimuli have been done both in humans [203, 210] and in animals [200]. When a monopolar recording electrode is placed directly on the surface of the cochlear nucleus in humans the response to a transient sound has an initial positive-negative deflection (P_1 and N_1 in Fig. 7.14) [210]. These components represent the arrival of the neural volley from the auditory nerve in the CN. They are followed by a slower deflection on which peaks are often riding. It is assumed that this component is generated by dendrites in the nucleus and its polarity depends on the placement of the recording electrode (Fig. 7.15).

The source of the slow potential can be described by a dipole with a certain orientation.

Since the activity of nerve cells may be regarded as a dipole source (Fig. 7.15), a reversal of the polarity occurs when a recording electrode is passed

---

BOX 7.5

SEPARATION OF AUDITORY NERVE GENERATED FFR FROM COCHLEAR POTENTIALS

Studies of the FFR from the auditory nerve in response to low frequency pure tones in animals are hampered by the contamination from cochlear microphonics. Snyder and Schreiner [276] reduced the contamination of the neural response from potentials generated in the cochlea by using a bipolar recording technique. The fact that the auditory nerve is longer in humans than in the cat makes it possible to record the FFR with a monopolar recording electrode without any noticeable contamination from cochlear potentials. That the FFR recorded from the human auditory nerve with a monopolar electrode is the result of propagated neural activity is supported by the finding

that the recorded potentials appear with a certain latency and are shifted in time when the recording electrode is moved along the eighth cranial nerve (Fig. 7.11).

The responses to low frequency tones recorded from the human auditory nerve have two components, a frequency following response and a slow component (Fig. 7.12) [214]. When the responses to tones of opposite phase were added, the frequency following response was canceled and the slow potential was seen alone. When the responses to tones of opposite phase were subtracted the slow potential was canceled and only the frequency following response remained.

BOX 7.5   (cont'd)

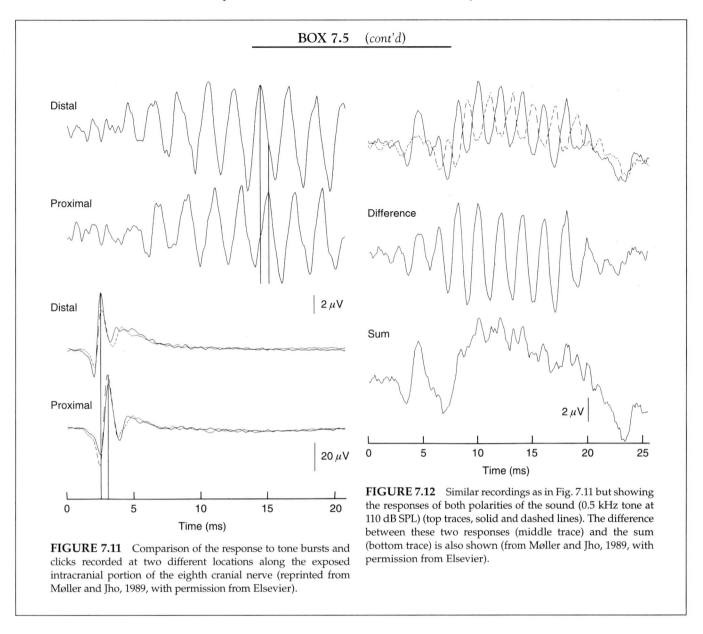

FIGURE 7.11 Comparison of the response to tone bursts and clicks recorded at two different locations along the exposed intracranial portion of the eighth cranial nerve (reprinted from Møller and Jho, 1989, with permission from Elsevier).

FIGURE 7.12 Similar recordings as in Fig. 7.11 but showing the responses of both polarities of the sound (0.5 kHz tone at 110 dB SPL) (top traces, solid and dashed lines). The difference between these two responses (middle trace) and the sum (bottom trace) is also shown (from Møller and Jho, 1989, with permission from Elsevier).

through the nucleus [5]. When the recording electrode is placed in between the two recording locations where the slow potential is positive and negative it will record only a very small slow potential because the positive contribution is equal to the negative contribution.

The peaks that are seen riding on this slow wave are assumed to be generated by discharges of cells in the nucleus (somaspikes). The latency of the sharp negative peak (N_2) in the response from the auditory nerve that follows the initial positive-negative deflection (P_1, N_1) is approximately 1 ms longer than that of the positive deflection (Fig. 7.14) which can be explained

by synaptic delay assuming that the N_2 response is generated by cells in the cochlear nucleus. The response from the cochlear nucleus shown in Fig. 7.14 was obtained from a monopolar electrodes placed in the lateral recess of the fourth ventricle (Fig. 7.16) [119, 203, 216].

The cochlear nucleus consists of three major subdivisions with different response characteristics as judged from recordings from single nerve cells (see Chapter 6). Therefore, the evoked responses recorded from the surface of the cochlear nucleus are likely to be different dependent on the subdivision from which they are recorded.

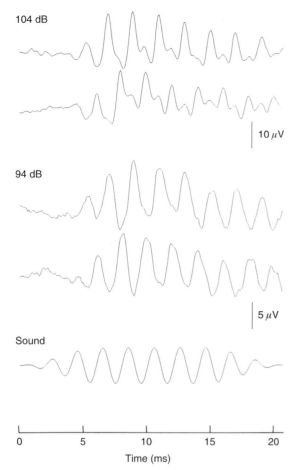

104 dB

10 μV

94 dB

5 μV

Sound

0   5   10   15   20

Time (ms)

FIGURE 7.13 Examples of responses to 0.5 kHz tones recorded from the exposed intracranial portion of the eighth cranial nerve to show distortion of the waveform. The responses to tones of two different intensities are shown. The two curves at each intensity are the responses to stimulation of opposite polarity (reprinted from Møller and Jho, 1989, with permission from Elsevier).

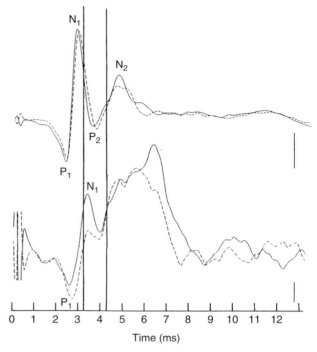

0   1   2   3   4   5   6   7   8   9   10   11   12

Time (ms)

FIGURE 7.14 Recordings from the exposed eighth nerve (top tracings) and the surface of the cochlear nucleus (bottom tracings). The response from the cochlear nucleus was obtained by placing an electrode in the lateral recess of the fourth ventricle. Solid lines are the responses to rarefaction clicks and the dashed lines are the responses to condensation clicks. Amplitude scales are 0.2 mV for the auditory nerve recording and 0.1 mV for the cochlear nucleus recording (reprinted from Møller et al., 1994, with permission from Elsevier).

The interpretation of the sources of the different components of the waveform of the response from a nucleus is based on studies of nuclei of the somatosensory system, done early in the history of neurophysiology. Experiments in a dog showed that a slow wave that followed after these initial waves gradually disappeared during anoxia [67]. The initial positive-negative complex was only affected by prolonged

---

BOX 7.6

ANATOMY OF THE LATERAL RECESS OF THE FOURTH VENTRICLE

The caudal portion of the floor of the lateral recess is the (dorsal) surface of the dorsal cochlear nucleus and the rostral portion of the floor of the lateral recess is the dorsal surface of the ventral cochlear nucleus [119]. When the lateral side of the brainstem is viewed in operations using a retromastoid craniectomy, the foramen of Luschka that leads to the lateral recess of the fourth ventricle is found dorsally to the exit of the ninth and tenth cranial nerves. Often a portion of the choroid plexus is seen to protrude from the foramen of Luschka and may have to be reduced by coagulation in order to place a recording electrode in the lateral recess of the fourth ventricle.

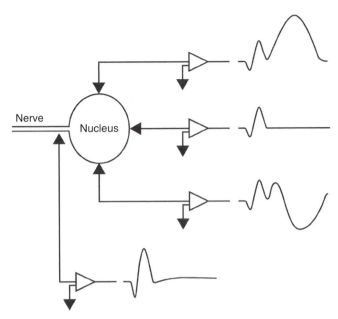

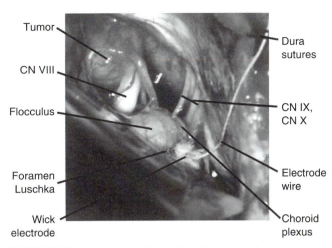

FIGURE 7.16 Placement of recording electrode in the lateral recess of the fourth ventricle (reprinted from Møller, A.R. 2006. *Intraoperative Neurophysiologic Monitoring*, 2nd Edition, Humana Press Inc, with permission from Humana Press Inc; modified from Møller AR, Jho HD, Jannetta PJ. Preservation of hearing in operations on acoustic tumors: An alternative to recording ABR. *Neurosurgery* 1994; 34: 688–693, with permission by Lippincott Williams and Wilkins).

FIGURE 7.15 Schematic illustration of the potentials that may be recorded from the surface of a sensory nucleus in response to transient stimulation such as a click sound for the auditory system. The three waveforms shown refer to recordings at opposite locations on the nucleus and in between to illustrate the dipole concept for describing the potentials that are generated by a nucleus. The waveform of the response that can be recorded from the nerve that terminates in the nucleus is also shown (reprinted from Møller, A.R. 2006. *Intraoperative Neurophysiologic Monitoring*, 2nd Edition, Humana Press Inc. with permission from Humana Press Inc).

severe anoxia thus indicating that the slow component was dependent on synaptic transmission while the initial deflections were generated in a nerve or a fiber tract. Synaptic transmission is more sensitive to anoxia than propagation of neural activity in nerves and fiber tracts.

Recordings from the surface of the cochlear nucleus have found practical clinical use in intraoperative neurophysiologic monitoring because it offers a more stable electrode position compared with recordings from the exposed eighth cranial nerve [216]. The amplitude of the auditory evoked potentials obtained by recording from these two locations is sufficiently high to allow interpretation after only a few responses have been added (averaged) [183].

3.3. Recordings from More Central Parts of the Ascending Auditory Pathways

Reports on recordings from more central brainstem structure of the ascending auditory pathways in humans have been few compared with recordings from the auditory nerve and the cochlear nucleus and such recordings have not yet found practical use in intraoperative monitoring, but they have been important for identification of the neural generators of the ABR. Recordings from the inferior colliculus and its vicinity using chronically implanted electrodes have been done recently [329].

The waveform of the response to short tone bursts (Fig. 7.17) recorded from the surface of the contralateral inferior colliculus in humans [79. 80, 206] is typical of a nucleus. The earliest positive deflection is presumably generated when the volley of neural activity in the lateral lemniscus reaches its termination in the inferior colliculus and the slow negative deflection is likely a result of dendritic activity, thus similar to the cochlear nucleus. Recordings of the response from the inferior colliculus to ipsilateral stimulation results in responses with much smaller amplitudes and a different waveform and indicates that the input from the ipsilateral ear that reaches the inferior colliculus is small [217].

4. FAR-FIELD AUDITORY EVOKED POTENTIALS

Far-field evoked potentials are the responses that can be recorded from electrodes placed far from their source. Far-field potentials therefore have much smaller amplitudes than near-field potentials and it is

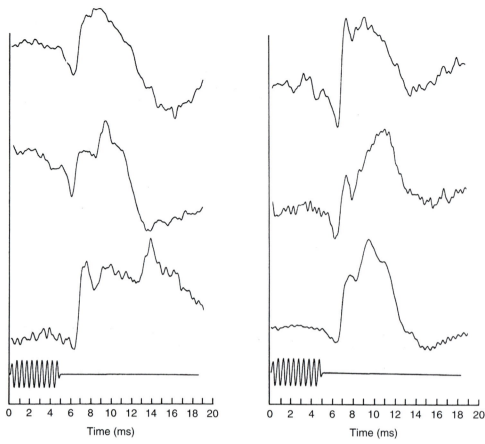

FIGURE 7.17   Responses recorded from the inferior colliculus in patients undergoing operations where the inferior colliculus was exposed or responses recorded from an electrode placed along the path of the fourth cranial nerve (reprinted from Møller and Jannetta, 1982, with permission from Elsevier).

necessary to add the responses to many stimuli in order to discern the various components of far-field potentials from the background of other biologic signals such as the spontaneous electroencephalographic (EEG) activity, potentials from muscles and electrical interference signals. Far-field evoked potentials could therefore not be studied before the development of the signal averager.

While evoked potentials can always be recorded from electrodes placed directly on nerves, fiber tracts and nuclei, such structures generate far-field potentials only when certain criteria are fulfilled. Thus, neural activity that propagates in a nerve or a fiber tract generates stationary peaks in the far-field when the electrical conductivity of the surrounding medium changes or when it is bent [94]. Neural activity that propagates in a straight nerve, the surrounding medium of which has uniform electrical conductivity, generates very little far-field potentials. A nucleus generates strong far-field potentials when its dendrites are organized uniformly whereas a nucleus where the dendrites are randomly

organized and point in all directions generates little far-field potentials. These two different types of nuclei are known to have an open and a closed field, respectively [142].

Far-field potentials are more complex than near-field potentials because they are likely to have contributions from sources with different anatomical locations. Neural structures activated sequentially by transient stimulation may generate a sequence of components, each of which occur with different latencies. The brainstem auditory evoked potentials (ABR) (Fig. 7.18) are examples of far-field-evoked potentials that are commonly used for clinical diagnosis and for intraoperative monitoring. The ABR is recorded from electrodes placed on the scalp and the earlobe (or mastoid). It is the most important functional test for detecting vestibular Schwannoma. The middle latency responses (MLR) are another kind of far-field auditory evoked potentials that can be recorded from electrodes placed on the scalp and which are used clinically to a lesser extent than the ABR. Proper interpretation of these auditory

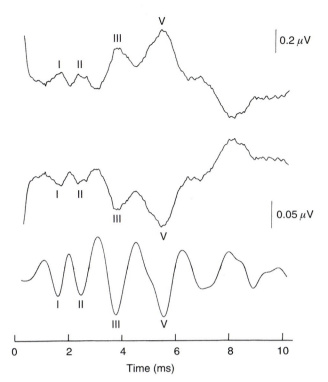

FIGURE 7.18 Typical recording of ABR obtained in a person with normal hearing. The curves are the average of 4,096 responses to rarefaction clicks, recorded from electrodes placed on the forehead at the hairline and the mastoid on the side where the stimuli were applied. The upper recording is shown with vertex positivity as an upward deflection, the middle curve is the same recording shown with positivity downward. These two curves are recordings that were filtered electronically with a band-pass of 10–3,000 Hz. The bottom curve is the same recording after digital filtering designed to enhance all five peaks of the ABR (from Møller, 1988).

evoked potentials for diagnostic purposes depends on knowledge about the anatomical origin of the different components of these potentials and how they are affected by pathologies. During the past two decades much knowledge about the neural generators of the ABR has accumulated but the neural generators of the MLR are not as well known and that has hampered the use of the MLR in diagnosis of neurologic disorders. The MLR is considerably more variable than the ABR and it is mainly used as an objective test of hearing threshold. The MLR has a potentially important role for diagnosis of disorders of the auditory nervous system, but insufficient knowledge about the origin of these potentials has so far prevented such use.

The far-field FFR to periodic sounds such as pure tones sounds can also be recorded from electrodes placed on the scalp. The modulation waveform of amplitude-modulated sounds likewise give rise to far-field potentials that can be recorded from electrodes placed on the scalp.

Electrodes placed on the scalp also record responses from muscles that are elicited by sound stimulation (myogenic evoked potentials). Muscle activity that is not related to the sound stimulation act as interference and prolong the time it takes to obtain an interpretable record.

4.1. Auditory Brainstem Responses

The human auditory brainstem response (ABR) consists of far-field evoked potentials from the auditory nervous system that occur during the first 10 ms after the presentation of a transient sound such as a click sound or a tone burst. The amplitudes of the ABR are small, less than 0.5 µV, and thus much smaller than the ongoing spontaneous activity of the brain (EEG). Therefore responses to many stimuli must be added to obtain a recording where the individual components can be discerned. The different components of the ABR are generated by neural activity in the ear, the auditory nerve and the nuclei and fiber tracts of the ascending auditory pathways.

When recorded differentially between two electrodes, one placed at the vertex and one at the mastoid or earlobe on the side where the stimulus sounds are presented, the ABR typically is characterized by five to seven vertex positive waves (Fig. 7.18). These waves (or peaks) are traditionally labeled with roman numerals. The first five of these peaks of the human ABR except peak IV can usually be discerned in individuals with normal hearing.

The labeling of the vertex positive waves by Roman numerals that Jewett and Williston [96] introduced is still the most common way to label the components of the ABR. This labeling is different from the way different components of other sensory evoked potentials are labeled. Usually, both positive and negative components of evoked potentials are labeled with the letter P and N respectively, followed by a number that gives the normal value of the latency of the respective component.

One of the consequences of only labeling the vertex positive peaks of the ABR has been that only the latency of these positive peaks have been used for diagnostic purposes and most studies of neural generators of the ABR have ignored the negative waves. At the time when this labeling was introduced it was not known which of the different components of the ABR were most important for diagnostic purposes. It would seem likely that the vertex negative waves would also be of diagnostic value as these negative waves also have distinct neural generators [217].

Some authors prefer to show the ABR with the vertex positivity as an upward deflection, whereas

---

BOX 7.7

HISTORY OF THE AUDITORY BRAINSTEM RESPONSE

It was probably Kiang [107] and his colleagues at the Eaton Peabody Laboratory in Boston who first demonstrated these potentials. Dr Kiang belonged to the group at MIT assembled by Professor Walter Rosenblith, who pioneered signal analysis of neuroelectric potentials and was in the forefront for developing signal averaging into a routine method for studies of neuroelectric potentials. Dr Kiang also predicted that these potentials might be useful in diagnosis of disorders of the auditory system and in intraoperative monitoring [107]. However, systematic studies of the ABR were not published until 10-15 years later at which time Jewett and his collaborators [95, 96] identified and described the different components of the ABR and introduced the placement of the recording electrodes that is now commonly used: one electrode

placed at the vertex and the other placed at the mastoid on the side where the stimuli are applied [96]. This placement, however, is not the ideal montage from a physiological point of view.

When evoked potentials are recorded differentially between two electrodes, one electrode is usually placed at a location where the potential to be recorded is large and the other electrode (reference electrode) is placed on a location where it records as little as possible of the evoked potentials that are studied. With the electrode placement commonly used for recording ABR both recording electrodes record auditory evoked potentials. Peak V has a larger amplitude in the vertex recording than in the recording from the mastoid while peaks I–III have larger amplitudes in the recordings from the mastoid or earlobe than from the vertex.

---

others display the ABR with the vertex positivity as a downward deflection (Fig. 7.18), the latter being in accordance with the common convention of displaying negative potentials as an upward deflection, assuming the vertex electrode to be the most active electrode. In this book, ABRs are always shown with vertex positivity as a downward deflection.

Many factors affect the waveform and the amplitude of the ABR. Recording parameters, filtering of the recorded potentials, individual variations some of which are related to age and size of the head, all influence the recorded potentials.

The main purpose of filtering the ABR is to reduce the number of responses that must be averaged in order to obtain an interpretable record. Filtering can also enhance specific components of the ABR and the appearance of the ABR depends on the way that the potentials are filtered (Fig. 7.19). Since it is the latencies of the different peaks that are important for diagnostic purposes, the filters used should enhance the peaks that are of diagnostic importance without shifting the peaks in time. When recorded with an open band pass (10–3,000 Hz), the ABR has the appearance of a series of three clear positive peaks (Fig. 7.19) followed by peak V. When low frequency components of the ABR are not attenuated by filtering (Fig. 7.19), peak V is seen to be followed by a broad negative peak, the SN_{10} component (Fig. 7.20). Electronic filters shift the peaks in time to an extent that depends on the spectrum of the peaks, the type of filters used, and their settings. Electronic filters may shift the different peaks of the ABR differently. It is possible to design electronic filters

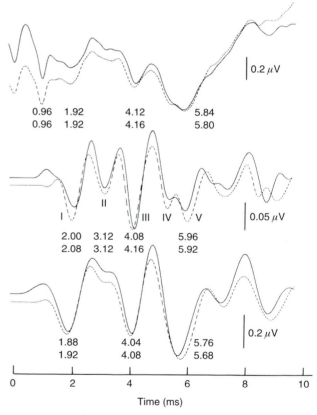

FIGURE 7.19 The same ABR, filtered in three different ways. Top traces: low pass filter with a digital filter with a triangular shaped weighting function with a base of 0.4 ms. Middle trace: digital band pass filter with a W shaped weighting function with a base of approximately 1 ms. Lower trace: digital filter with a W shaped weighting function with a base of approximately 2 ms (assuming a 40 μS sampling time).

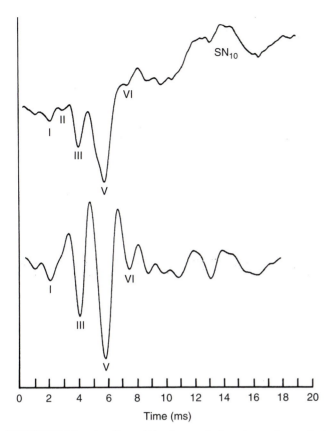

FIGURE 7.20 The effect of filtering on the broad negative peak at approximately 12 ms latency (SN₁₀). Top trace: unfiltered. Bottom trace: digitally band-pass filtered. Note that the time scale is 20 ms (reprinted from Møller and Jannetta, 1982, with permission from Elsevier).

with linear phase shift and such filters (known as Bessel filters [37]) will shift all components of the ABR with same amount but such filters are rarely available.

Digital filters are more flexible than electronic filters and optimal filtering can be obtained using digital filters whereas electronic filters have considerable limitations [20, 185]. Digital filters are computer programs that operate on the averaged waveform of the ABR. Zero-phase digital filters can be designed so that they do not shift any peak of the ABR at all. Aggressive filtering that is made possible using digital filters can enhance specific components of the ABR that are of interest (peaks). Peak II and peak IV of the ABR are often difficult to identify, but appropriate (digital) filtering can make these peaks appear clearly (Fig. 7.19). Such filtering also makes it possible to have computer programs identify the individual peaks and print their latencies automatically (Fig. 7.19). It is common in commercially available equipment to use digital filters that emulate electronic filters. However, such filters do not provide all the advantages of digital filters. So called zero phase finite impulse filters offer other

advantages that cannot be achieved with electronic filters (or digital filters that emulate electronic filters) such as reduced effect of stimulus artifacts [185].

The amplitude of the ABR decreases with decreasing stimulus intensity[2] and the latencies increase but the different components of the ABR are affected differently. Decrease in stimulus intensity cause the amplitude of early peaks (I, II, and III) to decrease more than that of peak V (Fig. 7.21). That means that at low stimulus intensities, practically only peak V is discernible.

For diagnosis of disorders of the auditory nervous system (or for excluding such disorders) the ABR is commonly elicited by click simulation presented to one ear at a time. Clicks can be either condensation clicks or rarefaction clicks. Condensation clicks move the tympanic membrane (initially) inward while rarefaction clicks cause movement of the tympanic membrane in the opposite direction. In attempts to reduce the stimulus artifact, it has been common to use alternating condensation and rarefaction clicks. While the ABR in individuals with normal hearing are nearly identical when elicited by condensation or rarefaction clicks, the ABR may differ considerably in response to these two types of clicks in individuals with cochlear hearing loss [212]. The use of alternating condensation and rarefaction clicks as stimuli is therefore not recommended. (For more details about the use of ABR in diagnostics, see Hall [75] or Jacobson [88].)

At a first approximation, the different components of the ABR are generated by neural activity in sequentially activated structures of the ascending auditory pathways. However, the classical auditory nervous system is not just a string of nuclei connected with fiber tracts but rather a complex series of nuclei with many interconnections, including a high degree of parallel processing, and that complicates the interpretation of the ABR (see Chapter 9).

The generation of ABR has been studied in animals and more recently, in humans during neurosurgical operations where the intracranial portion of the auditory

[2]The intensity of clicks used for recording ABR is often given in "peak equivalent SPL" (PeSPL), which is the sound pressure of a pure tone with the same peak sound pressure as the clicks. It is also common to give the intensity of click stimulation in dB above the normal hearing threshold (dB HL). (Hearing level [HL] is the sound level in dB above the average threshold of young individuals without disorders of the ear.) While the physical measure of click intensity (PeSPL) is independent of the rate at which the clicks are presented, the behavioral threshold (dB HL) decreases with increasing repetition rate because of temporal integration in hearing. Thus at a rate of 5 pps the threshold is approximately 37 dB PeSPL, at 20 pps it is 35 dB PeSPL and at a rate of 80 pps, the threshold is 32 dB PeSPL [283]. This means that the HL of clicks of (physical) intensity of which is 105 dB PeSPL is 70 dB when presented at 20 pps.

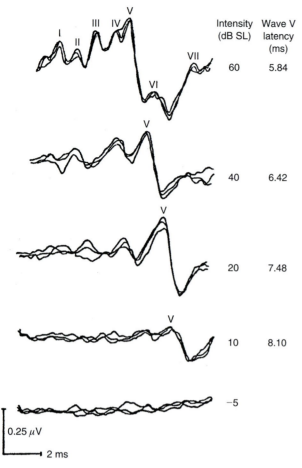

	Intensity (dB SL)	Wave V latency (ms)
	60	5.84
	40	6.42
	20	7.48
	10	8.10
	−5	

0.25 μV
2 ms

FIGURE 7.21 Effect of stimulus intensity on the ABR. ABRs in response to click stimulation presented at a rate of 30 pps at different sound intensities (given in dB SL). Two thousand responses were averaged and one repetition is shown. Note that vertex positivity is shown as an upward deflection (reprinted from Galambos and Hecox, 1977, with permission from Karger).

nerve and other structures of the ascending auditory pathways become exposed. Most studies of the generators of the ABR have focused on generators of the vertex positive peaks in the ABR and only a few studies have concerned the vertex negative waves of the ABR.

Much of our understanding of the contributions from the ascending auditory nervous system to the ABR has been gained from studies in animals but the specific anatomical differences between the auditory nerve in humans and the commonly used experimental animals including the monkey have caused misinterpretation of the neural generators of the ABR in humans. During the past 20–25 years extensive studies of the responses recorded directly from exposed structures of the human auditory nervous system during neurosurgical operations have contributed to the understanding of the generation of the human ABR.

The abnormalities of the ABR in patients with known pathologies such as tumors, etc., have also been used to identify the neural generators of the ABR. The use of pathologies presumes that the pathology in question affects specific parts of the ascending auditory pathways and that the anatomical location of the pathology is known. The disadvantages of using such methods are related to difficulties in assessment of the anatomical location and extent of the pathologies. Imaging studies such as the magnetic resonance imaging (MRI) scans can only detect changes in structure and not in function and it is uncertain how specific morphological abnormalities, as they appear in imaging studies, relate to functional abnormalities.

Reconstruction of the dipoles of the generators of sensory evoked potentials based on recordings of three-dimensional evoked potentials make it possible to obtain some information about the anatomical location of generators of sensory evoked potentials. The spatial resolution is, however, limited. Such recordings (three-dimensional Lissajous trajectories, 3-CLT) are made by placing three pairs of electrodes orthogonal on the scalp.

Perhaps the most successful method for identifying the neural generators of the ABR is the one that makes use of comparisons between the ABR and evoked potentials recorded directly from specific structures of the ascending auditory pathways in patients undergoing neurosurgical operations. Coincidence in time between the main components of the directly recorded potentials and the different (vertex positive) peaks of the ABR has been taken as an indication, but not a proof, that a specific structure is the generator of a certain component of the ABR. Such studies can be made in selected neurosurgical operations where it becomes possible to place a recording electrode directly on the intracranial portion of the auditory nerve, the cochlear nucleus or other structures of the ascending auditory pathways or in their immediate vicinity. In such studies, the ABR is recorded simultaneously before and during intracranially recordings to ensure that the surgical manipulations have not affected the function of the structures that contribute to the ABR [80, 147, 205, 253, 280]. Some investigators [79] have recorded directly from the auditory nervous system by inserting electrodes through burr holes in the skull and passing them through the brain to reach the desired location. Clicks have been the most commonly used stimuli in such studies, but some investigators have used tone bursts as stimuli.

Recordings from the intracranial portion of the eighth cranial nerve in operations where that portion of the nerve became exposed revealed that the negative peak of the response (CAP) occurs with approximately

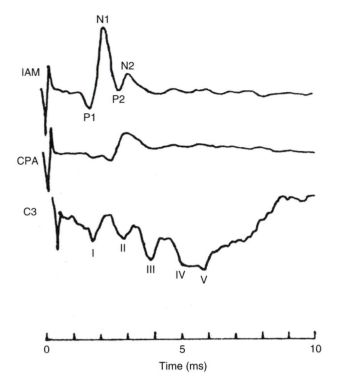

FIGURE 7.22 Comparison between intracranial recordings made from the exposed eighth nerve and the ABR. IAM = recording of the CAP from the intracranial portion of the eighth nerve where it exits the bony canal (porus acousticus). CPA = recordings of the CAP directly from the exposed eighth nerve in the cerebello pontine angle (CPA). C3 = ABR recorded between the ipsilateral earlobe and the C3 on the parietal ipsilateral scalp (international EEG recording nomenclature) (reproduced from Hashimoto et al., 1981, with permission from Oxford University Press).

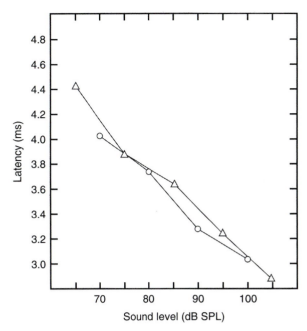

FIGURE 7.23 Latency of the negative peak in the CAP recorded from the intracranial portion of the eighth nerve as a function of the stimulus intensity (open triangles) and the latency of peak II of the ABR postoperatively (open circles). The sound stimuli were 5 ms long 2 kHz tone bursts (modified from Møller and Jannetta, 1981, with permission from Elsevier).

the same latency as the second vertex positive peak (peak II) in the ABR [80, 147, 205, 280]. This has been taken to indicate that peak II is generated by the central portion of the auditory nerve (Fig. 7.22). The relationship between the negative peak in the CAP recorded directly from the intracranial portion of the eighth nerve became even more convincing when the latencies of peak II of the ABR and the main negative peak of the CAP were compared over a large range of stimulus intensities (Fig. 7.23) [204].

Before these studies were published, comparison of the ECoG potentials with the ABR had shown that peak I of the ABR occurs with the same latency as the negative peak (N_1) in the ECoG [233]. The N_1 peak of the ECoG is generated by the most peripheral portion of the auditory nerve and therefore also peak I of the ABR is assumed to be generated in the most peripheral portion of the auditory nerve (in the ear). That means that in humans, peak I is generated by the distal portion of the auditory nerve and peak II is generated by the proximal (intracranial) portion of the auditory

nerve, thus the auditory nerve is the generator of two peaks in the ABR. This is different from animals used in auditory research where the auditory nerve is the generator of only one peak in the ABR. The reason for that is that the auditory nerve is much shorter in animals than in humans (0.8 cm in the cat [59] versus 2.5 cm in humans [124]).

The response recorded directly from the surface of the cochlear nucleus has a less clear relationship with the simultaneously recorded ABR than the CAP recorded directly from the auditory nerve. Comparison between the responses recorded from the exposed eighth cranial nerve, the cochlear nucleus and the ABR (Figs 7.24 and 7.25) show that the initial positive-negative deflection in the response from the cochlear nucleus has the same latency as peak II of the ABR and the sharp negative peak (N_2) that follows in the response from the cochlear nucleus has the same latency as peak III of the ABR. This large negative peak in the cochlear nucleus response is probably a result of firings of nerve cells.

Peak III is thus the earliest manifestation of neural activity in secondary neurons. The cochlear nucleus is also the main generator of the vertex negative wave between peak III and peak IV of the ABR. The fiber tract that leaves the cochlear nucleus may contribute to peak III of the ABR (including the negative component

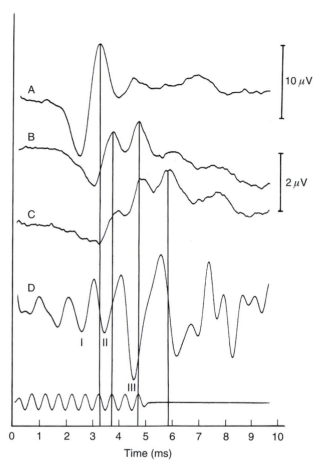

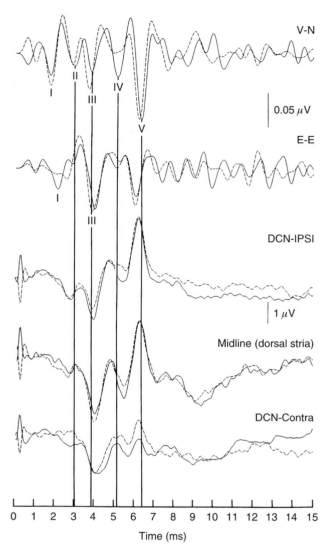

FIGURE 7.24 Comparison of the response (A) from the eighth nerve, (B) entrance of the eighth nerve into the brainstem, (C) the lateral side of the brainstem about 4 mm rostral to the entrance of the eighth nerve and digitally filtered ABR recorded differentially between the mastoid and vertex. Negativity of the near-field potentials is shown as an upward deflection and vertex positivity of the ABR is shown as a downward deflection. The stimuli were 2 kHz tone bursts at 90 dB SPL (reprinted from Møller and Jametta, 1984, with permission from Butterworth Publishers).

FIGURE 7.25 Recordings of the response from the floor of the fourth ventricle to click stimulation in a patient undergoing an operation where the floor of the fourth ventricle was exposed surgically. The stimulation was applied to the right ear. A monopolar recording electrode was used to record the responses from the ipsilateral, contralateral dorsal cochlear nucleus (DCN) and at the midline over the dorsal stria. Negativity is shown as an upward deflection. V–N = ABR recorded differentially between the vertex and the upper neck, vertex positivity is displayed as a downward deflection; E–E = ABR recorded between the two earlobes, ipsilateral earlobe negativity is displayed as a downward deflection. Solid lines: responses to rarefaction clicks (105 dB peSPL), dashed lines responses to condensation clicks (reproduced from Møller et al., 1994, with permission from Elsevier).

that follows peak III) [210]. Peak III may in addition receive contributions from (late) firings of auditory nerve fibers but peak III probably does not receive input from neurons of a higher order than the cochlear nucleus.

The timing of the neural activities in the three different divisions of the cochlear nucleus may be different and the different divisions may contribute to different peaks of the ABR. The three striae where they merge to form the lateral lemniscus may also contribute to peak III and the following negative deflection in the ABR.

The anatomical locations of the neural generators of peak IV are poorly understood and only few published studies have addressed the sources of peak IV of the ABR. Recordings of evoked potentials directly from the exposed lateral brainstem, rostral to the

entrance of the eighth cranial nerve near the fifth cranial nerve have revealed a distinct component with a latency value that is similar to that of peak IV of the ABR (Fig. 7.25). This recording location is anatomically close to the superior olivary complex, indicating that the superior olivary complex might generate these near field potentials and thus suggesting that peak IV of the ABR might be generated by the nuclei of the superior olivary complex [79, 209].

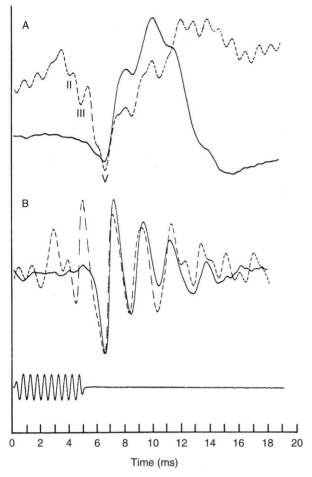

FIGURE 7.26 Responses recorded from the vicinity of the inferior colliculus with the reference electrode on the clavicle (solid lines) compared with the ABR recorded in the same operation between the vertex and a position immediately above the ipsilateral pinna (dashed lines). The top tracings: recordings with electronic filtering 0.003–10 kHz and digital low-pass filtering by a triangular weighting function with a 0.8 ms base. Bottom tracings: the same recordings after digital filtering with a triangular weighting function that has a band-pass characteristic and attenuates slow potentials. The stimuli were 5 ms long, 2 kHz tone bursts at 95dB SPL, presented at a rate of 7 pps (reprinted from Møller and Jannetta, 1983, with permission from Taylor & Francis).

Responses to contralateral stimulation that can be recorded from the floor of the fourth ventricle near the inferior colliculus [79, 80, 206, 217] reveal a sharp positive deflection that is followed by a broad negative wave. The sharp positive deflection occurs with nearly the same latency as peak V of the ABR (Fig. 7.25) [80, 206]. The response is larger and more distinct to contralateral stimulation than to ipsilateral stimulation (Fig. 7.25) [217]. That supports the hypothesis that the initial sharp positive deflection is generated by the terminations of the lateral lemniscus in the inferior colliculus on the side that is contralateral to the stimulation. Peak V is thus

generated mainly by structures that are activated by contralateral stimulation.

The slow deflection in the response from the inferior colliculus (Fig. 7.26) has a similar latency as the broad negative deflection seen to follow peak V in the ABR. This peak (SN_{10} [34]) is variable in humans and usually attenuated by the commonly used filtering of the ABR. When the responses recorded directly from the inferior colliculus or its close vicinity are filtered so that low frequency components are attenuated, a series of sharp waves appear after the initial positive peaks. These waves probably reflect firings of nerve cells in the inferior colliculus (somaspikes). Comparison between the ABR and such filtered recordings from the inferior colliculus indicate that these components may be the generators of peaks VI and VII of the ABR (Fig. 7.26).

Identification of the anatomical location of the generators of the ABR has been attempted by recording the ABR in three orthogonal planes [93, 147, 234, 262]. Such three-dimensional recordings, known as the three-channel Lissajous' trajectory (3 CLT), display evoked potentials as a line, each point of which represents the voltage at any given time after the stimulus (Fig. 7.27).

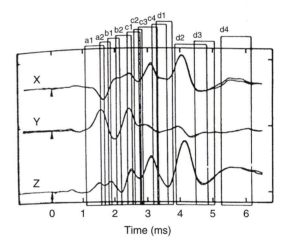

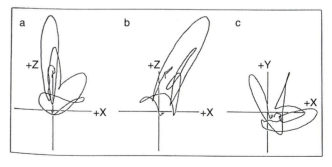

FIGURE 7.27 Illustration of the 3 CLT. Upper graph shows recordings of the ABR in three orthogonal planes. Lower graphs are two-dimensional plots (Lissajous' trajectories). (reprinted from Pratt et al., 1985, with permission from Elsevier).

The use of the 3 CLT recordings to identify the anatomical location of the generators of evoked potentials such as the ABR is based on the assumption that the head acts as a sphere. A dipole source that is located inside the sphere generates electrical potentials that can be recorded from electrodes placed on the surface of the sphere. These potentials can be calculated when the location of the dipole source is known. However, the opposite, namely calculating the location of generators when the potentials on the surface are known, can only be done when certain conditions are fulfilled. This is because a certain voltage distribution on the sphere can be caused by more than one location of a dipole source. Thus while such 3 CLT recordings provide a complete description of the potentials on the surface of the head, determination of the location of a source of the potentials on the basis of distribution of the electrical voltage on the surface of the head does not have a unique solution. Despite this deficiency, the 3 CLT method has yielded valuable results regarding identifying the anatomical location of individual generators of the components of the ABR [147, 235, 262, 309].

Scherg and von Cramon [262] showed that the generation of the ABR could be synthesized by six dipoles, approximately located in the coronal plane (a vertical plane that is perpendicular to the sagittal plane). Dipole I is nearly horizontally oriented and represents the auditory nerve (Fig. 7.28). The negative deflection that follows peak I (I-) has a slightly different orientation. Dipole III and III- are also horizontally oriented toward the contralateral ear, and located in the lower brainstem on the ipsilateral side at approximately the

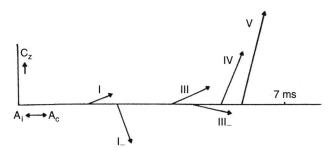

FIGURE 7.28   Orientation and strength of the six dipoles identified from recordings from electrodes placed in three planes. The horizontal line is a line between the two ears and it is also the time axis. The vertical axis is a line between the middle of that line and the vertex. The origin of the vectors is the latency of the first peak in the dipole and the length is the relative strength of the dipoles. Note the short distance between the two first dipoles (peak I and II of the ABR) and the third (peak III) (reprinted from Scherg and von Cramon, 1985, with permission from Elsevier).

same distance from the midline as the cochlear nucleus. The fifth and sixth dipole representing peaks IV and V are oriented vertically but the resolution of these vertical components did not allow determination of their exact location, nor was it possible to determine whether they were located ipsilaterally or contralaterally to the stimulated ear.

Studies of such 3-CLT recordings thus confirmed results obtained by other methods particularly regarding the generator of peak I and II and helped to explain how the recorded ABR depends on the electrode positions. Since the orientation of the dipoles of peaks I, II, and III are mostly along a line between the two ears, these peaks will appear with their highest amplitudes in recordings where the electrodes are placed at each earlobe (or mastoid), whereas peaks IV and V are best recorded from electrodes placed on the vertex and one earlobe or with a non-cephalic reference (such as the upper neck).

Studies of pathologies that affect the auditory nervous system have confirmed that peak II is generated by the intracranial portion of the auditory nerve.

Some studies of abnormalities of the ABR in individuals with known pathologies have produced results that at a glance contradict other studies of the neural generators of the ABR [146]. One such disagreement regards the question about laterality of the generators of the ABR.

While most investigators agree that peaks I, II, and III are generated on the ipsilateral side of the brainstem, some investigators disagree about the anatomical location of the generators of the later peaks [217]. Thus, Markand and coworkers [146] have interpreted available data and came to the conclusion that peak V is generated by brainstem structures on the side from which the ABR is elicited.

Determining the anatomical location of the neural generators of peak V of the ABR on the basis of abnormalities in the ABR in patients with lesions that affect the ascending auditory pathways has pitfalls that are often overlooked. The results of such studies are also more difficult to interpret than results from intracranial recordings. Nevertheless, such methods have the advantage that the results are directly related to the use of the ABR in diagnosis of disease processes that may affect the auditory nervous system.

Animal experiments make it possible to study the effect of inactivation (ablation) of specific neural structures on the ABR in addition to comparing the potentials recorded from specific structures of the ascending auditory pathways with the ABR. However, the ABR in the animals that have been studied differs from that of humans. Thus the ABR obtained from animals,

---

BOX 7.8

DIFFERENT INTERPRETATIONS OF ABR NEURAL GENERATORS BASED ON PATHOLOGY

The results of studies of the abnormalities of the ABR in patients with discrete intrinsic lesions of the brainstem [146, 222, 325] have been interpreted to show that the neural generators of all peaks, including peak V are located on the side from which the ABR is elicited. The results have been summarized by the widely cited statement: "When ABR abnormalities either occur exclusively on stimulation of one ear or are asymmetric on right and left ear stimulation, the responsible lesion in the brain stem is on the side of the ear eliciting maximal abnormality" [146].

Garg et al. [64] studied patients with a certain kind of hereditary motor-sensory neuropathy (type I) and found evidence that both peak I and peak II of the ABR were generated by the auditory nerve. That peak III is generated in the pontine region of the ipsilateral brainstem was supported by studies of the abnormalities of the ABR in patients with discrete lesions in the brainstem [64, 146].

Chiappa [26] in a recognized handbook on evoked potentials has extended the conclusions from lesion studies on the ABR to mean that the ABR does not reflect the parts of the ascending auditory pathways that are normally associated with hearing (the crossed pathways). This seems too strong a statement and the results from studies of pathologies can be explained in a more plausible way. As mentioned above, there is overwhelming evidence from physiological studies that peak V is mainly generated by structures located on the contralateral side with the main crossing occurring at the level of the superior olivary complex (see Chapter 5), but anatomical studies show uncrossed pathways as well. It is not known how many fibers cross and how many continue on the same side and the importance of the uncrossed pathways for hearing is unknown.

The findings that lesions (tumors, bleeding, etc.) verified by imaging techniques (such as MRI) affect components of the ABR elicited from the side of the lesion more than they affect (peak V) of the ABR elicited from the contralateral ear [146] does not need to be a contradiction to results of electrophysiologic studies that show that peak V of the ABR is generated mainly by structures located on the opposite side from which the ABR is elicited. These findings can be explained without resorting to such a drastic assumption that the uncrossed pathways are the (main) generator of the ABR [26].

First, recognize that changes at the peripheral levels are imposed on more centrally located structures. Lesions of peripheral structures will therefore also affect components of the ABR that are generated by more centrally located structures. Assume that peak V is generated when propagated neural activity in the lateral lemniscus (LL) halts, which normally occurs where the lateral lemniscus terminates in the inferior colliculus. Lesions to the LL from disease processes, such as a tumor, may cause a total or partial arrest of propagation of neural activity at the location of the lesion. Such a halt in propagation of neural activity in the LL can generate a stationary peak in the far field that is indistinguishable from a normal peak V. A lesion located at the LL anywhere between the ipsilateral cochlear nucleus and the contralateral inferior colliculus will have the same effect. The only obvious abnormalities in the ABR would be a slightly shorter latency of peak V that is unlikely to be noticeable. The SN_{10} would be abolished by interruption of the neural transmission in the LL but the SN_{10} is normally not detectable because the high-pass filter commonly used attenuates the SN_{10}. That means that a lesion of the LL in its contralateral course may not produce any noticeable change in the ABR. For the same reason, the effect of a tumor or other lesion on the inferior colliculus may not change the ABR noticeably.

The finding of changes in the ABR elicited from the ipsilateral side in patients with lesions in the brainstem may be explained by the fact that the anatomical extension of the lesions were poorly defined and lesions in the midbrain as determined by MRI scans may extend further caudally and affect the cochlear nucleus and superior olivary complex.

Studies that show that lesions that affect the midbrain may cause changes in peak V of the ABR elicited from the contralateral ear have often been overlooked. Thus, Zanette et al. [328] found that peak V was absent in the ABR of certain patients with brainstem hemorrhage when elicited from the ear contralateral to the bleeding. Fischer et al. [53] showed that wave V of the ABR was delayed and had a reduced amplitude when elicited from the opposite side in a patient with a lesion involving the inferior colliculus. It should also be noted that the study by Markand [146] indeed found changes in the ABR elicited from the side opposite to the lesion, but the changes were not noticeably larger than those in the ABR elicited from the side of the lesion.

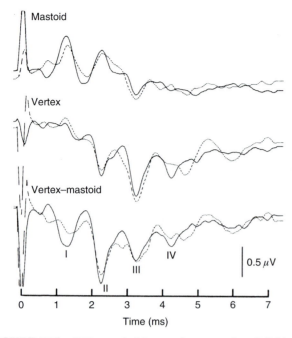

Mastoid

Vertex

Vertex–mastoid

0.5 μV

I

II

III

IV

0　1　2　3　4　5　6　7

Time (ms)

FIGURE 7.29 ABR recorded from a rhesus monkey. Solid lines: responses to rarefaction clicks, dashed lines: responses to condensation clicks. The reference electrodes were placed on the shoulder. In the mastoid and vertex recordings, negativity is shown as upward deflections and the vertex-mastoid recording vertex positivity is shown as a downward deflection (reprinted from Møller et al., 1986, with permission from Elsevier).

including the monkey, consists of only four constant vertex positive peaks (Fig. 7.29). The reason is that the auditory nerve is much shorter in animals used in auditory experiments than it is in humans. This causes the travel time in the auditory nerve to be

too short to generate two clearly separated peaks in animals. There may be other differences attributable to the differences in the ascending auditory pathways, mainly concerning the superior olivary complex (see [158, 159]).

A synthesis of the results using several different methods has provided the following general description of the neural generators of vertex positive peaks of the click evoked ABR recorded between electrodes placed on the vertex and the earlobe or mastoid on the side where the stimulation is applied (Fig. 7.30):

Peak
  I: Distal (peripheral) portion of the auditory nerve
  II: Proximal (central) portion of the auditory nerve
  III: Cochlear nucleus
  IV: Probably structures that are close to the midline (superior olivary complex?)
  V: Sharp vertex positive peak: The termination of the lateral lemniscus in the inferior colliculus on the contralateral side. The slow negative peak (SN_{10}): Dendritic potentials from the inferior colliculus.

It seems unlikely that any structure of the auditory system other than the auditory nerve can contribute to peaks I and II because cochlear nucleus cells would not fire earlier than 0.5 to 0.7 ms after the arrival of the neural volley in the auditory nerve. Peaks I and II are the only components that are generated by a single structure. Later components of the ABR are likely to receive contributions from several structures. All structures of the ascending auditory nervous

---

BOX 7.9

NEURAL GENERATOR OF PEAK II AND AUDITORY NERVE LENGTH

The results of early studies of the neural generators of the ABR in animals [23], which showed that peak II of the cat ABR was generated in the cochlear nucleus resulted in the erroneous assumption that also peak II of the human ABR was generated in the cochlear nucleus. The anatomical differences between the auditory nerve in humans and in the animals used in such experiments were not recognized and that caused the misinterpretation of the neural generators of the ABR in humans. The auditory nerve in humans is approximately 2.5 cm long [125, 126] compared with 0.5–0.8 cm in the cat [59]. The auditory nerve in humans is therefore sufficiently long to generate two well-separated

peaks in the ABR (I and II) while it generates only one peak in the ABR in animals. The auditory nerve in humans is longer than in small animals because humans have a larger head and a much larger sub-arachnoidal space in the cerebello-pontine angle than animals commonly used for studies of the auditory system, including the monkey.

It has been shown that under favorable circumstances two separate peaks that are generated by the auditory nerve could be identified in the ABR of small animals [1, 284]. The latency of the second peak was approximately 0.4 ms longer than that of peak I and it may correspond to peak II in humans.

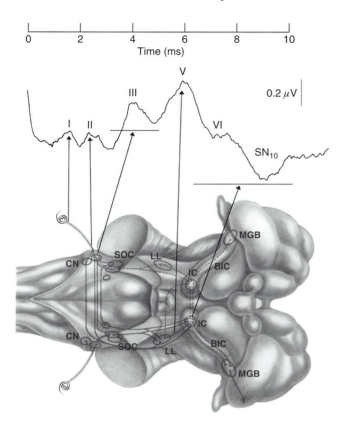

FIGURE 7.30 Schematic summary of the anatomical location of the neural generators of the ABR (reprinted from Møller, 2006, with permission from Cambridge University Press).

system other than the auditory nerve are likely to contribute to more than one peak of the ABR.

It is also important to consider that not only the vertex positive peaks are generated by specific structures of the auditory nervous system, but also the vertex negative waves have more or less specific neural generators [217, 329]. It is interesting to speculate what would have happened if it had been the vertex negative waves that were labeled and attention therefore drawn to these components instead of the vertex positive peaks.

It is not known with certainty whether the different components of the ABR are generated by fiber tracts (white matter) or cell bodies in nuclei (gray matter). Both of those two kinds of structures may contribute to far-field potentials. The fact that the auditory nerve is the sole generator of peaks I and II shows that a nerve can contribute to the ABR. This means that it is also reasonable to assume that fiber tracts can generate stationary peaks in the far field and thus contribute to the ABR. The contribution to the ABR that has been ascribed as coming from nuclei may in fact be

generated by fiber tracts that lead to and from the nuclei. An example is the sharp peak of peak V, which is generated by the termination of the LL in the IC. Whatever structures generate the sharp peaks in the far-field auditory potentials, their amplitude seems to depend on how well synchronized the neural activity is in the structure in question (nerve, fiber tracts or nuclei). Since these sources of the ABR can be regarded as dipoles that are oriented differently, the amplitude of the components of the ABR that they generate depends on the orientation of the pair of electrodes from which the ABR is recorded.

4.2. Middle Latency Responses

The middle latency response (MLR) consists of evoked potentials that occur in the interval between 10 and 80 ms (or 10–100 ms) after a sound stimulus. The MLR is commonly recorded in a similar way as the ABR, thus differentially between electrodes on the vertex and the earlobe on the side where the sound stimuli are applied. The different components of the MLR are generated by more central neural structures than the ABR, including the auditory cortex. The MLR were first described by Geisler et al. [65], and were later studied by many investigators such as Picton et al. [232] and by Nina Kraus and her co-workers [117].

The labeling of the components of the MLR is perhaps even more confusing than what is the case for the ABR. The most prominent components are labeled Na, Pa, and Nb, Pb, Nc, Pc, Nd with N for negative and P for positive waves (Fig. 7.31) [61]. The slow negative (SN_{10}) component of peak V of the ABR is usually visible in the MLR (as Na) because the MLR is

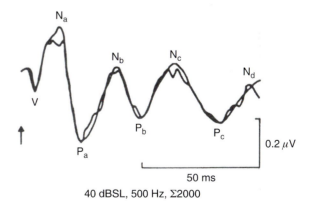

FIGURE 7.31 Middle latency responses (MLR) shown on a 100 ms time scale. The responses were elicited by click stimulation, presented at a rate of 10 pps (reprinted from Galambos et al., 1981).

recorded with preservation of low frequencies (low settings of the high pass filter in the amplifiers used). The Na component is followed by a large negative peak, Nb with a latency of approximately 35 ms. Two more negative peaks, Nc and Nd, can usually be identified.

Our knowledge about the origin of these potentials is sketchy at best. The fact that myogenic auditory evoked potentials (see p. 177) occur in the time window of the MLR has added to the difficulties in determining the neural generators of the MLR. Other factors, such as the much greater variability of the MLR compared with the ABR, make interpretation of studies of the neural generators of the MLR more difficult than that of the ABR. The degree of wakefulness affects the MLR and intraoperative studies are difficult to do because the MLR is suppressed by anesthesia. Thus, only a few studies have addressed the anatomical location of the neural generators of the MLR in humans.

It is a general rule that the amplitude of sensory evoked potentials such as auditory evoked potentials decreases with increasing repetition rate of the stimuli. However, Galambos and his co-workers [61], have shown that the amplitude of the response to repetitive sound stimuli has its highest value at stimulus repetition rates of approximately 40 Hz (or pps) when elicited by clicks or short tone bursts (Fig. 7.32).

The reason for the increased amplitude at a stimulus repetition rate of 40 pps is that peaks in the MLR occur with intervals of approximately 25 ms and they therefore add to each other when the interval between individual stimuli is 25 ms, thus a repetition rate of 40 pps (or Hz). The 40 Hz response elicited by bursts of pure tones seems to be useful for determining the hearing threshold in non-cooperative individuals.

4.3. Far-field Frequency Following Responses in Humans

Sound evoked potentials that reflect the waveform of the stimulus sounds are the frequency-following potentials. The latency of these responses is approximately 6 ms, indicating that the FFR is generated by structures that are rostral to the auditory nerve. It is probably the part of the cochlea that responds best to frequencies below 2 kHz, that generates the FFR [165, 166].

BOX 7.10

NEURAL GENERATORS OF MIDDLE LATENCY RESPONSE

Already by 1958, Geisler et al. [65] suggested that these potentials might be generated by the auditory cortex and later Lee et al. [133] found evidence in intracranial recordings in humans that the Pa component with normal latencies of 24–30 ms was generated by the auditory cortex. The neural generators of the MLR in humans may be different from those in animals and it is uncertain how the different components of the MLR in animals correspond to the components of the MLR in humans.

Recent studies in animals have revealed that some of the components of the MLR may be generated by the non-classical ascending auditory pathways. It has been shown that the ventral and the caudo-medial portions of the medial geniculate body in the guinea pig give specific contributions to the MLR. The ventral portion of the medial geniculate body is associated with the classical lemniscal auditory pathways and relays information to the primary cerebral auditory cortex where all information passes (see Chapter 5) while the caudo-medial portion of the medial geniculate body contains relay neurons for the non-classical ascending auditory system. McGee et al. [150] studied the contribution to the MLR recorded from each of these two parts of the medial geniculate body by inactivating one part at a time, by injecting Lidocaine in corresponding parts of the medial geniculate body. The MLR recorded from the scalp in guinea pigs overlying the temporal lobe was different from that recorded from a midline electrode position. Some of the components of the MLR recorded from a midline position were assumed to be generated by the non-classical system, whereas those recorded from the skull over the temporal lobe are generated by the classical system.

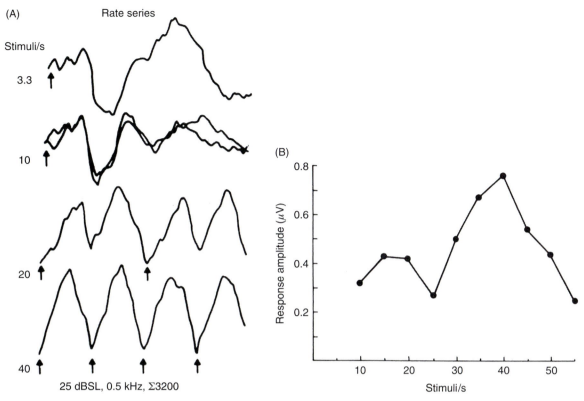

FIGURE 7.32   (A) The 40 Hz response. The change in the response when the repetition rate of the stimulation is changed. (B) Peak-to-peak amplitude of the response as a function of the repetition rate of the stimulation (from Galambos et al., 1981).

4.4. Myogenic Auditory Evoked Potentials

When sensory evoked potentials are concerned, it is usually assumed that the recorded potentials are generated by structures of the nervous system. Sensory stimuli can, however, also evoke motor responses that can be recorded as electromyographic (EMG) potentials.

Several investigators have described acoustically evoked muscle responses that occur with latencies between 10 and 30 ms in response to loud transient sounds [29, 65, 108].

Myogenic auditory evoked potentials are affected by attention, arousal, voluntary and involuntary muscle tension and other factors (Fig. 7.34). The variability of

---

BOX 7.11

GENERATORS OF FREQUENCY FOLLOWING RESPONSE

Moushegian and co-workers showed in 1973 [166] that FFR to low frequency tones could be recorded from electrodes placed on the scalp of human volunteers (Fig. 7.33). Masking studies confirmed that these potentials were of neural origin and not cochlear microphonics (CM) (Fig. 7.33A). The FFR recorded from electrodes placed on the scalp (vertex and mastoid) can be observed in the frequency range from 0.25 to 2 kHz and it is most pronounced for tones with frequencies in the range of 0.25 to 0.5 kHz (Fig. 7.33A). The amplitude of these potentials increases with increasing stimulus intensity (Fig. 7.33B) [166].

BOX 7.11   (*cont'd*)

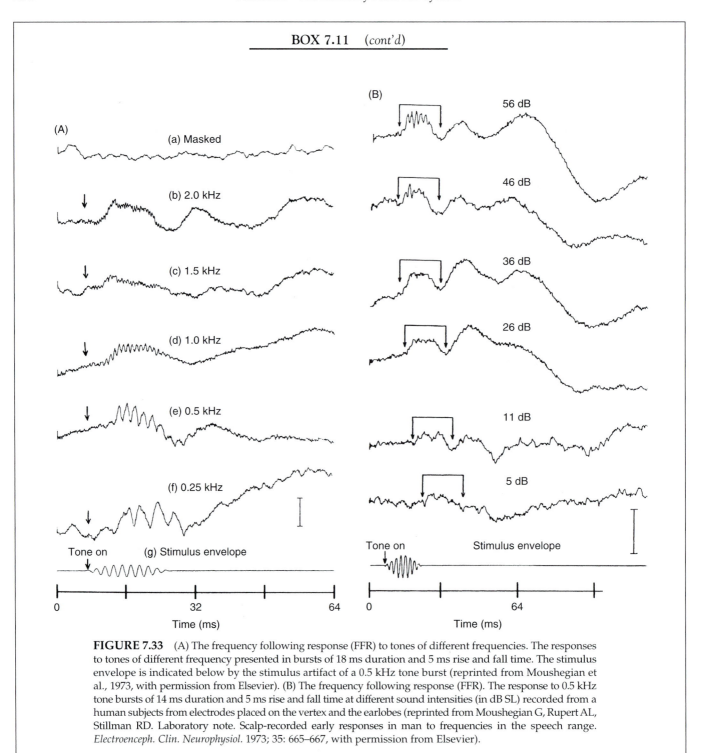

FIGURE 7.33   (A) The frequency following response (FFR) to tones of different frequencies. The responses to tones of different frequency presented in bursts of 18 ms duration and 5 ms rise and fall time. The stimulus envelope is indicated below by the stimulus artifact of a 0.5 kHz tone burst (reprinted from Moushegian et al., 1973, with permission from Elsevier). (B) The frequency following response (FFR). The response to 0.5 kHz tone bursts of 14 ms duration and 5 ms rise and fall time at different sound intensities (in dB SL) recorded from a human subjects from electrodes placed on the vertex and the earlobes (reprinted from Moushegian G, Rupert AL, Stillman RD. Laboratory note. Scalp-recorded early responses in man to frequencies in the speech range. *Electroenceph. Clin. Neurophysiol.* 1973; 35: 665–667, with permission from Elsevier).

the responses makes it difficult to interpret the results of such recordings and this is why myogenic evoked responses never gained clinical use [36]. The fact that the latency of the earliest components of these myogenic potentials are

between 10–30 ms makes such potentials sometimes occur at the end of the 10 ms recording window of the ABR and thus easily distinguishable from the components of the ABR that originate from auditory brainstem structures. Myogenic responses, however,

BOX 7.11   (cont'd)

(C)

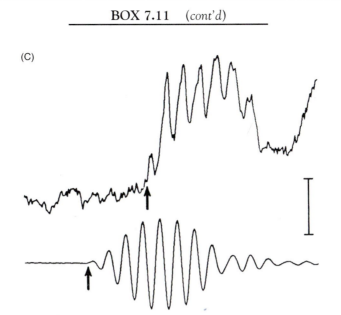

FIGURE 7.33 (C) The frequency following response (FFR). The latency of the FFR to 0.5 Hz tone bursts. The onset of the tone and the onset of the FFR are marked by arrows. The latency was 6.1 ms. The sound level was 60 dB SL (reprinted from Gerken et al., 1975, with permission from Elsevier).

BOX 7.12

AUDITORY EVOKED MYOGENIC POTENTIALS

Auditory evoked myogenic potentials can be recorded from an electrode placed behind the ear and from electrodes placed on the parietal region of the scalp (Fig. 7.34) [149]. It is not only muscles on the head that respond to sound stimulation but also extracranial muscles respond to strong click sounds (Fig. 7.35) [65].

P_I-C_Z Traction   Inion-ear

Backward 1.25 kg

None

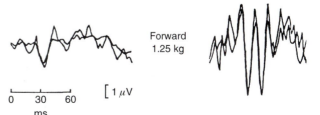

Forward 1.25 kg

FIGURE 7.34 Myogenic potentials recorded from an electrode placed on P1 and Cz (left column), and inion and Pz (right column). The different rows of records were obtained with different tension of neck muscles (reprinted from Mast, T.E. 1965. Short latency human evoked responses to clicks. *J. Appl. Physiol.* 20, 725–30, with permission from the American Physiological Society).

0   30   60
ms

[1 μV

BOX 7.12 (cont'd)

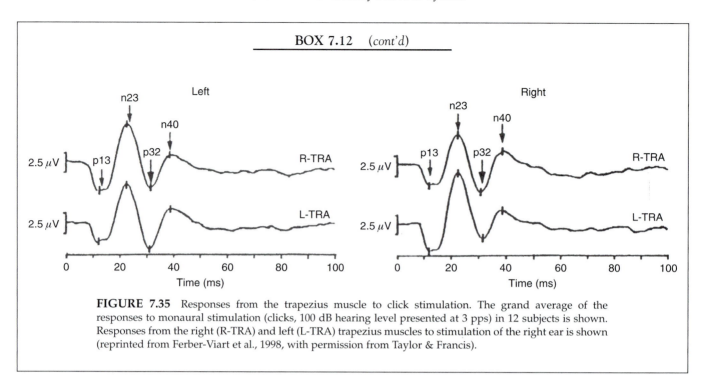

FIGURE 7.35 Responses from the trapezius muscle to click stimulation. The grand average of the responses to monaural stimulation (clicks, 100 dB hearing level presented at 3 pps) in 12 subjects is shown. Responses from the right (R-TRA) and left (L-TRA) trapezius muscles to stimulation of the right ear is shown (reprinted from Ferber-Viart et al., 1998, with permission from Taylor & Francis).

may interfere with the neural components of the MLR responses and that has been an obstacle in the clinical use of the MLR. Some of these responses recorded from extracranial muscles have been attributed to activation of the vestibular system and thus being parts of vestibular spinal reflexes [52]. Myogenic responses are not associated with visible contractions of muscles such as are the general sound evoked startle response which may occur in response to unanticipated loud sounds involving many muscles.

Acoustic Middle-ear Reflex

1. ABSTRACT

1. The acoustic middle-ear reflex involves only the stapedius muscle in humans but both the tensor tympani and the stapedius muscles contract as an acoustic reflex in animals commonly used in auditory research.
2. The acoustic middle ear reflex can be elicited by sounds of approximately 85 dB HL in individuals with normal hearing. The strength of the muscle contraction increases gradually with increasing stimulus intensity.
3. When elicited by sounds presented to one ear, the stapedius muscle in humans contracts in both ears but the response is slightly stronger in the ipsilateral ear and its threshold is lower.
4. The contraction of the stapedius muscle occurs with a latency that decreases from approximately 100 ms for sounds near the threshold of the reflex to approximately 25 ms for high intensity stimuli.
5. Contractions of the stapedius muscle decrease sound transmission to the cochlea, more for low frequencies than for high frequencies.
6. The acoustic middle ear reflex acts as a control system that makes the sounds that reaches the cochlea vary less than the sounds that reach the tympanic membrane.
7. Contraction of the stapedius muscle changes the ear's acoustic impedance.
8. Recording of changes in the ear's acoustic impedance is the most common method used for studies of the acoustic middle-ear reflex and it is used in clinical diagnosis.
9. The acoustic middle ear reflex response is reduced after intake of sedative (hypnotic) drugs such as barbiturates and alcohol.
10. The middle-ear muscles contract in other ways than in response to sounds. A few individuals can voluntarily contract their middle-ear muscles and the stapedius muscle contracts before vocalization.

2. INTRODUCTION

Involuntary muscle contractions that are elicited by sound are known as acoustic reflexes. The best known acoustic reflex is the acoustic middle-ear reflex, which involves one or both of the two middle-ear muscles. This reflex has been studied extensively, and recording of the reflex response plays an important role in modern audiological testing. The basic characteristics of the reflex will be described in this chapter, while its use in diagnosis of disorders of the middle ear, the cochlea and the auditory nervous system will be discussed in Chapters 9 and 10.

Other types of acoustic reflexes include the startle reflex where a loud and unexpected sound causes contraction of many skeletal muscles. The movement of the eyes toward the source of strong impulsive sounds may also be regarded as an acoustic reflex. Contractions of face and neck muscles commonly occur in response to loud sounds. Small invisible contractions can be detected by recording electromyographic (EMG) potentials in response to sounds over a wide range of intensities. The raise in heart rate in response to a strong sound is not commonly regarded as an acoustic reflex but it has most of the characteristics of an acoustic reflex. Whether or not seizures that under rare circumstances can be evoked by strong sounds (audiogenic seizures) may be called acoustic reflexes is a matter of definition. These acoustic reflexes are not discussed in this book.

The acoustic middle-ear reflex has been studied extensively in both humans and animals. The effector organ of the acoustic middle-ear reflex in humans is the stapedius muscle and in some animal species, the tensor tympani muscle also contracts in response to loud sounds. Thus, despite the fact that the stapedius and the tensor tympani muscles are innervated by two different cranial nerves (the facial and the trigeminal nerves, respectively, see Chapter 1) these two muscles contract together as an acoustic reflex, at least in the animals often used in auditory research, such as the cat, guinea pig, and rat. In humans a sound above the threshold of the reflex presented to one ear elicits a contraction of the stapedius muscle in both ears but the response is slightly larger when elicited from the ipsilateral ear [194].

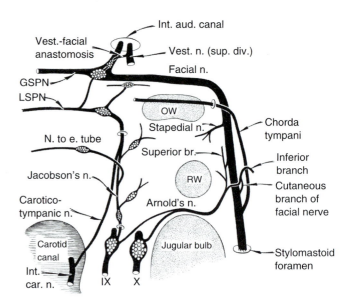

FIGURE 8.1 Schematic drawing of the course of the facial nerve in the skull. Notice the stapedius nerve (reprinted from Schucknecht, H.F. 1974. *Pathology of the Ear*. Cambridge, MA: Harvard University Press, with permission from Harvard University Press).

3. NEURAL PATHWAYS OF THE ACOUSTIC MIDDLE-EAR REFLEX

The stapedius muscle is innervated by the stapedius nerve, a branch of the facial nerve. The stapedius nerve takes off from the main trunk of the facial nerve in the facial canal peripheral to the petrosal nerve and the geniculate ganglion. In humans it branches off the facial nerve approximately 1 cm from the stylomastoid foramen (Fig. 8.1). The tensor tympani muscle is innervated by the mandibular branch (V3) of the fifth cranial nerve (the trigeminal nerve).

Studies in the rabbit [14] have shown that the auditory nerve and the ventral cochlear nucleus are the first part of the reflex arc of both the contralateral and the ipsilateral pathways of the acoustic stapedius reflex (Fig. 8.2). The tensor tympani reflex also uses

neurons in the ventral cochlear nucleus. The dorsal cochlear nucleus is not involved in the acoustic middle-ear reflex. Two main parallel ipsilateral pathways lead sound evoked neural activity to the facial motonucleus and subsequently activate the stapedius muscle (Figs 8.2 and 8.3). One pathway is a direct connection from the ventral cochlear nucleus to the facial motonucleus on the same side thus a two-synapse reflex arc. The other pathways connect the ventral cochlear nucleus to the facial motonucleus through a second synapse in nuclei of the superior olivary complex (SOC). Both a crossed and an uncrossed pathway connect neurons in the SOC to the facial motonucleus, thus a three-synapse reflex arc. The crossed pathways of the stapedius reflex (mediator of the contralateral response) have a synapse in the medial superior olivary (MSO) nucleus from where connections lead to the facial motonucleus. The neurons in the part of the facial motonucleus that

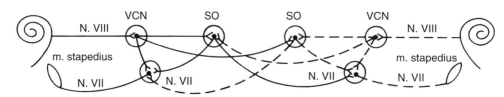

FIGURE 8.2 Schematic drawing of the reflex arc of the acoustic middle ear reflex (stapedius reflex). N.VIII = auditory nerve; N.VII = facial nerve; N.VII = facial motonucleus; VCN = ventral cochlear nucleus; and SO = superior olivary complex (reprinted from Møller, 1983, with permission from Elsevier).

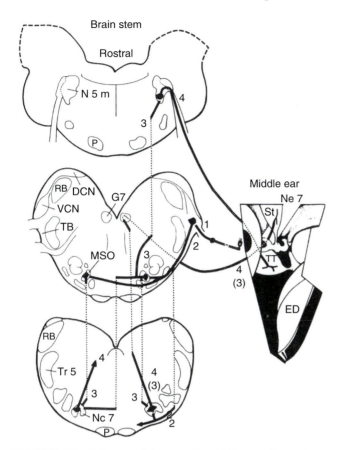

FIGURE 8.3 Pathways of the acoustic middle ear reflex as shown in three transverse sections through the brainstem of a rabbit. Solid lines represent nerve tracts; dotted lines show the connections between the sections. CR = restiform body; DCN = dorsal cochlear nucleus; ED = ear drum; G7 = internal geniculum of the seventh cranial nerve; MSO = medial superior olive; Nc 7 = nucleus of the seventh cranial nerve; N.VII = seventh cranial nerve; P = pyramidal tract; St. = stapedius; TB = trapezoid body; Tr 5 = spinal trigeminal tract; TT = tensor tympani; VCN = ventral cochlear nucleus. Numbers refer to the order of neurons in the stapedius reflex and the tensor tympani reflex (reprinted from Borg, 1973, with permission from Elsevier).

are anatomically close to the superior olivary complex innervate the stapedius muscle [103].

The direct pathways of the ipsilateral acoustic middle-ear reflex thus have two and three synapses, respectively, and the contralateral reflex has at least three synapses. A more diffuse pathway of the acoustic middle-ear reflex also exists [14]. These indirect pathways have many synapses. The reflex response mediated by the indirect pathways is slower and more sensitive to anesthesia than that mediated by the direct pathways, and it is probably also affected by the degree of wakefulness. It is not known in detail which neural structures comprise the indirect pathways but ablation of the inferior colliculus (IC) did not have any noticeable effect on the reflex response [14] nor did lesions in the pyramidal tract affect the response noticeably. Therefore, it may be assumed that the acoustic middle-ear reflex does not involve the cerebellum, nor the midbrain or the forebrain.

The reflex pathways for the tensor tympani are slightly different from that of the stapedius reflex. Axons from the MSO connect to the fifth cranial nerve motonucleus on both sides and these connections may be the most important ones for the tensor tympani reflex [14]. No connections from the trapezoidal body to the fifth cranial nerve motonucleus have been found.

4. PHYSIOLOGY

The response of the acoustic middle-ear reflex in humans has been studied using recordings of changes in the ear's acoustic impedance, displacement of the tympanic membrane and by recording of electromyographic potentials from the stapedius muscle.

BOX 8.1

TECHNIQUES FOR RECORDING THE CONTRACTIONS OF THE MIDDLE-EAR MUSCLES

Recordings of changes in the ear's acoustic impedance are a convenient and non-invasive method to record the contractions of the middle-ear muscles and it is the method commonly used in research as well as in clinical studies of the acoustic middle-ear reflex. Its use is based on the fact that contractions of the middle-ear muscles change the ear's acoustic impedance (Chapter 2).

Geffcken (1934) was probably the first to report that the ear's acoustic impedance changed when the middle ear muscles were brought to contraction by a loud sound. Metz [151, 152] was one of the first to use measurements of the ear's acoustic impedance for clinical purposes and he pioneered the use of measurement of changes in the ear's acoustic impedance to record the contractions of the middle-ear muscles. Since then recordings of the change of the ear's acoustic impedance have been used by numerous investigators for clinical studies of the acoustic middle-ear reflex [92, 296] and for research purposes [194]. While Metz [151] and Jepsen [92] used the Schuster bridge, the investigators who followed mainly used an electroacoustic method [33, 182, 194, 296] and that is also the principle used in the equipment that is presently used clinically. Most commercially available equipment that is designed for clinical recording the response of the acoustic middle ear reflex and for tympanometry use test tones of approximately 0.22 kHz but investigators of the function of the acoustic middle ear reflex have used a 0.8 kHz probe tone [194]. Another non-invasive method makes use of recordings of the displacement of the tympanic membrane as an indicator of contractions of the middle ear muscles but this method does not provide a reliable measure of the contraction of the stapedius muscle (see p. 38).

Recording electromyographic (EMG) potentials [19, 229] from the exposed stapedius muscle or recording the change in the cochlear microphonic (CM) potentials [177] has also been used to study the function of the acoustic middle ear reflex. Recording of EMG potentials makes it possible to discriminate between the contractions of the two muscles, which is not possible by recording of the ear's acoustic impedance. Recording CM makes it possible to measure the change in sound transmission through the middle ear that is caused by contractions of the middle-ear muscles [177]. Both the EMG and the CM methods are invasive and are not practical for use in humans except in special situations where the middle-ear cavity becomes exposed during a surgical operation [19].

4.1. Responses to Stimulation with Tones

The response amplitude of the acoustic middle ear reflex to sounds just above threshold of the reflex increases gradually after a brief latency and attains a plateau after approximately 500 ms. The response amplitude increases at a faster rate in response to sounds well above threshold (Fig. 8.4). The amplitude of the reflex response elicited by high frequency sounds decreases over time (adaptation) but normally the reflex response elicited by tones below 1.5 kHz shows little adaptation. The amplitude of the response is slightly larger when elicited from the ipsilateral ear, compared with the contralateral ear (Fig. 8.4) [169, 194]. The amplitude of the reflex responses increases with increasing stimulus intensity and reaches a plateau approximately 20 dB above the threshold (Fig. 8.5). The maximal response amplitude that can be obtained is higher when recorded from the ear from which the reflex is elicited than when recorded from the contralateral ear (Fig. 8.5). The rate of the increase in the response amplitude with increased stimulus intensity is similar for ipsilateral and contralateral stimulation (Fig. 8.5). The difference between the response to ipsilateral and contralateral stimulation is greater when the reflex response is elicited by low frequency tones than by tones above 0.5 kHz. When the stimulus tone is applied to both ears at the same time the response is larger than when only one ear is stimulated (Fig. 8.5) and the stimulus response curves are shifted approximately 3 dB relative to that of ipsilateral stimulation [169]. It is noteworthy that most studies of the acoustic middle-ear reflex, including its use in clinical diagnosis, have been restricted to studies of the contralateral responses.

The stimulus response curves are less steep for stimulation with short tones than for long tones (Fig. 8.6) and the difference between the response to bilateral, ipsilateral, and contralateral stimulation is greater when the reflex is elicited by short tones than by long tones. The response to short tones also reaches a plateau at a lower response amplitude than that to long tones, and the response to contralateral stimulation reaches a plateau at a lower response amplitude than for ipsilateral and bilateral stimulation.

Using recordings of changes in the ear's acoustic impedance, the threshold of the human acoustic middle-ear reflex is approximately 85 dB above normal hearing threshold [195] but there are considerable individual variations (Fig. 8.7). The threshold of the acoustic middle-ear reflex is poorly defined because small irregular responses are obtained in a large range of stimulus intensities near threshold (Fig. 8.8). The variability of these responses makes it difficult to accurately determine the absolute threshold of the acoustic middle-ear reflex. The "threshold" of the

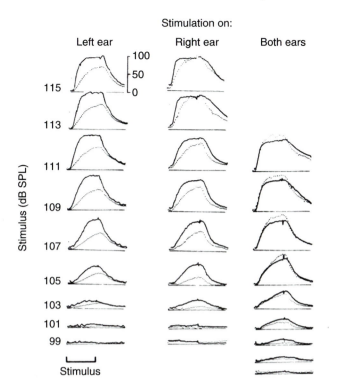

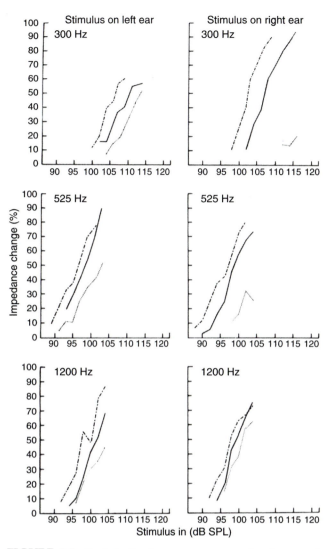

FIGURE 8.4 Change in the acoustic impedance recorded in both ears simultaneously as a result of contraction of the stapedius muscle elicited by tone bursts of different intensity. In the two left-hand columns, one ear was stimulated. The solid lines are the impedance change in the ipsilateral ear and the dashed lines are the impedance change in the contralateral ear. The right-hand columns show responses of both ears when both ears were stimulated simultaneously. The solid lines show contractions of the middle ear muscles in the ipsilateral ear and the dashed lines are the responses in the contralateral ear. The stimulus sound was 1.45 kHz pure tones presented in bursts of 500 ms duration. The intensity of the sound is given in dB SPL. The results were obtained in an individual with normal hearing (reprinted from Møller, 1962, with permission from the American Institute of Physics).

FIGURE 8.5 Typical stimulus response curves for the acoustic middle ear reflex in an individual with normal hearing. Dashes show the amplitude of the response to bilateral stimulation, solid lines are the response to ipsilateral stimulation and the dots are the contralateral response. Results from both ears are shown (right and left graphs). The stimuli were 500 ms tone bursts. In these experiments the stimulus intensity was first raised (in 2dB steps) from below threshold to the maximal intensity used and then lowered again (in 2 dB steps) to below threshold. The change in the ear's impedance given is the mean of two determinations, one when the stimulus was increased from below threshold and the other when the stimulus intensity was decreased from the maximal used intensity to the threshold. The change in the ear's acoustic impedance is given as a percentage of the maximally obtained response at any stimulus frequency and situation (usually bilateral stimulation) (reprinted from Møller, 1962, with permission from the American Institute of Physics).

acoustic middle ear reflex, defined as the sound intensity necessary to elicit a response the amplitude of which is 10% of the maximal response, is a more reproducible measure of the sensitivity of the reflex [195]. The threshold that is defined as the sound intensity needed to elicit a response with a small amplitude (for instance, 10% of the maximal response) has a high degree of reproducibility in the same individual when recorded at different times (Fig. 8.9). The reflex threshold, as defined here for stimulation of the contralateral ear, is approximately 85 dB above hearing threshold in young individuals with normal hearing. The reflex threshold shows considerable individual variations [195]. These large individual variations that are present even between young individuals with normal hearing and without history of middle-ear disorders (Fig. 8.7) should be considered when the threshold

of the acoustic middle-ear reflex is used for diagnostic purposes. The fact that the threshold in an individual person varies very little over time (Fig. 8.9) makes it possible to follow the progress of disorders of individual patients such as that of vestibular Schwannoma.

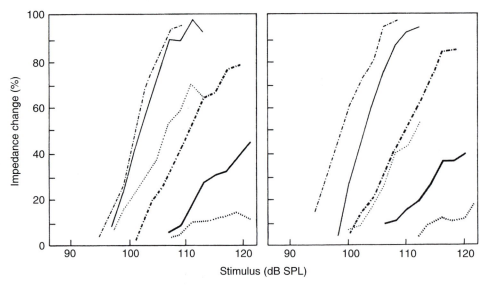

FIGURE 8.6 Stimulus responses curves similar to those in Fig. 8.5 showing the difference between the response to tones of 500 ms duration (thin lines) and the responses to shorter tones (25 ms duration, thick lines). Dots and dashes = bilateral stimulation; solid lines = ipsilateral stimulation; and dotted lines = contralateral stimulation. The stimulus frequency was 0.525 kHz. Left-hand graph: stimulation of the left ear; right-hand graph: stimulation of the right ear (reprinted from Møller, 1962, with permission from the American Institute of Physics).

It is not known how the threshold of the acoustic middle-ear reflex is set but it is interesting to note that individuals whose auditory nerve is injured have an elevated reflex threshold, and a poor growth of the reflex response amplitude with increasing stimulus intensity (see p. 291). Such injuries mainly affect the synchronization of neural activity in the auditory

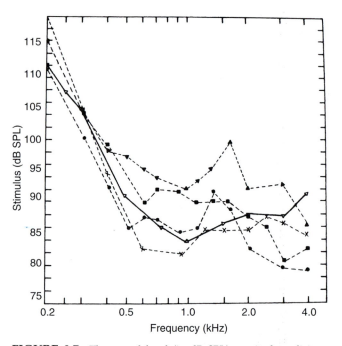

FIGURE 8.7 The sound level (in dB SPL) required to elicit an impedance change of 10% of the maximal obtainable response amplitude in the ear opposite to that which is stimulated is shown as a function of the frequency of the tones used for stimulation. The results were obtained in young individuals with normal hearing. The thick line shows the sound levels (in dB SPL) that are 80 dB above the threshold of hearing (80 dB HL) (reprinted from Møller, 1962, with permission from the Annals Publishing Company).

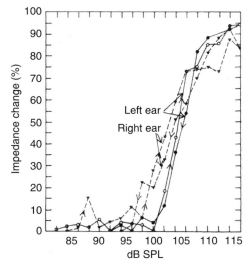

FIGURE 8.8 Similar graph as in Fig. 8.5 but showing the amplitude of the response to each stimulus. The stimulus was increased from below threshold to 115 dB SPL (in 2-dB steps and then reduced in a 2 dB steps to below threshold) (reprinted from Møller, 1961).

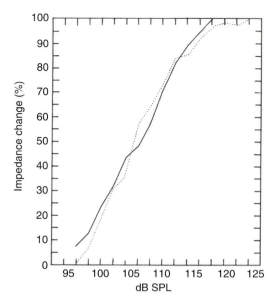

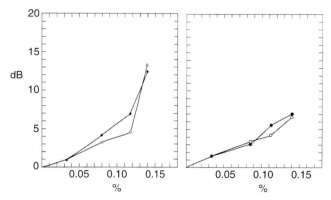

FIGURE 8.9 Illustration of the reproducibility of the responses of the acoustic middle ear reflex. The changes in the ear's impedance expressed in percentage of the maximally obtainable response amplitude are shown as a function in the stimulus intensity (dB SPL) at two occasions, 2 months apart. The stimulus sounds were 0.5 kHz tones applied to the contralateral ear (reprinted from Møller, 1961).

FIGURE 8.10 Mean value of the increase in stimulus intensity that is necessary to obtain a reflex response that is 10% of the maximally obtainable response as a function of blood alcohol concentration for two different frequencies of the stimulus tones. Left hand graph: stimulation with 0.5 kHz; right hand graph: stimulation with 1.45 kHz. Open circles are the ipsilateral response and closed circles the contralateral response (reprinted from Borg and Møller, 1967, with permission from Taylor & Francis).

nerve thus indicating that the function of the middle ear reflex may depend on synchronization (temporal coherence) of neural activity in many nerve fibers.

The latency of the earliest detectable response of the acoustic middle ear reflex (recorded as a change in the ear's acoustic impedance) decreases with increasing stimulus intensity. The shortest latency is approximately 25 ms and the longest is over 100 ms. The individual variation is large. The latency of the response to 1.5 kHz tones is shorter than the response to 0.5 kHz tones [182]. The latency of the ipsilateral and the contralateral responses are similar. The latency of the change in the acoustic impedance is the sum of the neural conduction time and the time it takes for the stapedius muscle to develop sufficient tension to cause a measurable change in the ear's acoustic impedance. Perlman and Case [229] recorded the EMG response to "loud" tones and found a mean latency of 10.5 ms based on recordings from several patients. This is a measure of the neural conduction time in humans. The latency of the EMG response is shorter than that of the change in the acoustic impedance, which involves the time it takes to build up strength of the contraction of the stapedius muscle.

The response of the acoustic middle-ear reflex is affected by drugs such as alcohol (Fig. 8.10), and sedative drugs such as barbiturates [16]. The threshold of the reflex response increases as a function of the concentration of alcohol in the blood. Blood alcohol

concentration of one tenth of one percent results in an elevation of the reflex threshold of an average of 5 dB. The individual variation is large.

4.2. Functional Importance of the Acoustic Middle-ear Reflex

Many hypotheses about the functional importance of the acoustic middle-ear reflex have been presented. Perhaps the most plausible hypothesis is that it keeps the input to the cochlea from steady sounds or sounds with slowly varying intensity nearly constant for sounds with intensities above the threshold of the reflex, while allowing rapid changes in the sound level to be preserved. The middle-ear reflex thus acts as a relatively slow automatic volume control that keeps the mean level of sound that reaches the cochlea within narrow limits (amplitude compression) [33, 194].

The functional importance of the acoustic middle-ear reflex for speech discrimination has been studied in individuals who have paresis of the stapedius muscle in one ear (Bell's Palsy [18]) and it was found that discrimination of speech at high sound levels is impaired when the acoustic middle-ear reflex is not active (Fig. 8.11). These studies indicate that the cochlea does not function properly at sound levels above the normal threshold for the acoustic reflex. Normally speech discrimination is nearly 100% in the range of speech sound intensities from 60 dB to 120 dB SPL but when the stapedius muscle is paralyzed, speech discrimination deteriorates when the sound intensity is above 90 dB SPL (Fig. 8.11).

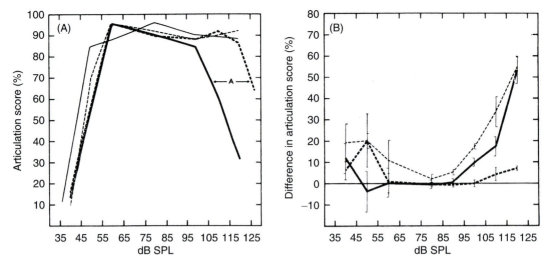

FIGURE 8.11   Effect of speech discrimination from paralysis of the stapedius muscle. (A) Speech discrimination's dependence on the function of the stapedius muscle (the average of results obtained in 13 patients). Speech discrimination scores (articulation scores in percentage) are shown as a function of the intensity for monosyllables (maximal levels, in dB SPL), during paralysis of the stapedius muscle (from Bell's Palsy) (thick continuous line), and after recovery of the paralysis (thin line). The thick interrupted line shows the discrimination scores in the opposite (unaffected) ear during the paralysis. (B) Average difference in articulation scores during and after paralysis of the stapedius muscle. The thick continuous line shows the difference between the articulation scores when the sound was led to the unaffected ear and obtained when the sounds were led to the affected ear at the time of paralysis. The thin interrupted line shows the difference between the articulation scores in the affected ear at the time of paralysis and after recovery for 6 of the subjects who participated in this study (reprinted from Borg and Zakrisson, 1973, with permission from the American Institute of Physics).

Since the acoustic middle-ear reflex attenuates the low frequency components of speech sounds more than high frequency components it may reduce masking from low frequency components of speech sounds that may impair discrimination of speech of high intensity. However, the high sound intensities (above 90 dB SPL) where speech discrimination without a functioning acoustic reflex becomes impaired do not normally exist. The acoustic middle-ear reflex therefore seems to have little importance under normal listening conditions.

When the acoustic middle-ear reflex is elicited by complex sounds such as speech sounds the contraction of the stapedius muscle will affect all low frequency components of the sound, independent of whether or not the spectral components contribute to activating the reflex. Thus high frequency components of broad band sounds will elicit contractions of the stapedius muscle when the intensities of these components are above the threshold of the reflex and that will cause attenuation of low frequency components of sounds even when these components are not sufficiently intense to activate the reflex.

Contraction of the stapedius muscle that attenuates low frequency sounds may help to separate specific sounds from a noise background and may reduce masking of high frequency components from strong low frequency components, including one's own vocalizing and sounds from chewing. The ability of the reflex to attenuate low frequency sounds of high intensity has been referred to as the perceptual theory of the action of the acoustic middle ear reflex [15], and it relates to the proposal by Simmons [273]. These features may have exerted evolutionary pressure to develop the acoustic middle-ear reflex.

Several studies have shown that the acoustic middle-ear reflex gives some protection against noise induced hearing loss. It is, however, questionable if reduced noise induced hearing loss could have played any role in the evolution of the acoustic middle-ear reflex. The type of noise it would protect against, i.e., long duration, high intensity sounds, are not common in nature.

The importance of being able to contract the middle-ear muscles voluntarily is unknown. The acoustic middle-ear reflex is well developed in mammals and the threshold of the reflex is generally lower in animals in which the acoustic reflex has been studied.

That the acoustic middle-ear reflex reduces the input to the cochlea has been supported by a study of the temporary threshold shift in response to exposure to loud noise. It was shown that the resulting

---

BOX 8.2

ACOUSTIC REFLEX AS A CONTROL SYSTEM

Contraction of the stapedius muscle reduces sound transmission through the middle ear (Chapter 2). The acoustic middle-ear reflex therefore functions as a control system that makes the input to the cochlea vary less than the sound that reaches the tympanic membrane, thus amplitude compression. The compression of the input to the cochlea is most effective for low frequency sounds and it occurs with a latency that is equal to the time it takes the stapedius muscle to contract after sound stimulation. That means that the latency of the reduction in sound transmission through the middle ear is at least 25 ms for sounds 20 dB or more above the threshold of the reflex and it takes in the order of 100 ms for the stapedius muscle to attain its full strength. The middle-ear reflex therefore does not affect fast changes in sound intensity and the amplitude compression is most effective for steady-state sounds or sounds with slowly varying amplitude.

The initial damped oscillation seen in the reflex response to low frequency tone bursts (Fig. 8.12) is a sign that the reflex regulates the input to the cochlea [194]. These oscillations occur because contractions of the stapedius muscle reduce the input to the cochlea. The attenuation caused by the stapedius muscle contraction decreases the input to the cochlea and thereby decreases the contraction of the stapedius muscle, and that in turn causes the input to the cochlea to again increase, and that increases the contraction of the stapedius muscle. This sequence of events repeats but the amplitude of the oscillations decay with time and the reflex response eventually becomes constant. The reflex response to tones above approximately 0.8 kHz do not show such oscillations, which is a sign that contraction of the stapedius muscle does not affect the sound transmission through the middle ear noticeably at that frequency, thus indicating that the acoustic middle-ear reflex is a less efficient control system for sounds at 0.8 kHz and above.

Studies of individuals with Bell's Palsy, in whom the stapedius muscle was paralyzed on one side, also indicated that low frequency sounds were more affected by the reflex than high frequency sounds [17]. When the

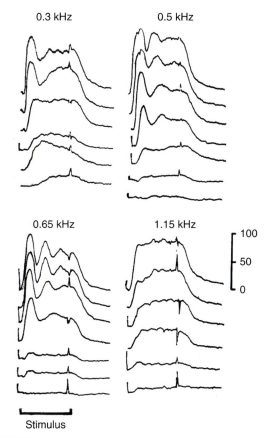

FIGURE 8.12 Recordings showing the change in the ear's acoustic impedance in response to stimulation of the ipsilateral ear with tones of different frequencies. The duration or the stimulus tones was 500 ms (reprinted from Møller, 1962, with permission from the American Institute of Physics).

reflex responses were elicited by stimulating the ear on the paralyzed side with a low frequency tone, the impedance change in the non-paralyzed side increased at a steeper rate as a function of the stimulus intensity than it did when the reflex was activated from the non-paralyzed side (Fig. 8.13). No such difference in the slope of the stimulus response curves was present when the reflex was elicited by a tone of a higher frequency (1.45 kHz).

BOX 8.2   *(cont'd)*

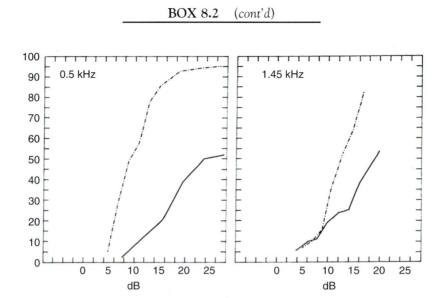

FIGURE 8.13 Stimulus response curves of the acoustic middle-ear reflex in an individual in whom the stapedius muscle was paralyzed, elicited from the side of the paralysis (Bell's Palsy) (reprinted from Borg, 1968, with permission from Taylor & Francis).

temporary threshold shift (TTS) was much greater in an ear where the stapedius muscle is paralyzed than it is in an ear with a normal functioning stapedius muscle (Fig. 8.14) [327]. These studies were performed in individuals with Bell's Palsy, in whom the stapedius muscle was paralyzed. The noise levels used caused little TTS in the ear with the normally functioning acoustic reflex. The TTS in the ear where the stapedius muscle was paralyzed increased as a nearly linear function of the level of the noise (Fig. 8.14). The individual variations were considerable. The TTS after exposure to noise centered at 2 kHz was not noticeably affected by the paralysis of the stapedius muscle [327] in agreement with the findings of other studies that have shown that the sound attenuation from contraction of the stapedius muscle is small at frequencies higher than 1 kHz.

Quantitative studies of the acoustic reflex as a control system [17, 33] have shown that above its threshold the reflex can keep the input to the cochlea nearly constant for low frequency sounds with slowly varying intensity despite the fact that the sound at the tympanic membrane may vary.

4.3. Non-acoustic Ways to Elicit Contraction of the Middle-ear Muscles

The tensor tympani muscle contracts normally during swallowing. It can be brought to contract by stimulating the skin around the eye, for instance by air puffs [133]. The response was elicited by stimulation of receptors in the skin that are innervated by the trigeminal nerve. (These investigators believed that it was the stapedius muscle that contracted while it in fact most likely was the tensor tympani muscle.) This response is similar to the blink reflex that is a natural protective reflex (see [187]), a test which is frequently used in neurologic diagnosis.

4.4. Stapedius Muscle Contraction May Be Elicited before Vocalization

Evidence that the stapedius muscle contracts a brief period before vocalization has been presented in studies in humans on the basis of EMG recording from the stapedius muscle [19] and in the flying bat where recordings of EMG potentials from the laryngeal muscles and the middle ear muscles have shown that contractions of the middle ear muscles are coordinated with the laryngeal muscles [90].

5. CLINICAL USE OF THE ACOUSTIC MIDDLE-EAR REFLEX

Recording of the acoustic middle-ear reflex response can provide information about the function of the

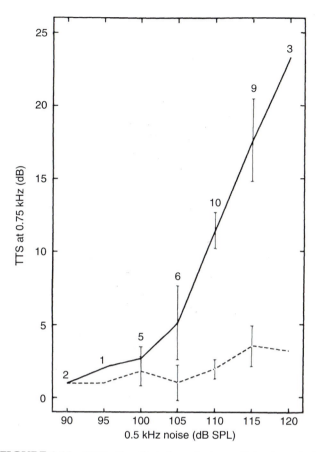

FIGURE 8.14 TTS in the affected ear during unilateral paralysis of the stapedius muscle compared with the TTS in the other ear (dashed line), as a result of exposure to band pass filtered noise (centered at 0.5 kHz, 0.3 kHz wide), for 5 min. Mean values from 18 subjects and standard error of the mean are shown as a function of the intensity of the noise. The TTS was measured 20 s after the end of the exposure. In this study the noise exposure consisted of a band of noise, centered at 0.5 kHz, and a width of 0.3 kHz. The exposure time was 5 or 7 min. Hearing threshold was measured at 0.75 kHz before exposure and 20 s after the end of the exposure using continuous pure tone (Békésy) audiometry (reprinted from Zakrisson, 1975, with permission from Taylor & Francis).

middle ear and it can help differentiate between hearing loss caused by cochlear injury and that caused by injury of the auditory nerve. The use of the acoustic middle-ear reflex in diagnosis of middle-ear disorders is based on the fact that contraction of the stapedius muscle does not cause any noticeable change in the ear's impedance if the stapes is immobilized or if the ossicular chain is interrupted (see Chapter 9). The threshold of the acoustic middle-ear reflex is elevated in patients with injuries to the auditory nerve but it is nearly normal in patients with hearing loss of cochlear origin (see Chapter 9). The acoustic middle-ear reflex is therefore a valuable aid in diagnosis of tumors of the auditory–vestibular nerve such as in vestibular Schwannoma or other forms of injuries to the auditory nerve (auditory nerve neuropathy) (see Chapter 10). Testing the acoustic middle-ear reflex may also help to identify malingering because it is an objective test that does not require the patient's cooperation. The response of the acoustic middle-ear reflex is now a routine test used in most clinics involved in diagnosis of the auditory system.

BOX 8.3

EMG ACTIVITY IN THE STAPEOIUS MUSCLE FOLLOWING VOCALIZATION

Recordings from the stapedius muscle in a patient in whom the tympanic membrane had been deflected as a part of a middle-ear operation have shown that EMG potentials are present before the start of vocalization (recorded by a microphone close to the patient's mouth) (Fig. 8.15). This means that the contractions of the stapedius muscle are not a result of an acoustic reflex but the muscle must have been brought to contract by activation of the facial motonucleus from the brain center that is involved in controlling vocalization. Studies in humans who have had laryngectomy do not show signs (change in acoustic impedance) of contraction of middle ear muscles during efforts to vocalize, thus contradicting the hypothesis that middle-ear muscles are controlled by CNS structures that are involved in generating commands to vocalize [106].

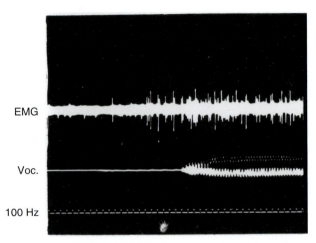

FIGURE 8.15 Electrical activity (electromyographic [EMG] potentials) recorded from the stapedius muscle during vocalization (upper trace). The sound of the vocalization (lower trace) was recorded near the patient's mouth. The intensity of the sound was 97 dB SPL. The timing impulses shown below have intervals of 10 ms (reprinted from Borg and Zakrisson, 1975, with permission from Taylor & Francis).

SECTION II REFERENCES

1. Achor L and Starr A. Auditory brain stem responses in the cat: I. Intracranial and extracranial recordings. *Electroenceph Clin Neurophysiol* 48: 154–173, 1980.

2. Agmon-Snir H, Carr CE, and Rinzel J. The role of dendrites in auditory coincidence detection. *Naure* 393: 268–272, 1998.

3. Aitkin LM. *The auditory midbrain, structure and function in the central auditory pathway.* Clifton, NJ: Humana Press, 1986.

4. Aitkin LM, Tran L, and Syka J. The responses of neurons in subdivisions of the inferior colliculus of cats to tonal, noise and vocal stimuli. *Exp Brain Res* 98: 53–64, 1994.

5. Andersen P, Eccles JC, Schmidt RF, and Yokota T. Slow potential wave produced by the cunate nucleus by cutaneous volleys and by cortical stimulation. *J Neurophys* 27: 71–91, 1964.

6. Arthur RM, Pfeiffer RR, and Suga N. Properties of "two tone inhibition" in primary auditory neurons. *J Physiol (Lond)* 212: 593–609, 1971.

7. Békésy von G. *Experiments in hearing.* New York: McGraw Hill, 1960.

8. Bieser A and Muller-Preuss P. Auditory responsive cortex in the squirrel monkey: neural responses to amplitude–modulated sounds. *Exp Brain Res* 108: 273–284, 1996.

9. Blauert J. *Spatial hearing: The psychophysics of human sound localization.* Cambridge, MA: M.I.T, 1983.

10. Blauert J and Lindemann W. Auditory spaciousness: some further psychoacoustic analyses. *J Acoust Soc Am* 80: 533–542, 1986.

11. Blum PS, Abraham LD, and Gilman S. Vestibular, auditory, and somatic input to the posterior thalamus of the cat. *Exp Brain Res* 34: 1–9, 1979.

12. Boer de E. Correlation studies applied to the frequency resolution of the cochlea. *J Aud Res* 7: 209–217, 1967.

13. Borg E. A quantitative study of the effect of the acoustic stapedius reflex on sound transmission through the middle ear in man. *Acta Oto–laryng (Stockh)* 66: 461–472, 1968.

14. Borg E. On the neuronal organization of the acoustic middle ear reflex. A physiological and anatomical study. *Brain Res* 49: 101–123, 1973.

15. Borg E, Counter SA, and Roesler G. *Theories of middle ear muscle function.* Orlando, FL: Academic Press, 1984.

16. Borg E and Møller AR. Effect of ethylalcohol and pentobarbital sodium on the acoustic middle ear reflex in man. *Acta Otolaryngol (Stockh)* 64: 415–426, 1967.

17. Borg E and Møller AR. The acoustic middle ear reflex in unanesthetized rabbit. *Acta Otolaryngol (Stockh)* 65: 575–585, 1968.

18. Borg E and Zakrisson JE. Stapedius reflex and speech features. *J Acoust Soc Am* 54: 525–527, 1973.

19. Borg E and Zakrisson JE. The activity of the stapedius muscle in man during vocalization. *Acta Otolaryng (Stockh)* 79: 325–333, 1975.

20. Boston JR and Ainslie PJ. Effects of analog and digital filtering on brain stem auditory evoked potentials. *Electroenceph Clin Neurophysiol* 48: 361–364, 1980.

21. Brodal P. *The central nervous system.* New York: Oxford Press, 1998.

22. Brownell WE. Observation on the motile response in isolated hair cells. In: *Mechanisms of hearing,* edited by Webster WR and Aitken LM. Melbourne: Monash University Press, 1983, p. 5–10.

23. Buchwald JS and Huang CM. Far field acoustic response: Origins in the cat. *Science* 189: 382–384, 1975.

24. Casseday JH, Ehrlich D, and Covey E. Neural tuning for sound duration: Role of inhibitory mechanisms in the inferior colliculus. *Science* 264: 847–850, 1994.

25. Celesia GG and Puletti F. Auditory cortical areas of man. *Neurology* 19: 211–220, 1969.

26. Chiappa K. *Evoked potentials in clinical medicine, 3rd ed.* Philadelphia: Lippincott-Raven, 1997.

27. Clarke SF, Ribaupierre de F, Bajo VM, Rouiller EM, and Kraftsik R. The auditory pathway in cat corpus callosum. *Exp Brain Res* 104: 534–540, 1995.

28. Code RA and Winer JA. Commissural connections in layer III of cat primary auditory cortex (AI): pyramidal and non-pyramidal cell input. *J Comp Neurol* 242: 485–510, 1985.

29. Cody DT and Bickford RG. Averaged evoked myogenic responses in normal man. *Laryngoscope* 79(3): 400–416, 1969.

30. Cooper NP, Robertson D, and Yates GK. Cochlear nerve fiber responses to amplitude-modulated stimuli: variations with spontaneous rate and other response characteristics. *J Neurophysiol* 70: 370–386, 1993.

31. Covey E and Casseday JH. The monaural nuclei of the lateral lemniscus in an echolocating bat: parallel pathways for analyzing temporal features of sound. *J Neurosci* 11: 3456–3470, 1991.

32. Dallos P and Cheatham MA. Compound action potential tuning curves. *J Acoust Soc Am* 59: 591–597, 1976.

33. Dallos PJ. Study of the acoustic reflex feedback loop. *IEEE Trans Bio-Med Eng* 11: 1–7, 1964.

34. Davis H and Hirsh SK. A slow brain stem response for low-frequency audiometry. *Audiology*: 441–465, 1979.

35. Diamond IT. The subdivisions of neocortex: a proposal to revise the traditional view of sensory, motor, and association areas. In: *Progress in psychobiology and physiological psychology,* edited by Sprague JM and Epstein AN. New York: Academic Press, 1979, p. 2–44.

36. Douek EE, Ashcroft PB, and Humphries KN. The clinical value of the postauricular myogenic (crossed acoustic) response in neuro-otology. In: *Disorders of auditory function II,* edited by Stephens SDG. London: Academic Press 1976, p. 139–144.

37. Doyle DJ and Hyde ML. Bessel filtering of brain stem auditory evoked potentials. *Electroenceph Clin Neurophysiol* 51: 446–448, 1981.

38. Druga R, Syka J, and Rajkowska G. Projections of auditory cortex onto the inferior colliculus in the rat. *Physiol Res* 46: 215–222, 1997.

39. Dudley H. Remaking speech. *J Acoust Soc Am* 11: 169–177, 1939.

40. Eggermont JJ. Temporal modulation transfer functions for AM and FM stimuli in cat auditory cortex. Effects of carrier type, modulating waveform, and intensity. *Hear Res* 74: 51–66, 1994.

41. Eggermont JJ. Wiener and Volterra analyses applied to the auditory system. *Hear Res* 66: 177–201, 1993.

42. Eggermont JJ, Johannesma PIM, and Aertsen AMH. Reverse-correlation methods in auditory research. *Quart Rev Biophys* 16: 341–414, 1983.

43. Ehret G and Romand R. *The central auditory pathway.* New York: Oxford University Press, 1997.

44. Engineer ND, Percaccio CR, Pandya PK, Moucha R, Rathbun DL, and Kilgard MP. Environmental enrichment improves response strength, threshold, selectivity, and latency of auditory cortex neurons. *J Neurophys* 92: 73–82, 2004.

45. Erulkar SD, Nelson PG, and Bryan JS. *Experimental and theoretical approaches to neural processing in the central auditory pathway.* New York: Academic Press, 1968.

46. Evans EF. Frequency selectivity at high signal levels of single units in cochlear nerve nucleus. In: Psycho physics and physiology of heamy, edited by Evans EF and Wilson JP. New York: Academic Press, 1977.

47. Evans EF. Normal and abnormal functioning of the cochlear nerve. *Symp Zool Soc Lond* 37: 133–165, 1975.

48. Evans EF. The frequency response and other properties of single fibers in the guinea pig cochlear nerve. *J Physiol* 226: 263–287, 1972.

49. Evans EF. The sharpening of cochlear frequency selectivity in the normal and abnormal cochlea. *Audiology* 14: 419–442, 1975.

50. Evans EF and Nelson PG. The responses of single neurons in the cochlear nucleus of the cat as a function of their location and the anaesthetic state. *Exp Brain Res* 17: 402–427, 1973.

51. Feddersen WE, Sandel TT, Teas DC, and Jeffress LA. Localization of high frequency tones. *J Acoust Soc Am* 29: 988–991, 1957.

52. Ferber-Viart C, Soulier N, Dubreuil C, and Duclaux R. Cochleovestibular afferent pathways of trapezius muscle responses to clicks in humans. *Acta Otolaryng (Stockh)* 118: 6–10, 1998.

53. Fischer C, Bognar L, Turjman F, and Lapras C. Auditory early- and middle-latency evoked potentials in patients with quadrigeminal plate tumors. *Neurosurgery* 35: 45–51, 1994.

54. Fishman KE, Shannon RV, and Slattery WH. Speech recognition as a function of the number of electrodes used in the SPEAK cochlear implant speech processor. *J Speech Lang Hear Res* 40: 1201–1215, 1997.

55. Fitzpatrick DC, Kanwal JS, Butman JA, and Suga N. Combination-sensitive neurons in the primary auditory cortex of the mustached bat. *J Neurosci* 13: 931–940, 1993.

56. Frisina RD, Karcich KJ, Tracy TC, Sullivan DM, Walton JP, and Colombo J. Preservation of amplitude modulation coding in the presence of background noise by chinchilla auditory-nerve fibers. *J Acoust Soc Am* 99: 475–490, 1996.

57. Frisina RD, Smith RL, and Chamberlain SC. Encoding of amplitude modulation in the gerbil cochlear nucleus. I. A hierarchy of enhancement. *Hear Res* 44: 99–122, 1990.

58. Frisina RD, Smith RL, and Chamberlain SC. Encoding of amplitude modulation in the gerbil cochlear nucleus. II. Possible neural mechanisms. *Hear Res* 44: 123–142, 1990.

59. Fullerton BC, Levine RA, Hosford Dunn HL, and Kiang NYS. Comparison of cat and human brain stem auditory evoked potentials. *Hear Res* 66: 547–570, 1987.

60. Galambos R and Hecox K. Clinical application of the brainstem auditory evoked potentials. In: *Auditory evoked potentials in man. Psychopharmacology correlates of EPs prog clin neurophysiol Vol 2*, edited by Desmedt JE. Basel: Karger, 1977, p. 1–19.

61. Galambos R, Makeig S, and Talmachoff PJ. A 40 Hz auditory potentials recorded from the humans scalp. *Proc Natl Acad Sci USA* 78: 2643–2647, 1981.

62. Galambos R, Myers R, and Sheatz G. Extralemniscal activation of auditory cortex in cats. *Am J Physiol* 200: 23–28, 1961.

63. Gardner MB. Historical background of the Haas/ or precedence effect. *J Acoust Soc Am* 43: 1243–1248, 1968.

64. Garg BP, Markand ON, and Bustion PF. Brainstem auditory evoked responses in hereditary motor sensory neuropathy : site of origin of wave II. *Neurology* 32: 1017–1019, 1982.

65. Geisler CD, Frishkopf LS, and Rosenblith WA. Extracranial responses to acoustic clicks in man. *Science* 128: 1210–1211, 1958.

66. Geisler CD, Rhode WS, and Kennedy DT. The responses to tonal stimuli of single auditory nerve fibers and their relationship to basilar membrane motion in the squirrel monkey. *J Neurophysiol* 37: 1156–1172, 1974.

67. Gelfan WR and Tarlov IM. Differential vulnerability of spinal cord structures to anoxia. *J Neurophysiol* 18: 170–188, 1955.

68. Gerken GM, Moushegian G, Stillman RD, and Rupert AL. Human frequency-following responses to monaural and binaural stimuli. *Electroencephalogr Clin Neurophysiol* 38: 379–386, 1975.

69. Goldberg JM and Brown PB. Response of binaural neurons of dog superior olivary complex to dichotic tonal stimuli: some physiological mechanisms of sound localization. *J Neurophysiol* 32: 613–636, 1969.

70. Goldman-Rakic PS. Modular organization of prefrontal cortex. *Trends Neurosci* 7: 419–429, 1984.

71. Graybiel AM. Some fiber pathways related to the posterior thalamic region in the cat. *Brain Behavior Evol* 6: 363–393, 1972.

72. Guinan Jr. JJ, Warr WB, and Norris BE. Topographic organization of the olivocochlear projections from the lateral and medial zones of the superior olivary complex. *J Comp Neurol* 226: 21–27, 1984.

73. Guinan Jr. JJ, Norris BE, and Guinan SS. Single auditory units in the superior olivary complex. II. Location of unit categories and tonotopic organization. *Int J Neurosci* 4: 147–166, 1972.

74. Gummer M, Yates GK, and Johnstone BM. Modulation transfer function of efferent neurons in the guinea pig cochlea. *Hear Res* 36: 41–52, 1988.

75. Hall JW. *Handbook of auditory evoked responses*. Boston, MA: Allyn and Bacon, 1992.

76. Harrison JM and Howe ME. Anatomy of the descending auditory system in auditory system. In: *Handbook of sensory physiology*, edited by Keidel WD and Neff WD. Berlin: Springer-Verlag, 1974, p. 363–388.

77. Harrison RV, Aran J-M, and Erre JP. AP tuning curves in normal and pathological human and guinea pig cochlea. *J Acoust Soc Am* 69: 1374–1385, 1981.

78. Hart HC, Palmer AR, and Hall DA. Different areas of human non-primary auditory cortex are activated by sounds with spatial and nonspatial properties. *Hum Brain Mapp* 21: 178–190, 2004.

79. Hashimoto I. Auditory evoked potentials from the humans midbrain: Slow brain stem responses. *Electroenceph Clin Neurophysiol* 53: 652–657, 1982.

80. Hashimoto I, Ishiyama Y, Yoshimoto T, and Nemoto S. Brainstem auditory evoked potentials recorded directly from human brain stem and thalamus. *Brain* 104: 841–859, 1981.

81. Haykin S and Chen Z. The cocktail party problem. *Neural Comput* 17: 1875–1902, 2005.

82. Heil P and Irvine DR. Functional specialization in auditory cortex: responses to frequency-modulated stimuli in the cat's posterior auditory field. *J Neurophysiol* 79: 3041–3059, 1998.

83. Heil P, Rajan R, and Irvine DRF. Sensitivity of neurons in cat primary auditory cortex to tones and frequency-modulated stimuli: I. Organization of response properties along the "isofrequency" dimension. *Hear Res* 63: 135–156, 1992.

84. Hellige J. *Hemispheric asymmetry: What's right and what's left*. Cambridge, MA: Harvard University Press, 1993.

85. Honrubia V and Ward PH. Longitudinal distribution of the cochlear microphonics inside the cochlear duct (guinea pig). *J Acoust Soc Am* 44: 951 958, 1968.

86. Irvine DRF. A comparison of two methods for measurement of neural sensitivity to interaural intensity differences. *Hear Res* 30: 169–180, 1987.

87. Irvine DRF. Interaural intensity differences in the cat: Changes in sound pressure level at the two ears associated with azimuthal displacements in the frontal horizontal plane. *Hear Res* 26: 267–286, 1987.

88. Jacobson JT. *Principles and applications in auditory evoked potentials*. Boston: Allyn & Bacon, 1994.

89. Jeffress LA. A place theory of sound localization. *J Comp Physiol Psychol* 41: 35–39, 1948.

90. Jen PH and Suga N. Coordinated activities of middle-ear and laryngeal muscles in echolocating bats. *Science* 91: 950–952, 1976.

91. Jenkins WM, Merzenich MM, Ochs MT, Allard T, and Guic-Robles E. Functional reorganization of primary somatosensory cortex in adult owl monkeys after behaviorally controlled tactile stimulation. *J Neurophysiol* 63: 82–104, 1990.

92. Jepsen O. *Studies of the acoustic stapedius refelx in man.* Aarhus: Universitetsforlaget 1955.

93. Jewett DL. The 3 channel Lissajous' trajectory of the auditory brain stem response. IX. Theoretical aspects. *Electroenceph Clin Neurophysiol* 68: 386–408, 1987.

94. Jewett DL, Deupree DL, and Bommannan D. Far field potentials generated by action potentials of isolated frog sciatic nerves in a spherical volume. *Electroenceph Clin Neurophys* 75: 105–117, 1990.

95. Jewett DL, Romano MN, and Williston JS. Human auditory evoked potentials: possible brain stem components detected on the scalp. *Science* 167: 1570–1571, 1970.

96. Jewett DL and Williston JS. Auditory evoked far fields averaged from scalp of humans. *Brain* 94: 681–696, 1971.

97. Johnstone BM, Patuzzi R, and Yates GK. Basilar membrane measurements and the traveling wave. *Hear Res* 22: 147–153, 1986.

98. Joris PX, Carney LH, Smith PH, and Yin TC. Enhancement of neural synchronization in the anteroventral cochlear nucleus. I. Responses to tones at the characteristic frequency. *J Neurophysiol* 71: 1022–1036, 1994.

99. Joris PX, Schreiner CE, and Rees A. Neural processing of amplitude-modulated sounds. *Physiol Rev* 84: 541–577, 2004.

100. Joris PX, Smith PH, and Yin TCT. Coincidence detection in the auditory system: 50 years after Jeffress. *Neuron* 21: 1235–1238, 1998.

101. Joris PX and Yin TC. Envelope coding in the lateral superior olive. III. Comparison with afferent pathways. *J Neurophysiol* 79: 253–269, 1998.

102. Joris PX and Yin TCT. Responses to amplitude modulated tones in the auditory nerve of the cat. *J Acoust Soc Am* 91: 215–232, 1992.

103. Joseph MP, Guinan JJ, Fullerton BC, Norris BE, and Kiang NYS. Number and distribution of stapedius motoneurons in cats. *J Comp Neurol* 232: 43–54, 1985.

104. Kaas JH and Hackett TA. Subdivisions of auditory cortex and processing streams in primates. *Proc Nat Acad Sci USA* 97: 11793–11799, 2000.

105. Katsuki Y, Sumi T, Uchiyama H, and Watanabe T. Electric responses of auditory neurons in cat to sound stimulation. *J Neurophysiol* 21: 569–588, 1958.

106. Kawase T, Ogura M, Kakehata S, and Takasaka T. Measurement of stapedius contraction during vocalization effort in patients after laryngectomy or tracheostomy. *Hear Res* 149: 248–252, 2000.

107. Kiang NY-S. The use of computers in studies of auditory neurophysiology. *Trans Am Acad Ophthal Otolaryngol* 65: 735–747, 1961.

108. Kiang NY-S, Christ AH, French MA, and Edwards AG. Postauricular electric response to acoustic stimuli in humans. *MIT Quart Prog Rep Research Laboratory of Electronics* 68: 218–225, 1963.

109. Kiang NY-S and Peake WT. Cochlear Responses to Condensation and Rarefaction Clicks. *Biophys J* 2: 23–34, 1962.

110. Kiang NYS, Watanabe T, Thomas EC, and Clark L. *Discharge patterns of single fibers in the cat's auditory nerve.* Cambridge, MA: MIT Press, 1965.

111. Kilgard MP and Merzenich MM. Cortical map reorganization enabled by nucleus basalis activity. *Science* 279: 1714–1718, 1998.

112. Kilgard MP and Merzenich MM. Distributed representation of spectral and temporal information in rat primary auditory cortex. *Hear Res* 134: 16–28, 1999.

113. Klockhoff I and Anderson H. Recording of the stapedius reflex elicited by cutaneous stimulation. *Acta Otolaryngol (Stockh)* 50: 451–454, 1959.

114. Knudsen EI, du Lac S, and Esterly SD. Computational maps in the brain. *Annu Rev Neurosci* 10: 41–65, 1987.

115. Konishi M and Nielsen DW. The temporal relationship between motion of the basilar membrane and initiation of nerve impulses in auditory nerve fibers. *J Acoust Soc Am* 53: 325, 1973.

116. Kral A, Hartmann R, Tillrin J, Heid S, and Klinke R. Congenital auditory deprivation reduces synaptic activity within the auditory cortex in layer specific manner. *Cerebral Cortex* 10: 714–726, 2000.

117. Kraus N, Odzamar O, Hier D, and Stein L. Auditory middle latency responses (MLRs) in patients with cortical lesions. *Electroencephalogr Clin Neurophys* 54: 275–287, 1982.

118. Kreiman J and Van Lancker D. Hemispheric specialization for voice recognition: evidence from dichotic listening. *Brain & Language* 34: 246–252, 1988.

119. Kuroki A and Møller AR. Microsurgical anatomy around the foramen of Luschka with reference to intraoperative recording of auditory evoked potentials from the cochlear nuclei. *J Neurosurg:* 933–939, 1995.

120. Kuwada S and Batra R. Coding of sound envelopes by inhibitory rebound in neurons of the superior olivary complex in the unanesthetized rabbit. *J Neurosci* 19: 2273–2287, 1999.

121. Kvasnak E, Popelar J, and Syka J. Discharge properties of neurons in subdivisions of the medial geniculate body of the guinea pig. *Physiol Res* 49: 369–378, 2000.

122. Kvasnak E, Suta D, Popelar J, and Syka J. Neuronal connections in the medial geniculate body of the guinea-pig. *Exp Brain Res* 132: 87–102, 2000.

123. Lang J. Anatomy of the brainstem and the lower cranial nerves, vessels, and surrounding structures. *Am J Otol* Suppl, Nov: 1–19, 1985.

124. Lang J. *Clinical anatomy of the posterior cranial fossa and its foramina.* Stuttgart: Thieme Verlag, 1981.

125. Lang J. Facial and vestibulocochlear nerve, topographic anatomy and variations. *The cranial nerves,* edited by Samii M and Jannetta P. New York: Springer-Verlag, 1981, p. 363–377.

126. Lang JJ, Ohmachi N, and Lang JS. Anatomical landmarks of the rhomboid fossa (floor of the 4th ventricle), its length and its width. *Acta Neurochir (Wien)* 113: 84–90, 1991.

127. Langner G and Schreiner CE. Periodicity coding in the inferior colliculus of the cat. I. Neuronal mechanisms. *J Neurophysiol* 60: 1799–1822, 1988.

128. Langner GA. Review: periodicity coding in the auditory system. *Hear Res* 60: 115–142, 1992.

129. Lazorthes G, Lacomme Y, Ganbert J, and Planel H. La constitution du nerf auditif. *Presse Medicine* 69: 1067–1068., 1961.

130. Le Prell CG, Dolan D, Schacht J, Miller JM, Lomax MI, and Altschuler RA. Pathways for protection from noise induced hearing loss. *Noise Health* 5: 1–17, 2003.

131. Leake PA, Snyder RL, Rebscher SJ, Moore CM, and Vollmer M. Plasticity in central representation in the inferior colliculus induced by chronic single- vs. two-channel electrical stimulation by cochlear implant after neonatal deafness. *Hear Res* 147: 221–241, 2000.

132. LeDoux JE. Brain mechanisms of emotion and emotional learning. *Curr Opin Neurobiol* 2: 191–197, 1992.

133. Lee YS, Lueders H, Dinner DS, Lesser RP, Hahn J, and Klem G. Recording of auditory evoked potentials in man using chronic subdural electrodes. *Brain Res Brain Res Rev* 107: 115–131, 1984.

134. Lepore F, Poirier P, Provencal C, Lassonde M, Miljours S, and Guillemot JP. *Cortical and callosal contribution to sound localization.* New York: Plenum, 1997.

135. Liberman MC. Auditory-nerve response from cats raised in low-noise chamber. *J Acoust Soc Am* 63: 442–455, 1978.

136. Liberman MC and Mulroy MJ. Acute and chronic effects of acoustic traume: cochlear pathology and auditory nerve pathology. In: *New perspectives in noise-induced hearing loss*, edited by Hamernik RP, Henderson D and Salvi R. New York: Raven Press, 1982.

137. Licklider JCR. *Three auditory theories*. New York: McGraw-Hill, 1959.

138. Liegeois-Chauvel C, Lorenzi C, Trebuchon A, Regis J, and Chauvel P. Temporal envelope processing in the human left and right auditory cortices. *Cereb Cortex* 14: 731–740, 2004.

139. Loizou PC. Introduction to cochlear implants. *IEEE Signal Processing Magazine* September: 101–130, 1998.

140. Loizou PC. On the number of channels needed to understand speech. *J Acoust Soc Am* 106: 2097–2103, 1999.

141. Loizou PC, Dorman M, and Fitzke J. The effect of reduced dynamic range on speech understanding: implications for patients with cochlear implant. *Ear Hear* 21: 25–31, 2000.

142. Lorente de No R. Action potentials of the motoneurons of the hypoglossus nucleus. *J Cell Comp Physiol* 29: 207–287, 1947.

143. Lorente de No R. Analysis of the distribution of action currents of nerve in volume conductors. *Studies of the Rockefeller Institute for Medical Research* 132: 384–482, 1947.

144. Lorente de No R. Anatomy of the eighth nerve III. General plan of structure of the primary cochlear nuclei. *Laryngoscope* 43: 327–350, 1933.

145. Malhotra S, Hall, A.J., and Lomber, S.G. Cortical control of sound localization in the cat: unilateral cooling deactivation of nineteen cortical areas. *J Neurophysiol* 92: 1625–1643, 2004.

146. Markand ON, Farlow MR, Stevens JC, and Edwards MK. Brain stem auditory evoked potential abnormalities with unilateral brain stem lesions demonstrated by magnetic resonance imaging. *Arch Neurol* 46: 295–299, 1989.

147. Martin WH, Pratt H, and Schwegler JW. The origin of the human auditory brainstem response wave II. *Electroenceph Clin Neurophysiol* 96: 357–370, 1995.

148. Mast TE. Binaural interaction and contralateral inhibition in dorsal cochlear nucleus of chinchilla. *J Neurophysiol* 62: 61–70, 1973.

149. Mast TE. Short latency human evoked responses to clicks. *J Appl Physiol* 20: 725–730, 1965.

150. McGee T, Kraus N, Littmanm T, and Nicol T. Contributions of medial geniculate body subdivisions to the middle latency response. *Hear Res* 61: 147–154, 1992.

151. Metz O. Studies of the contraction of the tympanic muscles as indicated by changes in the impedance of the ear. *Acta Otolaryng (Stockh)* 39: 397–405, 1951.

152. Metz O. The acoustic impedance measured on normal and pathological ear. *Acta Otolaryng (Stockh)* 63, 1946.

153. Middlebrooks JC. Binaural mechanisms of spatial tuning in the cat's superior colliculus distinguished using monaural occlusion. *J Neurophysiol* 57: 688–701, 1987.

154. Mishkin M, Ungerleider LG, and Macko KA. Object vision and spatial vision: two cortical pathways. *Trends Neurosci* 6: 415–417, 1983.

155. Mitani A and Shimokouchi M. Neural connections in the primary auditory cortex: and electrophysiologic study in the cat. *J Comp Neurol* 235: 417–429, 1985.

156. Moore BC. Psychoacoustics of normal and impaired hearing. *Br Med Bull* 63: 121–134, 2002.

157. Moore JK. The human auditory brain stem as a generator for auditory evoked potentials. *Hear Res* 29: 33–44, 1987.

158. Moore JK. The human auditory brain stem: a comparative view. *Hear Res* 29: 1–32, 1987.

159. Morest DK. Structural organization of auditory pathways. In: *The nervous system*, edited by Eagles EL. New York: Raven Press, 1975, p. 19–29.

160. Morest DK. The neuronal architecture of the medial geniculate body of the cat. *J Anat (Lond)* 98: 611–630, 1964.

161. Morest DK and Oliver DL. The neuronal architecture of the inferior colliculus in the cat. *J Comp Neurol* 222: 209–236, 1984.

162. Mountain DC and Cody AR. Multiple modes of inner hair cell stimulation. *Hear Res* 132: 1–14, 1999.

163. Mountain DC, Geisler C, D., and Hubbard AE. Stimulation of efferents alters the cochlear microphonic and sound-induced resistance changes measured in the scala media of the guinea pig. *Hear Res* 3: 231–240, 1980.

164. Mountcastle VB. Modality and topographic properties of single neurons of cat's somatic cortex. *J Neurophysiol* 20: 408–434, 1957.

165. Moushegian G, Rupert AL, and Stillman RD. Evaluation of frequency following potentials in man: masking and clinical studies. *Electroenceph Clin Neurophysiol* 45: 711–718, 1978.

166. Moushegian G, Rupert AL, and Stillman RD. Laboratory note. Scalp-recorded early responses in man to frequencies in the speech range. *Electroenceph Clin Neurophysiol* 35: 665–667, 1973.

167. Møller A. Unit responses in the cochlear nucleus of the rat to sweep tones. *Acta Physiol Scand* 76: 503–512, 1969.

168. Møller AR. *Auditory physiology*. New York: Academic Press, 1983.

169. Møller AR. Bilateral contraction of the tympanic muscles in man, examined by measuring acoustic impedance-change. *Ann Otol Rhinol Laryngol* 70: 735–753, 1961.

170. Møller AR. Bilateral contraction of the tympanic muscles in man. *Royal Institute of Technology (KTH), Div of Telegraphy-Telephony, Report No 18, Speech Transmission Laboratory* 1–51, 1961.

171. Møller AR. Coding of amplitude and frequency modulated sounds in the cochlear nucleus of the rat. *Acta Physiol Scand* 86: 223–238, 1972.

172. Møller AR. Coding of increments and decrements in stimuli intensity in single units in the cochlear nucleus of the rat. *J Neurosci Res* 4: 1–8, 1979.

173. Møller AR. Coding of sounds with rapidly varying spectrum in the cochlear nucleus. *J Acoust Soc Am* 55: 631–640, 1974.

174. Møller AR. Dynamic properties of excitation and inhibition in the cochlear nucleus. *Acta Physiol Scand* 93: 442–454, 1975.

175. Møller AR. Dynamic properties of primary auditory fibers compared with cells in the cochlear nucleus. *Acta Physiol Scand* 98: 157–167, 1976.

176. Møller AR. Dynamic properties of the responses of single neurons in the cochlear nucleus. *J Physiol* 259: 63–82, 1976.

177. Møller AR. Effect of tympanic muscle activity on movement of the eardrum, acoustic impedance, and cochlear microphonics. *Acta Otolaryngol (Stockh)* 58: 525–534, 1965.

178. Møller AR. *Evoked potentials in intraoperative monitoring*. Baltimore, MD: Williams and Wilkins, 1988.

179. Møller AR. Frequency selectivity of phase–locking of complex sounds in the auditory nerve of the rat. *Hear Res* 11: 267–284, 1983.

180. Møller AR. Frequency selectivity of single auditory nerve fibers in response to broadband noise stimuli. *J Acoust Soc Am* 62: 135–142, 1977.

181. Møller AR. Frequency selectivity of the basilar membrane revealed from discharges in auditory nerve fibers. In: *Psychophysics and physiology of hearing*, edited by Evans EF and Wilson JP. London: Academic Press, 1977, p. 197–205.

182. Møller AR. Intra-aural muscle contraction in man, examined by measuring acoustic impedance of the ear. *Laryngoscope* LXVIII: 48–62, 1958.

183. Møller AR. *Intraoperative neurophysiologic monitoring*. Luxembourg: Harwood Academic Publishers, 1995.

184. Møller AR. Intraoperative neurophysiologic monitoring in neurosurgery: Benefits, efficacy, and cost–effectiveness.

In: *Clinical neurosurgery, proceedings of the congress of neurological surgeons' 1994 meeting, Chapter 12.* Baltimore: Williams & Wilkins, 1995, p. 171–179.

185. Møller AR. *Intraoperative neurophysiologic monitoring, 2nd edition*: Humana Press Inc., 2006.

186. Møller AR. Latency of unit responses in the cochlear nucleus determined in two different ways. *J Neurophysiol* 38: 812–821, 1975.

187. Møller AR. *Neural plasticity and disorders of the nervous system.* Cambridge: Cambridge University Press (in press), 2006.

188. Møller AR. Origin of latency shift of cochlear nerve potentials with sound intensity. *Hear Res* 17: 177–189, 1985.

189. Møller AR. Responses of units in the cochlear nucleus to sinusoidally amplitude modulated tones. *Exp Neurol* 45: 104–117, 1974.

190. Møller AR. Review of the roles of temporal and place coding of frequency in speech discrimination. *Acta Otolaryngol (Stockh)* 119: 424–430, 1999.

191. Møller AR. *Sensory systems: anatomy and physiology.* Amsterdam: Academic Press, 2003.

192. Møller AR. Similarities between severe tinnitus and chronic pain. *J Amer Acad Audiol* 11: 115–124, 2000.

193. Møller AR. Statistical evaluation of the dynamic properties of cochlear nucleus units using stimuli modulated with pseudorandom noise. *Brain Res* 57: 443–456, 1973.

194. Møller AR. The acoustic reflex in man. *J Acoust Soc Am* 34: 1524–1534, 1962.

195. Møller AR. The sensitivity of contraction of the tympanic muscles in man. *Ann Otol Rhinol Laryngol* 71: 86–95, 1962.

196. Møller AR. Unit responses in the cochlear nucleus of the rat to pure tones. *Acta Physiol Scand* 75: 530–541, 1969.

197. Møller AR. Unit responses in the rat cochlear nucleus to repetitive transient sounds. *Acta Physiol Scand* 75: 542–551, 1969.

198. Møller AR. Unit responses in the rat cochlear nucleus to tones of rapidly varying frequency and amplitude. *Acta Physiol Scand* 81: 540–556, 1971.

199. Møller AR. Use of stochastic signals in evaluation of the dynamic properties of a neuronal system. *Scand J Rehab Med* 3: 37–44, 1974.

200. Møller AR and Burgess JE. Neural generators of the brain stem auditory evoked potentials (BAEPs) in the rhesus monkey. *Electroenceph Clin Neurophysiol* 65: 361–372, 1986.

201. Møller AR, Colletti V, and Fiorino F. Click evoked responses from the exposed intracranial portion of the eighth nerve during vestibular nerve section: bipolar and monopolar recordings. *Electroenceph Clin Neurophysiol* 92: 17–29, 1994.

202. Møller AR, Colletti V, and Fiorino FG. Neural conduction velocity of the human auditory nerve: Bipolar recordings from the exposed intracranial portion of the eighth nerve during vestibular nerve section. *Electroenceph Clin Neurophysiol* 92: 316–320, 1994.

203. Møller AR and Jannetta PJ. Auditory evoked potentials recorded from the cochlear nucleus and its vicinity in man. *J Neurosurg* 59: 1013–1018, 1983.

204. Møller AR and Jannetta PJ. Comparison between intracranially recorded potentials from the human auditory nerve and scalp recorded auditory brainstem responses (ABR). *Scand Audiol (Stockholm)* 11: 33–40, 1982.

205. Møller AR and Jannetta PJ. Compound action potentials recorded intracranially from the auditory nerve in man. *Exp Neurol* 74: 862–874, 1981.

206. Møller AR and Jannetta PJ. Evoked potentials from the inferior colliculus in man. *Electroenceph Clin Neurophysiol* 53: 612–620, 1982.

207. Møller AR and Jannetta PJ. Interpretation of brainstem auditory evoked potentials: results from intracranial recordings in humans. *Scand Audiol (Stockh)* 12: 125–133, 1983.

208. Møller AR and Jannetta PJ. Monitoring auditory functions during cranial nerve microvascular decompression operations by direct recording from the eighth nerve. *J Neurosurg* 59: 493–499, 1983.

209. Møller AR and Jannetta PJ. Neural generators of the brainstem auditory evoked potentials. In: *Evoked potentials II: the second international evoked potentials symposium*, edited by Nodar RH and Barber C. Boston, MA: Butterworth Publishers, 1984, p. 137–144.

210. Møller AR, Jannetta PJ, and Jho HD. Click–evoked responses from the cochlear nucleus: a study in human. *Electroenceph Clin Neurophysiol* 92: 215–224, 1994.

211. Møller AR, Jannetta PJ, and Sekhar LN. Contributions from the auditory nerve to the brainstem auditory evoked potentials (BAEPs): results of intracranial recording in man. *Electroenceph Clin Neurophysiol* 71: 198–211, 1988.

212. Møller AR and Jho HD. Effect of high frequency hearing loss on compound action potentials recorded from the intracranial portion of the human eighth nerve. *Hear Res* 55: 9–23, 1991.

213. Møller AR and Jho HD. Late components in the compound action potentials (CAP) recorded from the intracranial portion of the human eighth nerve. *Hear Res* 45: 75–86, 1990.

214. Møller AR and Jho HD. Response from the exposed intracranial human auditory nerve to low-frequency tones: Basic characteristics. *Hear Res* 38: 163–175, 1989.

215. Møller AR and Jho HD. Responses from the exposed human auditory nerve to pseudorandom noise. *Hear Res* 42: 237 252, 1989.

216. Møller AR, Jho HD, and Jannetta PJ. Preservation of hearing in operations on acoustic tumors: an alternative to recording BAEP. *Neurosurgery* 34: 688–693, 1994.

217. Møller AR, Jho HD, Yokota M, and Jannetta PJ. Contribution from crossed and uncrossed brainstem structures to the brainstem auditory evoked potentials (BAEP): A study in human. *Laryngoscope* 105: 596–605, 1995.

218. Møller AR, Kern JK, and Grannemann B. Are the non-classical auditory pathways involved in autism and PDD? *Neurol Res* 27: 625–629, 2005.

219. Møller AR, Møller MB, and Yokota M. Some forms of tinnitus may involve the extralemniscal auditory pathway. *Laryngoscope* 102: 1165–1171, 1992.

220. Møller AR and Rollins P. The non-classical auditory system is active in children but not in adults. *Neurosci Lett* 319: 41–44, 2002.

221. Müller M, Robertson D, and Yates GK. Rate-versus-level functions of primary auditory nerve fibres: evidence of square law behavior of all fibre categories in the guinea pig. *Hear Res* 55: 50–56, 1991.

222. Müsiek FE and Geurkink, N.A. Auditory brainstem response and central auditory test findings for patients with brain stem lesions: a preliminary report. *Laryngoscope* 92: 891–900, 1982.

223. Neff WD, Diamond JT, and Casseday JH. Behavioral studies of auditory discrimination: central nervous system: central nervous system. In: *Handbook of sensory physiology*, edited by Keidel WD and Neff WD. Berlin: Springer-Verlag, 1975, p. 307–400.

224. Oertel D and Young ED. What's a cerebellar circuit doing in the auditory system? *Trends Neurosci* 27: 104–110, 2004.

225. Oliver DL and Morest DK. The central nucleus of the inferior colliculus in the cat. *J Comp Neurol* 222: 237–264, 1984.

226. Ota CY and Kimura RS. Ultrastructural study of the human spiral ganglion. *Acta Otolaryngol (Stockh)* 89: 53–62, 1980.

227. Palmer AR. Encoding of rapid amplitude fluctuations by cochlear nerve fibers in the guinea pig. *Arch Otorhinolaryngol* 236: 197 202, 1982.

228. Peake WT, Goldstein MH, and Kiang NYS. Responses of the auditory nerve to repetitive stimuli. *J Acoust Soc Am* 34: 562–570, 1962.

229. Perlman HB and Case TJ. Latent period of the crossed stapedius reflex in man. *Ann Otol Rhin Laryngol* 48: 663–675, 1939.

230. Pfeiffer RR. Classification of response patterns of spike discharges for units in the cochlear nucleus: tone burst stimulation. *Exp Brain Res* 1: 220–235, 1966.

231. Pickles JO. *An introduction to the physiology of hearing, 2nd edition*. London: Academic Press, 1988.

232. Picton TW, Hillyard SA, Krausz HI, and Galambos R. Human auditory evoked potentials. I. Evaluation of components. *Electroenceph Clin Neurophysiol* 36: 176–190, 1974.

233. Portmann M, Cazals Y, Negrevergne M, and Aran JM. Transtympanic and surface recordings in the diagnosis of retrocochlear lesions. *Acta Otolaryngol (Stockh)* 89: 362–369, 1980.

234. Pratt H, Bleich N, and Martin WH. Three channel Lissajous' trajectory of humans auditory brain stem evoked potentials. I. Normative measures. *Electroenceph Clin Neurophysiol* 61: 530–538, 1985.

235. Pratt H, Martin WH, Schwegler JW, Rosenwasser RH, and Rosenberg SJ. Temporal Correspondence of intracranial, cochlear and scalp recorded humans auditory nerve action potentials. *Electroenceph Clin Neurophysiol* 84: 447–455, 1992.

236. Rao KR and Ben-Arie J. Optimal head related transfer functions for hearing and monaural localization in elevation: a signal processing design perspective. *IEEE Trans Biomed Eng* 43: 1093–1105, 1996.

237. Rauschecker JP. Cortical processing of complex sounds. *Curr Opinion Neurobiol* 8: 516–521, 1998.

238. Rauschecker JP and Harris LR. Auditory and visual neurons in the cat's superior colliculus selective for the direction of apparent motion stimuli. *Brain Res* 490: 56–63, 1989.

239. Rauschecker JP and Tian B. Mechanisms and streams for processing of "what" and "where" in auditory cortex. *Proc Nat Acad Sci USA* 97: 11800–11806, 2000.

240. Razak KA and Fuzessery ZM. Functional organization of the pallid bat auditory cortex: emphasis on binaural organization. *J Neurophysiol* 87: 72–86, 2002.

241. Razak KA, Fuzessery ZM, and Lohuis TD. Single cortical neurons serve both echolocation and passive sound localization. *J Neurophysiol* 81: 1438–1442, 1999.

242. Rees A and Møller AR. Responses of neurons in the inferior colliculus of the rat to AM and FM tones. *Hear Res* 10: 301–330, 1983.

243. Rees A and Møller AR. Stimulus properties influencing the responses of inferior colliculus neurons to amplitude-modulated sounds. *Hear Res* 27: 129–144, 1987.

244. Rhode WS and Greenberg S. Encoding of amplitude modulation in the cochlear nucleus of the cat. *J Neurophysiol* 71: 1797–1825, 1994.

245. Rodrigues–Dagaeff C, Simm G, Ribaupierre de Y, Villa A, Ribaupierre de F, and Rouiller EM. Functional organization of the ventral division of the medial geniculate body of the cat: evidence for a rostro–caudal gradient of response properties and cortical projections. *Hear Res* 39: 103–126, 1989.

246. Roeder KD. Acoustic alerting mechanisms in insects. *Ann New York Acad Sci* 188: 63–79, 1971.

247. Roeder KD and Treat AE. Detection and evasion of bats by moths. *American Scientist* 49: 135–148, 1961.

248. Romanski LM, Tian B, Fritz J, Mishkin M, Goldman-Rakic PS, and Rauschecker JP. Dual streams of auditory afferents target multiple domains in the primate prefrontal cortex. *Nature Neurosci* 2: 1131–1136, 1999.

249. Rose JE, Galambos R, and Hughes JR. Microelectrode studies of the cochlear nuclei in the cat. *Bull Johns Hopkins Hosp* 104: 211–251, 1959.

250. Rose JE, Gross NB, Geisler CD, and Hind JE. Some neural mechanisms in the inferior colliculus of the cat which may be relevant to localization of a sound source. *J Neurophysiol* 29: 288–314, 1966.

251. Rose JE, Hind JE, Anderson DJ, and Brugge JF. Some effects of stimulus intensity on response of auditory fibers in the squirrel monkey. *J Neurophysiol* 34: 685–699, 1971.

252. Rouiller EM. Functional organization of the auditory system. In: *The central auditory system*, edited by Ehret G and Romand R. New York: Oxford University Press, 1997, p. 3–96.

253. Rouiller EM, Ribaupierre de Y, Toros-Morel A, and Ribaupierre de F. Neural coding of repetitive clicks in the medial geniculate body of cat. *Hear Res* 5: 81–100, 1981.

254. Rouiller EM, Rodrigues-Dagaeff C, Simm G, Ribaupierre de Y, Villa A, and Ribaupierre de F. Functional organization of the medial division of the medial geniculate body of the cat: tonotopic organization, spatial distribution of response properties and cortical connections. *Hear Res* 39: 127–142, 1989.

255. Ruben RJ and Walker AE. The VIIIth nerve action potential in Ménière's disease. *Laryngoscope* 11: 1456–1464, 1963.

256. Ruggero MA and Rich NC. *Peak splitting: intensity effects in cochlear afferent responses to low frequency tones*: New York: Plenum, 1989.

257. Ryugo DK. The auditory nerve: Peripheral innervation, cell body morphology, and central projections. In: *The mammalian auditory pathway: neuroanatomy*, edited by Webster DB, Popper AN and Fay RR. New York: Springer-Verlag, 1992.

258. Sachs MB and Kiang NYS. Two tone inhibition in auditory nerve fibers. *J Acoust Soc Am* 43: 1120–1128, 1968.

259. Sachs MB and Young ED. Encoding of steady-state vowels in the auditory nerve: representation in terms of discharge rate. *J Acoust Soc Am* 66: 470–479, 1979.

260. Sachs MB, Young ED, and Miller MI. Speech encoding in the auditory nerve: implications for cochlear implants. *Cochlear Prostheses: An International Symposium* 405: 94–113, 1983.

261. Scharf B, Magnan J, and Chays A. On the role of the olivocochlear bundle in hearing: 16 case studies. *Hear Res* 103: 101–122, 1997.

262. Scherg M and von Cramon D. A new interpretation of the generators of BAEP waves I V: results of a spatio temporal dipole. *Electroenceph Clin Neurophysiol* 62: 290–299, 1985.

263. Schmidt RF and Thews G. *Human physiology*. Berlin: Springer-Verlag, 1983.

264. Schreiner CE and Urbas JV. Representation of amplitude modulation in the auditory cortex of the cat. I. The anterior auditory field (AAF). *Hear Res* 21: 227–242, 1986.

265. Schroeder M. Vocoders: Analysis and synthesis of speech. *Proc IEEE* 54: 720–734, 1966.

266. Schucknecht HF. *Pathology of the ear*. Cambridge, MA: Harvard University Press, 1974.

267. Sellick PM, Patuzzi R, and Johnstone BM. Measurement of basilar membrane motion in the guinea pig using the Mossbauer technique. *J Acoust Soc Am* 72: 131–141, 1982.

268. Sellick PM, Patuzzi R, and Johnstone BM. Modulation of responses of spiral ganglion cells in the guinea pig cochlea to low frequency sound. *Hear Res* 7: 199–221, 1982.

269. Shannon RV, Zeng F-G, Kamath V, Wygonski J, and Ekelid M. Speech recognition with primarily temporal cues. *Science* 270: 303–304, 1995.

270. Shore SE, El Kashlan H, and Lu J. Effects of trigeminal ganglion stimulation on unit activity of ventral cochlear nucleus neurons. *Neuroscience* 119: 1085–1101, 2003.

271. Shore SE, Vass Z, Wys NL, and Altschuler RA. Trigeminal ganglion innervates the auditory brainstem. *J Comp Neurol* 419: 271–285, 2000.

272. Silverstein H, Norrell H, Haberkamp T, and McDaniel AB. The unrecognized rotation of the vestibular and cochlear nerves from

the labyrinth to the brain stem: its implications to surgery of the eighth cranial nerve. *Otolaryngol Head Neck Surg* 95: 543–549, 1986.

273. Simmons FB. Perceptual theories of middle ear muscle function. *Ann Otol Rhinol Laryngol* 73: 724–739, 1964.

274. Sinex DG and Geisler CD. Auditory-nerve fiber responses to frequency-modulated tones. *Hear Res* 4: 127–148, 1981.

275. Snyder RL, Rebscher SJ, Cao K, and Leake PA. Effects of chronic intracochlear stimulation in the neonatally deafened cat: I. Expansion of central spatial representation. *Hear Res* 50: 7–33, 1990.

276. Snyder RL and Schreiner CE. Forward masking of the auditory nerve neurophonic (ANN) and the frequency following response (FFR). *Hear Res* 20: 45–62, 1985.

277. Snyder RL and Schreiner CE. The auditory neurophonic: basic properties. *Hear Res* 15: 261–280, 1984.

278. Sokolich WG, Hamernick RP, Zwislocki JJ, and Schmiedt RA. Inferred response polarities of cochlear hair cells. *J Acoust Soc Am* 59: 963–974, 1976.

279. Spangler KM, Cant NB, Henkel CK, Farley GR, and Warr WB. Descending projections from the superior olivary complex to the cochlear nucleus of the cat. *J Comp Neurol* 259: 452–465, 1987.

280. Spire JP, Dohrmann GJ, and Prieto PS. *Correlation of brainstem evoked response with direct acoustic nerve potential.* New York: Raven Press, 1982.

281. Spoendlin H and Schrott A. Analysis of the human auditory nerve. *Hear Res* 43: 25–38, 1989.

282. Stainsby TH, Moore BC, and Glasberg BR. Auditory streaming based on temporal structure in hearing-impaired listeners. *Hear Res* 192: 119–130, 2004.

283. Stapells DR, Picton TW, and Smith AD. Normal hearing thresholds for clicks. *J Acoust Soc Am* 72: 74–79, 1982.

284. Starr A and Zaaroor M. Eighth nerve contributions to cat auditory brainstem responses (ABR). *Hear Res* 48: 151–160, 1990.

285. Stevens SS. The relation of pitch to intensity. *J Acoust Soc Am* 6: 150–154, 1935.

286. Suga N. Auditory neuroetology and speech processing: complex-sound processing by combination-sensitive neurons. In: *Auditory function*, edited by Edelman GM, Gall WE, and Cowan WM. New York: John Wiley & Sons, 1988, p. 679–720.

287. Suga N. Multi-function theory for cortical processing of auditory information: implications of single-unit and lesion data for future research. *J Comp Physiol* 175: 135–144, 1994.

288. Suga N. Parallel-hierarchical processing of complex sounds for specialized auditory function. In: *Encyclopedia of acoustics*, edited by Crocker MJ. New York: John Wiley & Sons, 1997, p. 1409–1418.

289. Suga N. Sharpening of frequency tuning by inhibition in the central auditory system. Tribute to Yasuji Katsuki. *Neurosci Res* 21: 287–299, 1995.

290. Suga N, Yan J, and Zhang Y. Cortical maps for hearing and egocentric selection for self-organization. *Trends Cogn Sci* 1: 13–19, 1997.

291. Suga N, Zhang Y, and Yan J. Sharpening of frequency tuning by inhibition in the thalamic auditory nucleus of the mustached bat. *J Neurophysiol* 77: 2098–2114, 1997.

292. Sullivan WE. Possible neural mechanisms of target distance coding in auditory system of the echolocating bat Myotis lucifugus. *J Neurophysiol* 48: 1033–1047, 1982.

293. Sunderland S. A classification of peripheral nerve injuries producing loss of function. *Brain* 74: 491–516, 1951.

294. Syka J, Popelar J, and Kvasnak E. Response properties of neurons in the central nucleus and external and dorsal cortices of the inferior colliculus in guinea pig. *Exp Brain Res* 133: 254–266, 2000.

295. Szczepaniak WS and Møller AR. Interaction between auditory and somatosensory systems: a study of evoked potentials in the inferior colliculus. *Electroencephologr Clin Neurophysiol* 88: 508–515, 1993.

296. Terkildsen K. Acoustic reflexes of the human musculus tensor tympani. *Acta Otolaryng (Stockh)* Suppl. 158, 1960.

297. Tian B and Rauschecker JP. Processing of frequency-modulated sounds in the lateral auditory belt cortex of the rhesus monkey. *J Neurophysiol* 95: 2993–3013, 2004.

298. Tian B, Reser D, Durham A, Kustov a, and Rauschecker JP. Functional specialization in rhesus monkey auditory cortex. *Science* 292: 290–293, 2001.

299. Tobias JV and Zerlin.S. Lateralization threshold as a function of stimulus duration. *J Acoust Soc Am* 31: 1591–1594, 1959.

300. Tollin DJ. The lateral superior olive: a functional role in sound source localization. *Neuroscientist* 9: 127–143, 2003.

301. Ungerleider LG and Mishkin M. Analysis of visual behavior. In: *Analysis of visual behavior*, edited by Ingle DJ, Goodale MA, and Mansfield RJW. Cambridge MA: MIT Press, 1982.

302. Wallach H, Newman EB, and Rosenzweig MR. The precedence effect in sound localization. *Am J Psychol* 62: 315–336, 1949.

303. Warr WB. Organization of olivocochlear systems in mammals. In: *The mammalian auditory pathway: neuroanatomy*, edited by Webster DB, Popper AN, and Fay RR. New York: Springer-Verlag, 1992.

304. Warren EH and Liberman MC. Effects of contralateral sound on auditory-nerve responses. I. Contributions of cochlear efferents. *Hear Res* 37: 89–104, 1989.

305. Webster DB. An overview of mammalian auditory pathways with an emphasis on humans. In: *The mammalian auditory pathway: neuroanatomy*, edited by Webster DB, Popper AN, and Fay RR. New York: Springer-Verlag, 1992, p. 1–22.

306. Weinberg RJ and Rustioni A. A cuneocochlear pathway in the rat. *Neuroscience* 20: 209–219, 1987.

307. Wever EG. *Theory of hearing.* New York: John Wiley & Sons, 1949.

308. Whitfield IC and Evans EF. Responses of auditory cortical neurons to stimuli of changing frequency. *J Neurophysiol* 28: 655–672, 1965.

309. Williston JS, Jewett DL, and Martin WH. Planar curve analysis of three–channel auditory brain stem response: a preliminary report. *Brain Res* 223: 181–184, 1981.

310. Winer JA. The functional architecture of the medial geniculate body and the primary auditory cortex. In: *The mammalian auditory pathway: neuroanatomy*, edited by Webster DB, Popper AN, and Fay RR. New York: Springer-Verlag, 1992, p. 222–409.

311. Winer JA, Chernock ML, Larue DT, and Cheung SW. Descending projections to the inferior colliculus from the posterior thalamus and the auditory cortex in rat, cat, and monkey. *Hear Res* 168: 181–195, 2002.

312. Winer JA, Diehl JJ, and Larue DT. Projections of auditory cortex to the medial geniculate body of the cat. *J Comp Neurol* 430: 27–55, 2001.

313. Winer JA, Kelly JB, and Larue DT. Neural architecture of the rat medial geniculate body. *Hear Res* 130: 19–41, 1999.

314. Winer JA, Larue DT, Diehl JJ, and Hefti BJ. Auditory cortical projections to the cat inferior colliculus. *J Comp Neurol* 400: 147–174, 1998.

315. Winer JA and Prieto JJ. Layer V in cat primary auditory cortex (AI): cellular architecture and identification of projection neurons. *J Comp Neurol* 434: 379–412, 2001.

316. Winer JA, Sally SL, Larue DT, and Kelly JB. Origins of medial geniculate body projections to physiologically defined zones of rat primary auditory cortex. *Hear Res* 130: 42–61, 1999.

317. Winguth SD and Winer JA. Corticocortical connections of cat primary auditory cortex (AI): laminar organization and identification of supragranular neurons projecting to area AII. *J Comp Neurol* 248: 36–56, 1986.

318. Wise LZ and Irvine DRF. Topographic organization of interaural intensity difference sensitivity in deep layers of cat superior colliculus: implications for auditory spatial representation. *J Neurophysiol* 54: 185–211, 1985.

319. Xiao Z and Suga N. Reorganization of the cochleotopic map in the bat's auditory system by inhibition. *Proc Natl Acad Sci* 99: 15743–15748, 2002.

320. Yan J and Suga N. Corticofugal modulation of time-domain processing of biosonar information in bats. *Science* 273: 1100–1103, 1996.

321. Yan J and Suga N. The midbrain creates and the thalamus sharpens echo-delay tuning for cortical representation of target-distance information in the mustached bat. *Hear Res* 93: 102–110, 1996.

322. Yin TCT. Neural mechansims of encoding binaural localization cues in the auditory brainstem. In: *Integrative functions in the mammalian auditory pathway*, edited by Oertel D, Fay RR, and Popper AN. New York: Springer, 2002, p. 99–159.

323. Yin TCT and Chan JCK. Interaural time sensitivity in medial superior olive of cat. *J Neurophysiol* 64: 465–488, 1990.

324. Yin TCT and Kuwada S. *Neuronal mechanisms of binaural interaction*. New York: Wiley & Sons, 1984.

325. York DH. Correlation between a unilateral midbrain–pontine lesion and abnormalities of the brain–stem auditory evoked potential. *Electroenceph Clin Neurophysiol* 65: 282–288, 1986.

326. Young ED and Sachs MB. Representation of steady-state vowels in the temporal aspects of the discharge patterns of populations of auditory nerve fibers. *J Acoust Soc Am* 66: 1381–1403, 1979.

327. Zakrisson JE. The role of the stapedius reflex in poststimulatory auditory fatigue. *Acta Otolaryngol (Stockh)* 79: 1–10, 1975.

328. Zanette G, Carteri A, and Cusumano S. Reappearance of brainstem auditory evoked potentials after surgical treatment of a brain-stem hemorrhage: contributions to the question of wave generation. *Electroencephalogr Clin Neurophys* 77: 140–144, 1990.

329. Zappia M, Cheek JC, and Luders H. Brain-stem auditory evoked potentials (BAEP) from basal surface of temporal lobe recorded from chronic subdural electrodes. *Electroencephalogr Clin Neurophys* 100: 141–151, 1996.

330. Zeki S. *A vision of the brain*. London: Blackwell Scientific Publications, 1993.

331. Zhang M, Suga N, and Yan J. Corticofugal modulation of frequency processing in bat auditory system. *Nature* 387: 900–903, 1997.

332. Zurek PM. The precedence effect. In: *Directional hearing*, edited by Yost W and Gourevitch G. New York: Springer-Verlag, 1987.

333. Zwicker E. On a psychoacoustical equivalent of tuning curves. In: *Facts and models in hearing*, edited by Terhardt E. Berlin: Springer-Verlag, 1974, p. 132–141.

334. Zwislocki JJ. Are nonlinearities observed in firing rates of auditory–nerve afferents reflections of a nonlinear coupling between the tectorial membrane and the organ of Corti? *Hear Res* 22: 217–222, 1986.

335. Zwislocki JJ. What is the cochlear place code for pitch? *Acta Otolaryngol (Stockh)* 111: 256–262, 1991.

336. Zwislocki JJ and Sokolich WG. Velocity and displacement responses in auditory nerve fibers. *Science* 182: 64–66, 1973.

III

DISORDERS OF THE AUDITORY SYSTEM AND THEIR PATHOPHYSIOLOGY

The signs and symptoms of disorders of the auditory system are decreased function and abnormal function. Decreased function includes elevated threshold and decreased speech discrimination, generally known as impairment of hearing. The most common abnormal function is tinnitus, which is a sign of hyperactivity. Other examples of abnormal function are recruitment of loudness, hyperacusis, distortion of sound, and phonophobia. Such symptoms are mostly caused by changes in the function of the auditory nervous system. Impairment of conduction of sound to the cochlea mainly causes elevation of the hearing threshold with speech discrimination being unaffected provided that sound is amplified to compensate for the threshold elevation. The symptoms and signs from pathologies of the cochlea are similar but may in addition include symptoms such as recruitment of loudness. While disorders of the ear normally are associated with morphological abnormalities, such as loss of outer hair cells, disorders of the auditory the nervous system often occur without any detectable morphological abnormalities. Injuries to the auditory nerve and tumors, bleeding and ischemia are examples of morphological changes in the auditory nervous system.

Speech discrimination can be predicted relatively well on the basis of the elevation of the hearing threshold in disorders of the middle ear and the cochlea but the effect on the function of the nervous system may affect speech discrimination. Impairment of speech discrimination from injuries to the auditory nerve is less predicable from the elevation in hearing threshold. Speech discrimination is typically more impaired than it is in disorders of the ear with similar audiograms. Disorders of more central structures of the auditory system are rare and give complex symptoms and signs.

Earlier, disorders of the auditory system have been divided into two broad groups: conductive hearing loss and sensorineural hearing loss. Conductive hearing loss was defined as hearing loss caused by disorders of the apparatus that conduct sound to the cochlea (or rather to the sensory cells), and sensorineural hearing loss was hearing loss that was caused by pathologies of the cochlea (sensory cells) and the auditory nervous system. While these broad divisions of causes of hearing impairment still serve as a clinical useful division, it has become evident that the anatomical location of the physiological abnormalities that cause hearing impairment is not localized only to the structures that have detectable morphological abnormalities. The old concept that hearing loss that occurs in connection with impairment of sound conduction to the cochlea is caused only by the effect of the morphological abnormalities in the conductive apparatus is no longer valid, and symptoms of such disorders cannot be completely described by the pure tone audiogram. In a similar way, injuries to cochlear hair cells that impair the function of the cochlea are often associated with abnormal function of the auditory nervous system. Deficits in neural processing of sound may therefore affect individuals with pathologies of the conductive apparatus and the cochlea. This means that the distinction between peripheral and central causes of symptoms is blurred and it is no longer valid to divide disorders of the auditory system according to the anatomical location of the detectable morphological abnormalities.

It is only relatively recently that it has become evident that the function of the auditory nervous system can change without detectable changes in morphology. Such changes occur as a result of expression of neural plasticity, which means that the function of specific parts of the nervous system changes more or less permanently as a result of how it is activated.

The anatomical location of the physiological abnormalities that cause hyperactive symptoms (tinnitus, hyperacusis and phonophobia) is the auditory nervous system. Various forms of retraining can ameliorate symptoms of tinnitus and hyperacusis. Even presbycusis may be affected by sound stimulation, and that opens a possibility for reducing the risk of hearing loss with age.

Normal development of the auditory nervous system depends on appropriate stimulation early in life. Deprivation from stimulation such as occurs from any form of hearing deficit can severely affect the normal development of the auditory nervous system during childhood and even change the function of the mature nervous system. Since deficits from insufficient stimulation during childhood development are difficult to reverse by sound exposure later in life, it is imperative that hearing of neonates is tested and any hearing loss that is detected be compensated for in an early stage of life so that appropriate stimulation of the auditory system is established.

Expression of neural plasticity plays a role in creating symptoms of disease. A typical example is some forms of tinnitus, but expression of neural plasticity may also be involved in creating other abnormal functions such as hyperacusis, phonophobia and perhaps even depression that occurs together with tinnitus. This has added importance to understanding the pathophysiology of hearing disorders in general. Disorders that are thought to have no biological basis are sometimes referred to as "functional disorders", meaning that the disorder is either psychological or psychiatric in nature. We now believe that such disorders are also caused by biological changes, although they may not have any detectable morphologic or physiological correlates.

While hearing loss that is caused by pathologies of the conductive apparatus (ear canal and middle ear) can be successfully treated, little treatment is available for treating hearing loss caused by pathologies of the cochlea and the nervous system, but such patients can be helped by hearing aids and cochlear implants. Cochlear implants now offer the possibility to restore some forms of hearing in people with profound hearing loss due to injuries to cochlear hair cells, provided that the auditory nerve is intact. More recently cochlear nucleus implants have been introduced to aid people in whom the auditory nerve is severely injured or surgically removed such as is often the case after operations for vestibular Schwannoma. Cochlear implants and auditory brainstem implants are used for individuals with bilateral hearing deficits only. Restoration of cochlear hair cells has not yet been done but extensive research efforts are presently devoted to that task. Treatment of disorders that are caused by expression of neural plasticity is in its infancy but the possibilities are there and future development most likely will provide adequate treatment of such disorders as tinnitus and hyperacusis.

This section will discuss the underlying pathologies of hearing impairment (Chapter 9) and hyperactive hearing disorders (Chapter 10). The design and function of cochlear and brainstem implants are discussed in Chapter 11.

9

Hearing Impairment

1. ABSTRACT

1. Disorders that affect conduction of sound to the sensory cells in the cochlea cause elevation of the hearing threshold and affect speech discrimination in a similar way as reducing the intensity of the sound that reaches the ear.

2. Common causes of conductive hearing loss are obstruction of the ear canal by cerumen and accumulation of fluid in the middle ear or the air pressure in the middle-ear cavity being different from the ambient pressure.

3. Various disorders and trauma can cause interruption of the ossicular chain or perforation of the tympanic membrane, resulting in conductive hearing loss. Otosclerosis is a disease that impairs sound conduction through the middle ear by bone growth around the stapes footplate, which ultimately becomes immobilized.

4. Absence of the tympanic membrane and/or the ossicles cause severe hearing loss (as much as 60 dB) because of loss of the transformer action of the middle ear and because the sound reaches both windows of the cochlea with nearly the same intensity.

5. Diagnosis of conductive hearing loss is made from pure tone audiometry, tympanometry and recordings of the acoustic middle-ear reflex response.

6. Some forms of conductive hearing loss reverse without treatment. Other forms are treatable by medicine or by surgery.

7. The most common cause of cochlear hearing loss is injury to outer hair cells, which impairs the cochlear amplifier. Injury to outer hair cells causes elevation of the hearing threshold and it is often accompanied by recruitment of loudness and tinnitus. High frequencies are normally affected more than low ones, and the hearing loss rarely exceeds 50 dB.

8. Besides an elevation of the threshold of hearing, injuries to outer hair cells causes the cochlear filter to become broader and that may increase masking and impair temporal coding of broadband sounds such as vowels.

9. Amplitude compression is impaired from injuries to outer hair cells causing recruitment loudness.

10. Brainstem auditory evoked potentials (ABR) and the acoustic middle-ear reflex are little affected by injuries to outer hair cells.

11. Age-related changes are the most common cause of cochlear hearing loss. Exposure to loud sounds can cause injuries to cochlear hair cells, as can drugs such as diuretics and aminoglycoside antibiotics, trauma, and diseases. Some forms of cochlear hearing loss are hereditary, and some forms of hearing loss worsen in the first year of life and may become severe.

12. Episodal cochlear hearing loss that is one of the triad of symptoms that defines Ménière's disease is probably caused by an imbalance of pressure (or rather volume) in the compartments of the cochlea. In early stages of the disease, hearing loss mostly affects low frequencies.

13. Impaired function of hair cells can reverse totally or partially, such as occurs after noise exposure (temporary threshold shift [TTS]) and after administration of some ototoxic substances.

14. Permanent injury to hair cells cannot be restored medically or surgically but can often be compensated for by wearing a hearing aid.

15. Cochlear implants offer a possibility to restore useable hearing in people with severe cochlear damages as long as the auditory nerve is intact.

16. Disorders of the central auditory nervous system are of two kinds, one that is associated with detectable morphological changes and one that is not associated with detectable morphologic changes.

17. The most common disorder that affects the neural conduction in the auditory nerve and which is associated with detectable morphological changes is vestibular Schwannoma. Injuries to the auditory nerve may also be caused by surgically induced injuries, by viral infections, and by vascular compression.

18. Disorders of the auditory nerve cause hearing loss with greater impairment of speech discrimination than cochlear hearing loss of the same magnitude, and the impairment cannot be predicted from the threshold elevation for pure tones. The audiogram often has irregular peaks and dips, the ABR is abnormal and the threshold of the acoustic middle ear reflex is elevated or absent.

19. Patients with disorders of the auditory nerve often have tinnitus (hearing meaningless sounds).

20. Disorders caused by lesions of brainstem structures are rare and auditory signs are complex.

21. Lesions of the auditory cerebral cortex often cause minimal threshold elevation and the speech discrimination is often normal when tested using standard audiological tests but such lesions can be diagnosed by using low-redundancy speech tests and by imaging techniques.

22. Disorders of sound conduction to the cochlea and injuries to hair cells are often accompanied by changes in the function of the auditory nervous system that have no detectable morphological correlates.

2. INTRODUCTION

Hearing impairment is a broad concept that comprises many disorders. It is mainly identified by pure tone audiometry and speech discrimination tests but there are other more specific tests in common use for diagnosis of disorders that cause impaired hearing. Hearing impairment had earlier been divided into two large groups: conductive hearing loss and sensorineural hearing loss. Conductive hearing loss was a group of disorders with morphological changes in the middle ear including obstruction of the ear canal. These disorders have similar effects on hearing as reducing the intensity of a sound (turning the volume of a loudspeaker down). Many forms of disorders of the conductive apparatus will resolve on their own, as they often do in the case of otitis media, or they can be successfully treated by surgery.

The definition of sensorineural hearing loss included disorders where morphological changes could be shown in the cochlea (mostly injuries to outer hair cells) and many kinds of disorders where the morphological abnormalities are located in the central auditory nervous system. This means that the old distinction between causes of hearing impairment was based the anatomical location of detectable morphological abnormalities.

Diagnosis of disorders of the conductive apparatus requires knowledge about the normal function of the middle ear and what changes occur in function in different kinds of pathologies. Recording otoacoustic emission can assess the nature of injuries to outer hair cells of the cochlea. Recent research has shown that pathophysiology of hearing impairment is often far more complex than detectable morphological changes in the middle ear and the cochlea and functional and even morphological changes in the central auditory nervous system occur in such disorders. The changes in function of the auditory nervous system that occur concurrent with morphological changes in the middle ear and the cochlea are more difficult to assess quantitatively and most of our knowledge about such changes originates from studies in animals. Electrophysiological tests such as auditory brainstem responses are valuable in assessing hearing loss caused by morphological changes in the auditory nervous system such as that caused by vestibular Schwannoma. This chapter will discuss the pathophysiology of the middle ear, the cochlea, and the auditory nervous system in disorders that present with signs of impairment of hearing.

3. PATHOLOGIES OF THE SOUND CONDUCTING APPARATUS

The anatomical location of impairment of sound transmission to the cochlea can be the ear canal, the tympanic membrane, or the ossicular chain. Various audiological tests can determine the anatomical location of the pathology. Correct interpretation of such tests require knowledge and understanding about the normal function of the sound conducting apparatus (see Chapter 2) as well as knowledge about how various disease processes and trauma can alter the function of the sound conducting apparatus.

3.1. Ear Canal

A build up of cerumen that blocks the ear canal (impacted wax) causes the simplest and easiest treatable form of hearing loss. The obstruction of sound conduction to the tympanic membrane caused by total blockage of the ear canal results in a nearly flat hearing loss that varies between 20 and 30 dB (Fig. 9.1) Hearing is restored to normal by removal of the cerumen. The ear canal in frequent swimmers often narrows because of formation of new bone (exostosis). This makes it easier for the accumulation of cerumen to obstruct the ear canal and it makes it more difficult to clean the ear canal for cerumen. Cerumen may also cover part of tympanic membrane.

With age, the outer (cartilaginous) portion of the ear canal in many individuals changes from a nearly circular cross-section to an oval shape and consequently it may become totally occluded. If the earphone used for audiometry has a supra-aural cushion (AR/MX41) it may cause a nearly collapsed ear canal to become totally occluded due to assertion of pressure on the ear canal by the earphone, thus providing erroneous results of audiometry. The hearing loss is similar to that caused by impacted cerumen (approximately 25 dB). Placing a short plastic tube in the ear canal during hearing tests can solve the problem. These problems do not exist when insert earphones are used for audiometry.

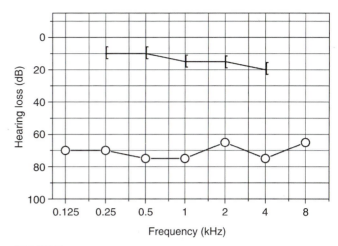

FIGURE 9.2   Hearing loss from congenital ear canal atresia (data from Lidén, 1985).

Ear-canal atresia is a condition where one or both ear canals have not opened during prenatal life. The mild form of this congenital malformation is characterized by a small ear canal and a nearly normal middle ear. In a more severe form the ear canal is totally occluded (or actually missing) and the ossicular chain is malformed. In the most severe form the middle ear space is small or absent in addition to the ear canal being occluded. Ear-canal atresia impairs transmission of airborne sound to the tympanic membrane and the function of the middle ear may be impaired. Hearing loss of 55–70 dB (Fig. 9.2) occurs in ear canal atresia. If the atresia is bilateral such hearing loss implies a listening distance of less than 10 cm (4 inches) (hearing loss of 60 dB in the speech frequency range results in a listening distance of 10 cm from the ear in order to obtain speech communication.) A person with such a condition will require a hearing aid. The bone conduction is little affected (Fig. 9.2) and therefore bone conduction hearing aids have been used to help such patients as an alternative to surgical intervention.

Ear-canal atresia on one side will allow a person to hear with one ear, but such a person will have difficulties in determining the direction of a sound source and have difficulties in the discrimination of speech in noisy environments and where more than one speaker is present.

3.2. Middle Ear

The middle ear is the site of most disorders that affect sound transmission to the cochlea. The air pressure in the middle ear cavity being different from the ambient pressure is probably the most common cause of impairment of sound transmission to the cochlea.

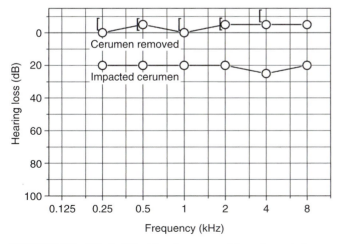

FIGURE 9.1   Effect of blocking the ear canal from impacted cerumen (data from Sataloff and Sataloff, 1993).

Accumulation of fluid in the middle-ear cavity and when the pressure in the middle-ear cavity is different from the ambient pressure are the causes of some of the most common disorders that can impair sound transmission to the cochlea. More serious pathologies of the middle ear include perforation of the tympanic membrane and interruption or fixation of the ossicular chain. Each one of these conditions affects sound transmission through the middle ear in specific ways.

Sound transmission to the cochlea is impaired when the air pressure in the middle-ear cavity is different from that in the ear canal (the ambient pressure) as was discussed in Chapter 2. The effect is a decrease in transmission that is greatest for low frequencies. A negative pressure in the middle-ear cavity causes larger hearing loss than the same value of positive pressure. Negative pressure in the middle-ear cavity can be caused by malfunction of the Eustachian tube and often occurs in connection with middle-ear infections. If the Eustachian tube does not open normally, oxygen absorption by the mucosa in the middle ear will cause the pressure to decrease. Positive pressure in the middle-ear cavity may occur in the ascending phase of flying because of a decrease in the ambient pressure but it usually equalizes spontaneously, even with a partly functioning Eustachian tube. Negative pressure that occurs during descent is more difficult to equalize because the higher ambient pressure exerts a closing pressure on the opening of the Eustachian tube in the pharynx. Therefore, a person is more likely to have problems equalizing pressure in the middle ear on landing than after take-off.

The mucosa of the middle-ear cavity has attracted attention because it is involved in a common disorder known as otitis media with effusion (OME), which is an inflammation of the lining of middle-ear cavity, the mastoid cell system and the Eustachian tube. It has been estimated that approximately 90% of children within the first 3 years of life acquire OME [324, 325] see also Bernstein [22]. OME at an early age may also disturb the pneumatization of mastoid air cells [324, 325]. Small mastoid cell systems promote middle-ear infections later in life. The incidence of middle-ear infections decreases rapidly with age and it is usually over around the age of 7 years. OME occurs rarely in adults.

The inflammation of the middle-ear mucosa prevents the Eustachian tube from opening normally, which creates a negative air pressure in the middle-ear cavity because of absorption of oxygen by the mucosa. Clear fluid may effuse from the mucosa of the middle ear and this fluid accumulates in the middle-ear cavity. Viscous fluid may accumulate as a result of inflammatory processes. The major reasons that OME is more frequent in children than adults are that the Eustachian tube is shorter in children up to age 5–6 years than in adults and that the direction of the Eustachian tube is nearly horizontal rather than pointing 45 degrees downwards as it does in adults (see Chapter 1, and Fig. 1.6B).

The hearing loss from fluid in the middle ear depends on how much air remains in the middle ear and the location of the air [265]. The presence of clear fluid in the middle-ear cavity affects sound conduction only when it covers the tympanic membrane. Fluid that covers the tympanic membrane in the middle-ear cavity impairs its movement. When the entire tympanic membrane is covered with fluid it resembles a situation where sound is transferred to a fluid, and is thus very inefficient, as was discussed in Chapter 2. Hearing, however, is likely to be essentially unaffected by fluid that fills the middle-ear cavity incompletely as long as there is air behind the tympanic membrane. Clear (low viscosity) fluid that covers the ossicles has minimal effect on their movement and fluid covering the round window of the cochlea will not affect the motion of the cochlear fluid noticeably. Since hearing loss depends on how large a portion of the tympanic membrane is covered with fluid, the resulting hearing loss will depend on the head position, provided that the fluid has a low viscosity so that it can move freely in the middle-ear cavity. Hearing loss may only be evident when a patient is lying down with the head turned to the side of the fluid in the ear. In that body position hearing loss may become evident even when the amount of fluid in the middle-ear cavity is small. The air behind the tympanic membrane acts as a cushion that adds stiffness to the middle ear and impedes the motion of the tympanic membrane for low frequencies. The stiffness of that air cushion increases when the volume decreases but it causes only slight hearing loss at low frequencies and it therefore often escapes detection in audiometric testing. The tympanogram will have a small peak in an ear where fluid does not completely cover the backside of the tympanic membrane and it is unlikely that a response of the acoustic middle-ear reflex can be recorded in such an ear.

Hearing loss is independent of the head position in patients whose middle ear is totally fluid filled or in patients with highly viscous fluid. Fluid with the consistency of a gel, as often is present in the middle ear in patients with middle-ear infections, may impair hearing noticeably even without covering the tympanic membrane because it impedes the motion of the middle-ear ossicles. Such "glue ears" thus typically present with hearing loss that is independent of the position of the head.

When fluid fills the middle-ear cavity sound must be transferred to the fluid and it exerts approximately the same force on both the round and the oval window, thus a situation similar to hearing without a middle ear. However, statistics show that the average hearing loss is approximately 30 dB with nearly normal bone conduction thresholds (Fig. 9.3) [155, 265]. Only few patients have hearing losses of 50 dB. Fluid in the middle ear can be diagnosed by tympanometry. If the fluid covers the entire backside of the tympanic membrane, the tympanogram will be flat because the acoustic impedance of the ear does not change with changing air pressure in the ear canal. It is not possible to record the response of the acoustic middle-ear reflex in such an ear even when elicited from a normal opposite ear. The reflex response can be elicited from an ear with fluid in the middle ear cavity when recorded from the opposite ear provided that the middle ear is normal in that ear. The reflex threshold is elevated by the amount of hearing loss in the affected ear. The difference between air conduction threshold and bone conduction threshold (air–bone gap) can be used to diagnose a conductive hearing loss.

When hearing loss caused by fluid in the middle-ear cavity impairs speech perception in children it is important that hearing is restored sufficiently because exposure to speech and other natural sounds is important for the normal development of the auditory nervous system (see p. 248). This is one of the reasons that OME is routinely treated by placing a small tube (polyethylene tube [PE]) in the tympanic membrane so that the fluid can drain and the air pressure in the middle-ear cavity can equalize to that of the ambient pressure. The main purpose is to restore hearing. Although such

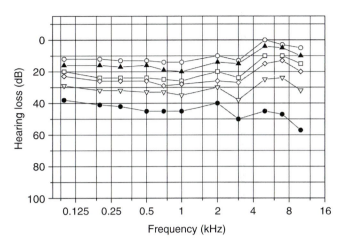

FIGURE 9.4   Effect of perforation of the tympanic membrane on hearing threshold (data from Payne, 1951; in Wever, 1954).

a tube acts as a hole in the tympanic membrane, it will not interfere noticeably with hearing because of its small opening (see Fig 9.4).

The fluid in the middle-ear cavity may contain toxic substances produced by bacteria and these substances may enter the cochlea by diffusion through the membranes of the round and the oval windows. Studies using evoked potentials in rats with middle-ear effusion showed signs of cochlear involvement in addition to the conductive hearing loss [305]. Whether or not these substances cause permanent injury to the cochlea is unknown.

Cholesteatomas are examples of other growths in the middle-ear cavity that may affect hearing. Cholesteatomas are benign growths that may develop in the middle ear after long-term recurrent or chronic middle-ear infections, or they may occur with no apparent (known) cause. When a cholesteatoma grows in the middle-ear cavity, the extent of the hearing loss it causes varies with the size of the growth and whether it is in contact with the ossicular chain or to what extent it may cause erosion of the ossicular chain and interruption of the ossicular chain.

It is the difference between the sound that reaches the front side and the backside of the tympanic membrane that causes it to move (vibrate). A hole in the tympanic membrane will allow some sound to reach the backside of the tympanic membrane, which reduces the difference between the intensity of the sound that is present on the two sides of the tympanic membrane. The result is that the force that causes the tympanic membrane to vibrate is reduced. The reduction of the vibration of the tympanic membrane caused by a perforation depends on the size of the hole in the tympanic membrane and the size of the middle-ear cavity.

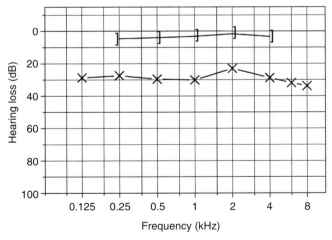

FIGURE 9.3   Hearing loss from chronic otitis media. Mean hearing loss in 95 patients (196 ears for air conduction and 122 ears for bone conduction) (data from Kokko, 1974).

BOX 9.1

EFFECT OF A HOLE IN THE TYMPANIC MEMBRANE

Acoustically, a small hole in the tympanic membrane acts in the same way as an electrical inductance in an electrical circuit and the middle-ear cavity acts as a capacitance (the mechanical analogy is a mass and a spring, respectively). A small hole in the tympanic membrane therefore acts as a low pass filter that lets sound of low frequencies reach the middle-ear cavity. The cut-off frequency of this low pass filter is lower for a small hole than for a large one and it is lower for larger middle-ear cavities.

Experimental studies of the effect of a perforation of the tympanic membrane in the cat [240, 333, 334] have confirmed that the effect of a hole in the tympanic membrane is largest at low frequencies. A small hole in the tympanic membrane causes hearing loss mainly at frequencies below 4 kHz, whereas a larger hole also affects higher frequencies.

A small hole in the tympanic membrane acts as a low pass filter and therefore causes hearing loss at low frequencies only. This is because only low frequencies will reach the middle-ear cavity and decrease the force that acts on the tympanic membrane. A larger hole will permit sounds within a larger frequency range to reach the middle-ear cavity and thus impair hearing over a larger range of frequencies (Fig. 9.4) [240].

The effect of a large perforation is not limited to the effect of sound reaching the backside of the tympanic membrane (Fig. 9.4). When large parts of the tympanic membrane are lost the perforation also affects the way the manubrium of malleus vibrates because some of the suspension of the malleus is lost. The effect on the hearing threshold depends not only on the size of the perforation but also on its location on the tympanic membrane. Animal experiments have shown that the largest effect of a hole of a certain size occurs when it is placed in the posterior or superior part of the tympanic membrane and the least effect occurs when it is located in the anterior, inferior portion of the tympanic membrane. Clinical experience is generally in good agreement with the results of animal experiments. Thus, a large hole in the tympanic membrane in humans commonly results in a 40–50 dB, mostly flat, hearing loss (45 dB hearing loss in the speech range corresponds to a listening distance of approximately 1 m, approximately 3 ft, for speech discrimination). When the tympanic membrane is totally missing the hearing loss can reach 60 dB (corresponding to a listening distance of approximately 10 cm) [241, 285].

The bone conduction threshold is normal in patients with a perforated tympanic membrane. The acoustic middle-ear reflex cannot be recorded in an ear with a perforated tympanic membrane because stiffening of the middle ear from contraction of the stapedius muscle will not affect the acoustic impedance of the ear at the low frequency at which it is usually measured (0.22 KHz) unless the perforation is very small. The acoustic reflex may be elicited from the affected ear and recorded in the opposite ear provided that the hearing loss in the affected ear is not excessive at the frequency of the stimulus tone and that the middle ear is normal in the opposite ear. The tympanogram is flat in an ear with a perforated tympanic membrane.

Interruption of the ossicular chain may occur as a result of trauma, or because of disease processes that erode the middle ear ossicles, such as cholesteatomas. The conductive hearing loss may exceed 60 dB when the ossicular chain is interrupted and the tympanic membrane is intact (Fig 9.5).

This hearing loss is thus greater than when sound reaches the two windows of the cochlea directly, such as occurs when the entire middle ear including the

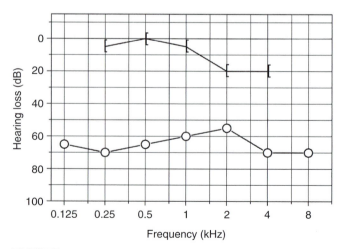

FIGURE 9.5 Effect on hearing threshold from interruption of the ossicular chain in an ear with intact tympanic membrane (data from Lidén, 1985).

tympanic membrane is missing (see p. 209). The greater hearing loss is caused by the attenuation of the sound that reaches the middle-ear cavity by the intact tympanic membrane.

Individuals with an interrupted ossicular chain hear better when the tympanic membrane is perforated. The hearing loss from an interrupted ossicular chain observed in humans (Fig. 9.5) is of the same order of magnitude as that obtained in animal experiments (cats) where interruption of the incudo-stapedial joint results in a hearing loss of 50–70 dB [241, 346].

Interruption of the ossicular chain can be diagnosed by tympanometry because the acoustic impedance of the ear is abnormally low. Tympanometry shows a larger than normal admittance (compliance). The acoustic middle-ear reflex response cannot be recorded in an ear in which the ossicular chain is interrupted because contraction of the stapedius muscle does not change the ear's acoustic impedance. Considering the large conductive hearing loss that results from interruption of the ossicular chain, it is also unlikely that the reflex can be elicited from the affected ear and recorded in the opposite ear even if that is normal. The bone conduction threshold is normal in individuals with an interrupted ossicular chain.

If a connection between the incus and the stapes is established by soft tissue hearing for low frequencies will improve because an elastic connection between the middle-ear bones transmits low frequency sounds but high frequencies are not conducted effectively. Growth of a cholesteatoma may result in the formation of a connection between the tympanic membrane and the stapes causing a paradoxical improvement of hearing despite a progression of the disorder. If left untreated, further growth of a cholesteatoma may cause sensorineural hearing loss due to erosion into the cochlea.

Such re-establishment of the connection between the middle-ear bones may also occur in otospongiosis that involves the ossicles. Other masses that may fill the middle-ear cavity, such as various kinds of tumors may have similar effects on hearing.

The normal motion of the stapes is impaired in a disorder known as otosclerosis in which the stapes footplate becomes fixated in the round window because new bone is constantly formed around the stapes footplate.

It is a complex disorder, or probably a group of disorders, that is associated with disorders of connective tissue. Genetic factors play a role in the occurrence of otosclerosis [3, 106]. The hearing loss in patients with otosclerosis is largest for low frequencies and it increases with the progression of the disease, usually over many years. Typically, hearing loss in patients

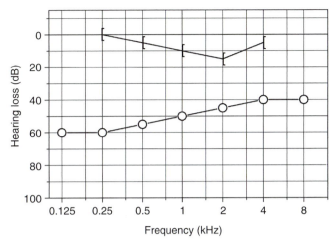

FIGURE 9.6   Effect of fixation of the ossicular chain (from otosclerosis) on hearing threshold (data from Lidén, 1985).

who have had otosclerosis for many years is 50 dB at low frequencies and less at high frequencies (Fig. 9.6). The bone conduction threshold is nearly normal but often has a dip (up to 30 dB) around 2 kHz. This dip, known as "Carhart's notch" is not a result of cochlear (sensorineural) involvement because it disappears after a successful operation. It is a sign that fixation of the stapes footplate affects sound conduction to the cochlea through bone conduction.

The response of the acoustic middle-ear reflex is absent in an ear in which the stapes is immobilized (e.g., in otosclerosis). The tympanogram has a small peak in an otosclerotic ear because the mobility of the middle ear is reduced by the fixation of the stapes.

Patients with otosclerosis may, over time, get a component of cochlear hearing loss because of formation of new bone inside the cochlea. The sensorineural hearing loss adds to the approximately 60 dB conductive hearing loss and the total hearing loss may reach 80–85 dB after 20–25 years of untreated otosclerosis. There is evidence that otosclerosis affects the function of the central auditory nervous system, manifest by the fact that many patients with otosclerosis have tinnitus which disappears after treatment of the otosclerosis (see Chapter 10).

The earliest treatments of otosclerosis involved making an artificial route for sound to the cochlea by making an opening in the bone of a semicircular canal. This operation, known as the fenestration operation was replaced by an operation in which the new bone around the stapes was removed, but that method has also been abandoned, because the relief was short due to formation of new bone.

Now, the common treatment of otosclerosis is replacement of the middle-ear bones by a prosthesis in an operation introduced by Shea about 1962. The stapes is removed and replaced by a prosthesis that connects the incus to an artificial stapes placed in the oval window. That has solved the problem of recurring hearing loss from bone growth and it is now the common treatment for otosclerosis. With modern microsurgical techniques and modern design of middle-ear prostheses, hearing can be restored nearly to its normal value. Techniques that are being developed include making a small hole in the stapes footplate to insert a prosthesis. This technique is less traumatic than replacing the entire stapes footplate and it works as well.

Other conditions of middle-ear pathologies also use prostheses as treatment. Such prostheses must have proper size and must be placed so that maximal sound transfer to the cochlea is accomplished. Middle ear prostheses do not replace the function of the stapedius muscle in regulating sound transmission through the middle ear and patients with middle ear-prostheses may have a higher susceptibility to noise induced hearing loss (see p. 226).

Understanding of the hearing deficits that result when the tympanic membrane or the entire middle ear is missing so that sound reaches both cochlear windows directly requires understanding of how the cochlear fluid is set into motion with and without the middle ear. Studies have shown that it is the difference between the pressure at the two windows that is the effective force that sets the cochlear fluid into motion [335]. The improvement of sound conduction to the cochlea by the transformer action of the middle ear was discussed in Chapter 2. However, that does not explain all ramifications on hearing from total or partial loss of the middle ear.

When the middle-ear transformer action is absent, the hearing loss exceeds that of the lost transformer gain (approximately 30 dB). This is because the sound then reaches both windows of the cochlea at approximately the same intensity. Since it is the difference between the force on the two cochlear windows that causes the cochlear fluid to move, the resulting hearing loss depends on the size of the differences between the sound that reaches the two cochlear windows. Normally, the force that acts on the oval window is much larger than that acting on the round window because of the gain of the middle ear transformer. If the sound pressure at the two windows is exactly the same there would be total deafness as there would be no motion of the cochlear fluid at all because it is only the difference in the forces at the two windows that can set the cochlear fluid into motion. However, in practice there will always be some difference in the amplitude and the phase of the sound that reaches the two windows and it is that difference that sets cochlear

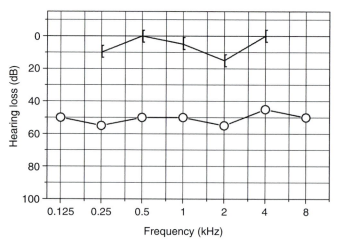

FIGURE 9.7   Hearing threshold in a patient without a middle ear (data from Lidén, 1985).

fluid in motion and makes it possible for individuals without a middle ear to hear, although at a much elevated threshold.

Without surgical restoration of hearing, individuals with an open middle-ear cavity and no middle-ear ossicles have hearing loss for air conducted sounds of 50–60 dB (Fig. 9.7), thus approximately 30 dB in addition to the approximately 30 dB loss from the transformer action of the middle ear. The hearing loss is often less for high frequencies, because the phase difference between the sound that reaches the two windows of the cochlea is larger for high frequencies than low frequencies and that creates a larger phase difference between the force that acts on the two windows for high frequencies than for low frequencies.

Individuals in whom restoring the function of the middle ear using a prosthesis is not possible can be helped by operations that aim at making the difference between the sound that reaches the two windows of the cochlea as large as possible. Since the distances in the middle ear in relation to the wavelength of the sound are small the sound level will be nearly uniform in the middle ear cavity in an ear without a tympanic membrane. This means that the difference between the sound that reaches the two windows of the cochlea will be small. The phase angle is 180° for 1/4 wavelength and at 1 kHz, 1/4 wavelength is 17 cm, at 10 kHz it is 1.7 cm. However, a smaller phase shift can cause a considerable motion of the cochlear fluid (Fig. 9.8). Thus, a difference of only 10 degrees produces a CM potential that is only 20 dB below its maximal value (i.e., a hearing loss of only 20 dB). A difference between the forces at the two cochlear windows of 10% (1 dB) will produce motion of the cochlear fluid that is only 20 dB less than its maximal value.

BOX 9.2

PRESSURE DIFFERENCES DRIVE COCHLEAR FLUID

Early animal experiments confirmed that it is the difference between the amplitude and phase of the pressure that acts on the two windows that drives the cochlear fluid [346]. In these experiments, pure tones were applied independently to the round window and the oval window of the cochlea of cats. The amplitude and the phase angle between the sound at the two windows were varied independently while the cochlear microphonic (CM) potentials were recorded. Recall that the CM is a valid measure of the volume velocity of the cochlear fluid (Chapters 2 and 3). The force that sets the cochlear fluid into motion is the vector difference between the forces at the two cochlear windows (Fig. 9.8). This means that the largest motion of the cochlear fluid is induced when the two sounds are precisely out of phase. Thus, a sinusoidal force (pure tone) produces the largest motion of the cochlear fluid when applied with a phase difference of 180° (opposite phase).

Voss et al. [335] determined the "common mode rejection" (CMR) as a measure of the efficiency of transfer of sound to the cochlear fluid. (The term CMR is borrowed from engineering to describe the properties of differential amplifiers.) In studies in cats, the values of the CMR were in the range of 35 dB for frequencies below 1 kHz. This means that when the forces on the two windows are exactly of opposite phase, the motion of the cochlear fluid is 35 dB larger than when the two forces are identical and in-phase. Sound that reaches the two windows of the cochlea in-phase will sound approximately 35 dB weaker than sounds that reach the two windows in opposite phase. The greater the difference (in amplitude or phase) between the two forces the greater the induced motion of the cochlear fluid. Adding that to the loss of the transformer action of the middle ear (approximately 30 dB) results in an estimated

hearing loss of 65 dB, a value that is close to that experienced in humans with an absent middle ear (Fig. 9.7).

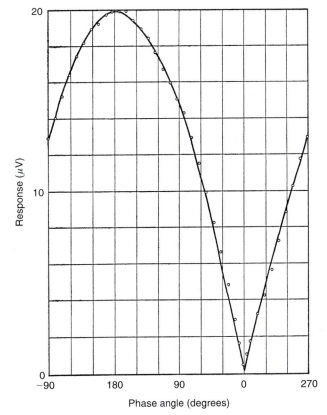

FIGURE 9.8 Illustration of the vectorial summation of sound that reaches both the oval and the round windows. The results were obtained in a cat using the cochlear microphonic potentials to measure the motion of the cochlear fluid (reprinted from Wever and Lawrence, 1954, with permission from Princeton University Press).

3.3. Impairment of Sound Conduction in the Cochlea

Sound conduction in the cochlea can be impaired for instance by bone growth similar to that in otosclerosis (known as cochlear otosclerosis). Some of the attributes of this kind of hearing loss are similar to impairment of sound conduction in the middle ear and some are different. Since this condition occurs together with immobilization of the stapes, tests such as tympanogram and the acoustic middle-ear reflex will be similar to that seen in common otosclerosis.

Patients with cochlear otosclerosis will in addition have signs of sensorineural hearing loss because the bone growth in the cochlea affects the nerve supply to the hair cells. The new bone formation in the cochlea may be slowed by treatment with fluor compounds.

3.4. Accuracy of Measurements of Conductive Hearing Loss

It is usually assumed that the pure tone audiograms are accurate measures of conductive hearing loss.

BOX 9.3

SHIELDING COCHLEAR ROUND WINDOW IN MIDDLE-EAR ABSENCE

Many different surgical methods have been tried to shield the two windows from each other. In the type IV tympanoplasty operation, a tissue graft (for example fascia from a muscle) is placed to shield the round window from sound while the stapes footplate resting in the oval window is exposed to sound. (If the stapes footplate is removed and replaced by a graft, the same operation is called tympanoplasty type V.) It is important that the cavity created over the round window, known as the cavum minor is kept aerated, and the shield should be as rigid as possible to reduce the sound that reaches the round window as much as possible. If a successful shielding between the two windows were accomplished, the expected hearing loss would be equal to that of the loss of the transformer action of the middle ear, thus 25–30 dB. That ideal situation cannot be achieved surgically and in practice the results of such operations are moderate hearing loss of 35–40 dB for low frequencies and 20–25 dB at 2 and 4 kHz. (The results of such operations are usually expressed as the air-bone gap, namely the difference between the air conduction threshold and the bone conduction threshold; the bone conduction threshold represents the threshold of the cochlea.) Recently the physiological basis for such operations has been studied extensively (193).

However, the earphones also conduct sound by bone conduction. It was shown in Chapter 2 that the average cross talk for earphones such as TDH 39 with MX41/AR was approximately 60 dB between 0.5 and 4 kHz and less for insert earphones (Fig. 2.8A in Chapter 2). This is the reason why it may be necessary to mask the ear with the better hearing when testing the hearing of a person with much larger hearing loss in one ear than in the other. If an earphone can stimulate the opposite ear by bone conduction, bone conduction must be equally effective in stimulating the cochlea on the side where the earphone is applied. In a patient with a conductive hearing loss that exceeds the conduction of cross talk the sound that reaches the cochlea by bone conduction will be stronger than the sound that reaches the cochlea by air conduction. Bone conducted sound generated by the earphones that are used therefore sets the limit for the maximal conductive hearing loss that can be measured (approximately 60dB for TDH39 and 70dB for modern insert earphones). That means that conductive hearing loss that is greater than that value cannot be assessed and conductive hearing losses that exceed these values may therefore be underestimated. The bone-conducted sound that earphones deliver varies from individual to individual (see Fig. 2.8A). The difference between the air conducted and the bone conducted sounds delivered by the TDH 39 earphone can be as small as 50 dB (Fig. 2.8A) and the bone conducted sounds delivered by insert earphones may be only 60 dB below the air-conducted sound (Fig. 2.8A).

These matters should be taken into account when interpreting hearing loss in individuals with large conductive components of hearing loss. The hearing loss associated with hearing without the middle ear, and that due to ear-canal atresia where the measured air conduction threshold exceeds 60 dB, may be higher than what it appears to be from conventional audiometry because of the bone-conducted sound. Especially old data obtained using TDH39 earphones should be viewed with that in mind.

3.5. Implications of Impairment of Conduction of Sound to the Cochlea

Middle-ear disorders can impair speech communication to an extent that it interferes with learning in school. Impairment of sound conduction to the cochlea causes deprivation of input to the auditory nervous system and that can cause expression of neural plasticity that changes processing of sounds in the auditory system. Input is thus important for the normal childhood development of the auditory system. Therefore disorders that cause impairment of sound conduction to the cochlea that occur early in life such as middle-ear infections can have lifelong consequences. These are all reasons why it is important to treat the hearing deficits in childhood caused by middle ear problems. It has been the topic of discussion how and if middle-ear infections should be treated, but there is little doubt about the benefit from treating the hearing deficit. This can be done independently of how the underlying middle ear problems are treated, using ventilation (PE) tubes inserted in the tympanic membrane.

4. PATHOLOGIES OF THE COCHLEA

The decline in hearing with age, known as presbycusis, is the most common form of cochlear pathology. Presbycusis is associated with detectable morphological changes in hair cells, mainly the outer hair cells [139] and subsequent reduced function of outer hair cells. Other causes of hearing impairment such as exposure to noise, administration of drugs such as certain antibiotics [94], certain diuretics, aspirin [234], quinine and many other substances cause similar morphological changes to hair cells [112]. Cochlear pathologies are also present in diseases such as Ménière's disease [199]. Distension of the basilar membrane that occurs as a result of cochlear hydrops is most likely involved in causing the low frequency fluctuating hearing loss in Ménière's disease that is typical for the disease, at least in its early stages.

The most obvious signs of such insults are reduced sensitivity, namely elevated pure tone threshold. Speech discrimination is not profoundly affected by moderate cochlear hearing loss and it is closely related to the pure tone hearing loss. Recruitment of loudness is common in people with cochlear hearing loss (see Chapter 10).

Earlier, hearing impairments from pathologies of the cochlea have primarily been regarded to be the effect of injuries to hair cells, mainly outer hair cells such as is typical for noise induced hearing loss, hearing loss from ototoxic antibiotics, Ménière's disease, etc. However, it has become evident that disorders of the cochlea can indirectly influence the function of the central nervous system and symptoms of, for instance, injuries to cochlear hair cells are likely to have components that originate in the central nervous system, causing deprivation and changed balance between inhibition and excitation. Such changes can in themselves cause symptoms but they may also promote expressions of neural plasticity, which in turn can cause symptoms such as hyperactivity, changed dynamic range, or redirection of information (discussed in Chapter 10). The fact that the morphological changes in the cochlea are so apparent has made these disorders to be known as cochlear disorders. The advances in our knowledge about the disorders of the auditory system have now blurred the distinction between cochlear and nervous system disorders.

It had earlier been assumed that such impairment of hearing as presbycusis was caused by impairment of the function of cochlear hair cells as a part of the normal aging process. However, recent studies have shown a more complex cause of such hearing impairment, involving changes in the function of the nervous system.

The impairment of speech perception, increased masking and recruitment of loudness are not only related to the impairments of the function of cochlear hair cells that are evident as a result of aging (presbycusis), exposure to loud noise, administration of ototoxic substances and various disease processes but it is also a result of the changes in the function (and morphology) of the central auditory nervous system. The anatomical location of the abnormal function that causes these symptoms is thus not only the cochlear hair cells, but changes in the function of the auditory nervous system also contribute to the symptoms.

The symptoms and signs (impaired hearing) that occur after insults such as exposure to traumatizing noise, ototoxic antibiotics [315], and the processes of normal aging causing decline in hearing sensitivity (presbycusis) are thus caused by a combination of deficits in the auditory periphery (cochlea) and the effect of changes in the central auditory nervous system [190, 213].

4.1. General Audiometric Signs of Cochlear Pathologies

Most forms of injury to cochlear hair cells affect the hearing threshold at high frequencies more than low frequencies. Typically, the hearing loss in cochlear pathologies begins at the highest frequencies that are tested by clinical audiometry (8 kHz) and progresses towards lower frequencies as it becomes more severe. The normal hearing range in humans extends to about 20 kHz but the hearing threshold is normally not tested at frequencies above 8 kHz. Therefore data about hearing loss in the frequency range above 8 kHz is sparse and the beginning of the progression of hearing loss therefore usually escapes detection. Hearing loss caused by exposure to noise is an exception because it normally affects the hearing threshold at 4 kHz more than other frequencies. Hearing loss from Ménière's disease mostly affects low frequencies. Certain forms of hereditary hearing loss mainly affect the mid-frequency range ("cookie-bite" audiogram).

Speech discrimination normally only becomes affected when the threshold elevation at frequencies below 2 kHz becomes noticeable, but the relationship between pure tone thresholds and speech discrimination varies considerably between different individuals. The average decrease in speech discrimination obtained in many individuals (Fig. 9.9) can therefore only serve as a guide in estimating speech discrimination in an individual. Only hearing loss of the high frequency type was included in the data shown in Fig. 9.9 because that is the commonly occurring type of hearing loss associated with injuries to cochlear hair cells.

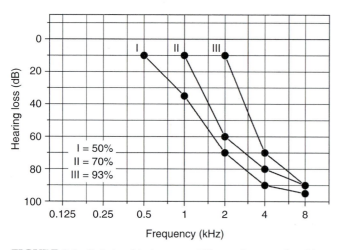

FIGURE 9.9   Relationship between different degrees of cochlear hearing loss and speech discrimination (data from Lidén, 1985).

The relationship between pure tone threshold and speech discrimination may be different in other forms of cochlear hearing pathologies.

The reason that speech discrimination is only affected to a small degree when the hearing loss is moderate is that only outer hair cells are affected and that only impairs the cochlear amplifier, leaving sensory transduction unaffected. Recall from Chapter 3 that the cochlear amplifier is mainly effective at low sound levels and its effect is small at physiological sound levels. Since speech tests are performed at physiologic sound levels, impairment of the cochlear amplifier has little effect on speech discrimination. Also, most speech tests are done in quiet conditions. If they were done in noisy conditions [303], deficits due to cochlear hearing loss would often appear more severe and perhaps more similar to what the individual experiences in normal everyday listening conditions. Individuals who have greater impairments of speech discrimination than the average person with the same hearing loss may have impairments of other structures in addition to cochlear impairments. Thus, changes to the auditory nerve that are present in some patients with presbycusis may be one such factor that is responsible for poor speech discrimination. The extent of injury or damage to cochlear hair cells from the same insult varies among individuals and that contributes to the individual variations in hearing impairment from seemingly identical conditions. The fact that cochlear pathology affects the function of the auditory nervous system is a further source of complexity and one which may contribute to the individual variations in hearing impairment. These changes in the function of the auditory nervous system are, however, not sufficient to cause detectable changes in auditory evoked potentials such as ABR. The ABR is within normal limits in patients with moderate degrees of cochlear pathologies and the threshold of the acoustic middle-ear reflex is within normal limits.

4.2. Age-related Hearing Loss (Presbycusis)

Presbycusis is commonly associated with degeneration of cochlear hair cells, mainly outer hair cells in the basal portion of the cochlea, and age-related hearing impairment was earlier assumed to be caused by the effect of these morphologic changes in cochlear hair cells. In that way, the changes are similar to those seen in other injuries to the cochlea such as those from noise exposure. Outer hair cells are affected most and the changes begin in the basal end of the cochlea, spreading toward the apex as the condition progresses. The individual variations are large (Fig. 9.11) [227, 228] and hereditary factors are important (more from mothers than fathers [98]).

Recent studies have indicated a much more complex pathology, involving the central auditory nervous system [349]. Age related changes may occur in more centrally located parts of the auditory nervous system because the function of the nervous system may change as a result of the change in input from the cochlea caused by loss of hair cells (through expression of neural plasticity, see [213]). Both human [149] and animal studies [128] have indicated that age-related hearing loss is preceded by changes in the middle olivocochlear pathway that connects cells in the superior olivary complex with outer hair cells.

Age-related changes often include morphological changes in the auditory nerve [309] in addition to pathologies of cochlear hair cells. Thus, it has been shown that the distribution of fiber diameters of auditory nerve fibers widens with age [309] (see also Chapter 5). This causes a wider distribution of conduction velocities of auditory nerve fibers and thus a decreased temporal coherence of the auditory nerve impulses that arrive at the cochlear nucleus. These morphologic changes in the auditory nerve may be one of the reasons that temporal processing in the auditory system deteriorates with age [313]. Since temporal coherence in the auditory nerve seems to be important for discrimination of complex sounds such as speech, the changes in the auditory nerve may contribute to the deteriorating ability to discriminate speech that often occurs with age. Both ABR and the acoustic middle-ear reflex may be abnormal.

Presbycusis appears as a gradually sloping hearing loss towards higher frequencies. It is often regarded as a part of the normal aging process. Several studies have analyzed hearing loss as a function of age [228, 310].

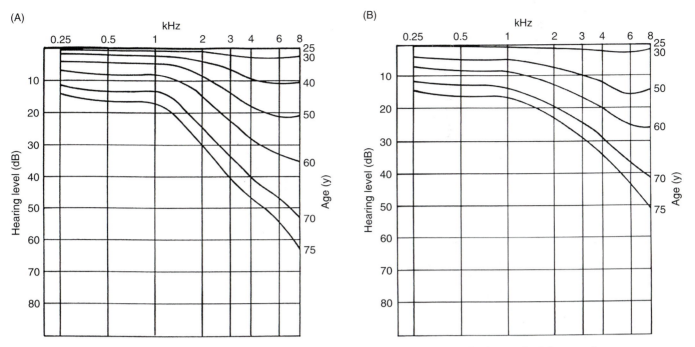

FIGURE 9.10   (A) Average hearing loss in different age groups of men. Results from eight different published studies based on a total of 7,617 ears. (B) Average hearing loss in different age groups of women. Results from eight different published studies based on 5,990 ears (reprinted from Spoor, 1967).

The results show that the high frequency hearing loss increases with age (Fig. 9.10). The data in Fig. 9.10 show the averages of eight published studies comprising data from more than 7,600 men (Fig. 9.10A) and almost 6,000 women (Fig 9.10B) (310). Such studies rarely define which criteria were used for inclusion in the studies and it is therefore possible that the results may reflect hearing loss that is caused by factors other than age. In large population studies such as those compiled by Spoor [310], many individuals have been exposed to noise, which results in greater hearing loss at 4 kHz than other frequencies (see p. 219).

A cross-sectional and longitudinal population study of hearing loss and speech discrimination scores in an unselected population of individuals aged 70 (Fig. 9.11) (228) showed that both these groups of individuals had high speech discrimination scores (Fig. 9.12), somewhat lower in men than women. Exposure to noise affected hearing in men more than in women and that appears as a slightly greater hearing loss for high frequencies. The reason for this gender difference may be that many men had noise induced hearing loss, but there may be other reasons related to hormonal influence on the progression of age-related changes in the cochlea and possibly differences in the age-related change in neural processing of sounds. M.B. Møller [228] also provided the distributions of hearing loss among the individuals of the study (Fig. 9.11) and

these data show that the hearing loss in the men and women studied is far from being normally (Gaussian) distributed. For low frequencies, the distributions are skewed with a long tail towards larger hearing loss while the distribution of hearing loss for higher frequencies is more symmetrical although it is far from being a normal distribution. The mean value and standard deviation are therefore not adequate descriptions of the hearing loss as a function of age. Despite that, mean and median values of hearing loss are commonly the only data provided in population studies of hearing [310].

Age related hearing loss (presbycusis) is associated with morphological changes in the cochlea in the form of loss of outer hair cells. As for other causes of cochlear impairments (noise exposure and ototoxic drugs), the loss of outer hair cells is more pronounced in the basal portion of the cochlea, thus affecting the cochlear amplifier for high frequency sounds more than for low frequency sounds. Loss of outer hair cells is the most obvious change, and it has received more attention than other changes, but there are also changes in the auditory nerve, and the variations in fiber diameter of the axons in the auditory nerve increases with age (Fig. 5.3).

Evidence of age-related changes in the function of the auditory nervous system such as changes in synthesis of inhibitory neurotransmitter such as gamma butyric acid (GABA) have been presented [44].

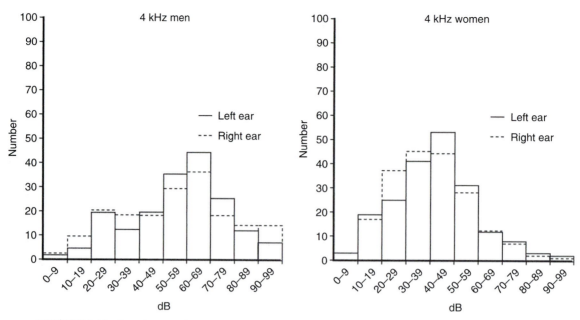

FIGURE 9.11 Distribution of hearing loss at different frequencies from a cross-sectional population study of hearing in people of age 70; men and women. Solid lines represent left ear and dashed lines represent right ears (data from Møller, 1981, with permission from Elsevier).

Expression of neural plasticity from reduced high frequency input from the cochlea may cause functional changes in the nervous system [190]. There is also evidence of changes in the function of the corpus callosum, affecting binaural hearing, and perhaps impairing the ability to fuse sound from the two ears [49, 137].

Some unexpected results of animal experiments have shown that the progression of age related hearing loss can be slowed by sound stimulation [328] (see p. 237).

That the progression of sensorineural hearing loss can be slowed has only been shown in a few studies because of the obvious difficulties in performing

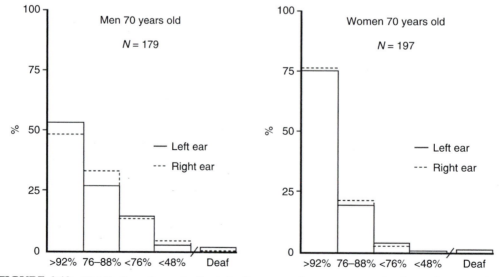

FIGURE 9.12 Distribution of speech discrimination scores from a cross-sectional population study of hearing in people of age 70. The speech discrimination scores were obtained using phonetically balanced word lists presented at 30 dB SL or at the most comfortable level. Solid lines represent left ears and dashed lines represent right ears (data from Møller, 1981, with permission from Elsevier).

BOX 9.4

EFFECT OF AAE ON AGE-RELATED HEARING LOSS

Experiments by Willott and co-workers [328, 349] in strains of mice that have early deterioration of hearing have shown that low level sound stimulation (augmented acoustic environment [AAE]) can reduce or slow the age related hearing loss in these animals. The mouse that these investigators used, DBA/2J, had progressive hearing loss from early adolescence.

---

controlled studies. This type of hearing loss is similar to presbycusis and is primarily a result of degeneration of cochlear hair cells.

The mechanisms for that reduction in hearing loss are unknown but several possibilities have been suggested [328], such as neural activity in the cochlear efferents that could affect outer hair cells, effects on neurotrophin action, effects on some unknown factors that are elicited by stimulation (excitotoxicity), regulation of certain genes, and possibly an effect of intracellular calcium concentration. It is important to point out that it is the progression of hearing loss that is affected (slowed) but the degenerative process does not seem to be reversed by such sound exposure. The fact that the progression of this kind of hearing loss can be reduced means that appropriate sound stimulation can actually affect cochlear degeneration. In the past it has been the negative aspects of exposure to sound that have been studied, and it is only recently that it has been shown in a few studies that there are also positive aspects of sound exposure. Thus as more knowledge about age related changes accumulate, it appears that presbycusis is more complex than just normal age related changes of cochlear hair cells.

4.3. Noise Induced Hearing Loss

Noise induced hearing loss (NIHL) is normally associated with noise exposure in industry and thus thought of as a product of modern civilization. It is mainly thought of as being caused by injury to cochlear hair cells but as our knowledge about disorders of the auditory system increases it has become evident that the effect of noise exposure is complex. It has been mainly the loss of hearing sensitivity that has been studied but NIHL has many other effects on hearing. Tinnitus may accompany any of the different forms of cochlear hearing deficits but it is more common in NIHL and in fact most incidences of tinnitus are associated with NIHL (see Chapter 10).

The effect on the cochlea in NIHL has been studied extensively and it was for a long time believed that the

morphological changes in the cochlea could explain the changes in hearing. However, it has more recently become evident that the effect of exposure to traumatic noise also causes both morphological and functional changes in the auditory nervous system. Expression of neural plasticity plays an important role in creating the symptoms from the auditory nervous system.

Exposure to a moderately loud noise causes hearing loss that decreases gradually after the end of the noise exposure. The hearing threshold may return to its normal value after minutes, hours or days depending on the intensity and duration of the noise exposure and the individual person's susceptibility to noise exposure. Exposure to noise above a certain intensity and duration results in hearing loss that does not fully recover to its pre-exposure level. This remaining hearing loss is known as permanent threshold shift (PTS). Hearing loss that resolves is known as temporary threshold shift (TTS).

Hearing loss caused by noise exposure affects high frequencies more than low frequencies. The audiogram of a person with noise induced hearing typically has a dip at 4 kHz (Fig. 9.13) and the hearing threshold at 8 kHz is better than it is at 4 kHz, at least for moderate

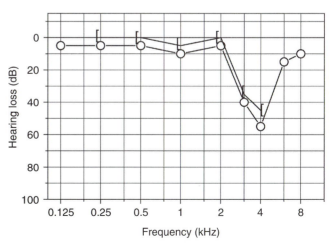

FIGURE 9.13 Typical audiogram for an individual who has suffered noise induced hearing loss (data from Lidén, 1985).

degrees of noise induced hearing loss. This distinguishes noise induced hearing loss from age related hearing loss (presbycusis), which results in threshold elevation that increases with the frequency (Fig. 9.10). The 4 kHz dip is more or less pronounced depending on the noise exposure and it is most pronounced in individuals who have been exposed to impulsive noise, thus noise with a broad spectrum.

The amount of acquired hearing loss depends not only on the intensity of the noise and the duration of exposure but also on the character of the noise (frequency spectrum and time pattern). The hearing loss from noise exposure is thus distinctly related to the physical characteristics of the noise exposure but great individual variations exist. The combination of noise level and duration of exposure is known as the immission level and it is used as a measure of the effectiveness of noise in causing PTS. However, the PTS caused by exposure to noise with the same immission level shows large individual variations (Fig. 9.14) [33].

Exposure to pure tones or sounds with a narrow spectrum causes the greatest hearing loss at about one half octave above the frequency of the highest energy of the sound. The reason for this half octave shift is most likely the shift of the maximal vibration of the basilar membrane towards the base of the cochlea with increasing sound intensity (see Chapter 3). Exposure to loud noise is expected to cause the most damage to hair cells at the location on the basilar membrane where the noise gives rise to the largest vibration amplitude. That means that the most damage is done at a location

that is tuned to the frequency of the maximal energy of the noise at the intensity of the noise. The location of maximal vibration amplitude is not the same for high intensity sounds as for sounds at the threshold used to measure the hearing loss. This is because the frequency to which a certain location along the basilar membrane is tuned shifts along the basilar membrane with increasing stimulus intensity.

The audiograms obtained in individuals who have been exposed to many different kinds of noise have similar shape, but the 4 kHz dip is probably most pronounced for exposure to impulsive noise. Studies have shown evidence that the enhancement of sound from the resonance of the ear canal [246] is the cause of the selective damage. Ear-canal resonance amplifies sounds in the region of 3 kHz (cf. Chapter 2). That the greatest hearing loss from exposure to sound with their highest energy around 3 kHz occurs near 4 kHz can be explained by the half octave shift discussed above. The point on the basilar membrane that was tuned to 3 kHz at a high sound intensity (e.g., 90 dB) will be tuned to a higher frequency when tested near the threshold. This is why the largest threshold shift from exposure to 3 kHz sound occurs at a higher frequency, approximately 4 kHz.

Individual variation is a characteristic feature of all forms of hearing impairment including NIHL, but the reason for this individual variation in acquired hearing loss from similar noise exposure is not well understood. It is characteristic of NIHL that the same noise exposure causes different degrees of hearing impairment in different individuals (see Fig. 9.14). This individual variation in susceptibility to noise induced hearing loss has many sources. Genetic variations are one [61], and age and health status are also important factors that affect injury to hair cells from noise exposure. Drugs of various kinds most likely also increase susceptibility to noise induced hearing loss. Hearing loss of conductive type also affects the risk of NIHL [237]. Absence or impairment of the acoustic middle ear reflex results in increased hearing loss from noise exposure [356]. Ingestion of alcohol and other drugs that impair the function of the acoustic middle ear reflex (cf. Chapter 8) may also affect susceptibility to NIHL.

Numerous hypotheses have been presented but published experimental evidence is rare. Besides variability in the exposure conditions, genetic differences, age, gender, pigmentation, differences in the sound conducting apparatus, blood supply and innervation of the cochlea have all been suggested as causes of the variability in NIHL to the same noise exposure. The hypothesis that age is a factor in the observed variations in susceptibility to NIHL has been supported by studies in mice [120]. Other factors that affect NIHL include a history of sound exposure, as discussed below.

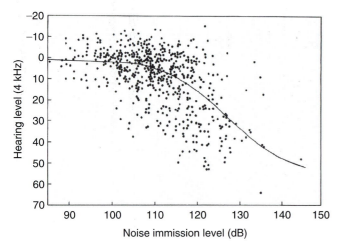

FIGURE 9.14   Hearing loss at 4 kHz as a function of noise exposure. Each dot represents the elevation in hearing threshold at 4 kHz for one ear. The solid line is the mean value. The horizontal axis represents both the sound level and the time of exposure (known as the noise immission level which is equal to the noise level (in dB) + 10 times the logarithm of the duration of exposure) (modified from Burns and Robinson, 1970, with permission from Her Majesty's Stationery Office).

BOX 9.5

FREQUENCY OF GREATEST NIHL DEPENDS ON EAR CANAL LENGTH

Studies of the correlation between the resonance frequency of the ear canal and the frequency of the greatest hearing loss in people with noise induced hearing loss [246] have shown that the mean resonance frequency of the ear canal in the group of people studied was 2.814 kHz and the maximal hearing loss occurred at 4.481 kHz. Assuming that the maximal energy of broad band noise occurred at the resonance frequency of the ear canal (2.814 kHz) and that the greatest hearing loss occurs at a frequency that is 1.5 times the frequency of the maximal energy of the noise exposure, then the maximal hearing loss would be expected to occur at 4.221 kHz. The mean frequency of maximal hearing loss was 4.481 kHz, thus very close to the expected value. This study also showed a high correlation between ear-canal resonance frequency and the frequency of the maximal hearing loss in individuals.

Earlier studies [38], showed that extending the ear canal by a tube that caused the resonance frequency to decrease caused a similar decrease in the frequency of the maximal TTS in volunteers who were exposed to broad band noise. The greatest hearing loss (TTS) occurred at frequencies about one half octave higher than the frequency of maximal sound energy.

These studies thus support the hypothesis that the typical 4 kHz dip in the audiograms of individuals who have suffered noise induced hearing loss is a result of the resonance of the ear canal. (It has been pointed out [270] that the maximal transfer of sound power to the cochlea does not necessarily occur at the frequency of the ear canal resonance but depends on other factors that are frequency dependent, such as the transformation ratio of the middle ear.)

Much of the individual variations in NIHL in humans can be explained by genetic differences, environmental factors, and inaccuracies in determination of the level and the duration of the noise to which they were exposed. The noise level and environmental facts can be controlled in animal experiments in the laboratory. Animals can be exposed to noise in the laboratory in a much more accurate way than humans. When normal guinea pigs are exposed to noise the acquired hearing loss varies considerably (Fig. 9.15) [186].

The fact that different animals are affected to different degrees from the same insults is an indication of individually different genetic makeup. This assumption

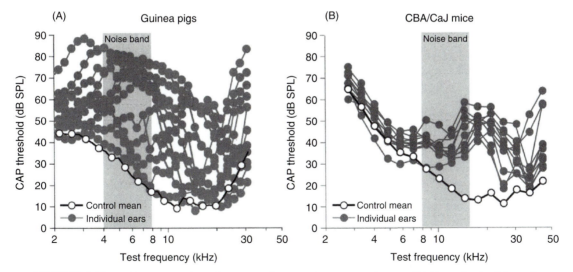

FIGURE 9.15 NIHL in animals with various degrees of genetic variations. (A) Data obtained in male guinea pigs (400–500 g); the exposure was a 2–4 kHz octave band of noise at 109 dB SPL for 4 h with a 1-week survival. The mean peak PTS was 35.1 dB at 7.6 kHz (SD of 21.33 dB) (reprinted from Maison, S.F. and Liberman, M.C. 2000. Predicting vulnerability to acoustic injury with a non-invasive assay of olivocochlear reflex strength. *J Neurosci* 20: 4701–4707, with permission from the Society for Neuroscience; Courtesy Charles Liberman. Copyright © 2000 Society for Neuroscience). (B) Inbred mice, males (23–29 g) exposed to octave band noise (8–16 kHz) at 100 dB for 2 h with a 1-week survival. The mean peak PTS was 38 dB at 17.5 kHz (SD of 4.06 dB) (reprinted from Yoshida and Liberman, 2000, with permission from Elsevier).

BOX 9.6

HYPERTENSIVE RATS ACQUIRE GREATER NIHL THAN
NORMOTENSIVE ANIMALS

Experiments in rats [26, 28] have shown that spontaneous hypertensive rats acquire more PTS from noise exposure than normotensive rats. However, hypertension caused by impairing blood supply to the kidney does not show such increased PTS [27]. Thus, hypertension in itself is probably not the cause of the higher susceptibility to noise induced hearing loss. The increase in susceptibility to noise induced hearing loss seen in the spontaneous hypertensive rats is probably related to factors that occur together with the predisposition for hypertension.

is supported by the finding that the variation is less when inbred animals are used in such experiments (Fig. 9.15) [353].

That genetics is important for acquiring NIHL is supported by the results of other animal experiments that have shown that animals with genetically related hypertension acquire more hearing loss than normotensive animal from the same noise exposure [26, 28].

The amount of hearing loss acquired by genetically identical animals from noise exposure under controlled laboratory conditions shows individual variation (Fig. 9.15). These variations in NIHL in genetically identical animals can be explained by difference in epigenetics[1] [140] or "noise in gene expressions" [260]. The variations that occur in the susceptibility to noise exposure between animals that are regarded to be genetically identical can be purely stochastic in nature or caused by differences in the internal states of a population of cells. Ongoing mutations are another source of variations that can manifest as differences in the physical characteristic of genetically identical organisms. Naturally, environmental factors can also affect the development of an animal.

These factors (epigentics and "noise" in gene expression) and perhaps other yet unknown ones, can explain the variations in the effect of insults such as noise exposure but it also explains why, for example, only one of two identical (homozogotic) twins acquires an inherited disease, despite both twins having exactly the same genetic set-up.

Other factors than genetics and epigentics may affect the susceptibility to noise exposure, such as hearing loss due to middle-ear pathologies. Middle-ear pathology acts as an ear protector and actually decreases the person's hearing loss from exposure to noise [237]. The conductive hearing loss does not affect hearing to any great extent at high frequencies but the protective effect from the low frequency conductive hearing loss against noise induced hearing loss is substantial. The result is that the acquired NIHL can be considerably less in the ear with conductive hearing loss than in the ear without conductive loss (Fig. 9.16).

NIHL has many similarities with presbycusis. It mainly affects outer hair cells and speech discrimination is little affected when the hearing loss is moderate and limited to frequencies around 4 kHz. It is mainly outer hair cells in the basal portion of the cochlea that are injured or totally destroyed, thus causing impairment of the cochlear amplifier. It is not known why hair cells located in the base of the cochlea are more susceptible to insults from noise exposure (and from ototoxic agents and aging, see pp. 216, 227) compared to hair cells in other parts of the cochlea. Pure tones or noise that has a narrow spectrum cause lesions within a restricted region of the basilar membrane.

Little damage to the stereocilia can be detected by light microscopic examination after noise exposure that produces 40–60 dB hearing loss [178]. In moderate degrees of cochlear hearing loss, inner hair cells are intact when examined by the light microscope. High resolution light microscopy (using Nomansky optics) and scanning electron microscopy (SEM) have shown that noise exposure causes a disarray of stereocilia on both inner and outer hair cells (Fig. 9.18) [178]. High-resolution light microscopy has revealed that the stereocilia of inner hair cells are altered to almost the same extent as were the stereocilia of outer hair cells after exposure to moderate levels of noise.

It has been shown that noise exposure causes disconnection between stereocilia of outer hair cells and the tectorial membrane. It should be noted that this is different from other types of insults to the cochlea such

[1]Epigentics: This term is used to describe activation and de-activation of genes. It is defined as the study of heritable changes in gene function that occur without a change in the DNA sequence. This mainly occurs in the uterus but can also occur after birth. It has become increasingly evident that epigenetic mechanisms such as DNA methylation, histone acetylation, and RNA interference, and their effects in gene activation and inactivation, are important factors in phenotype transmission and development [107].

(A)

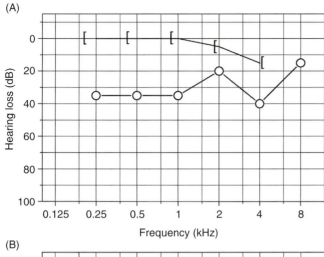

(B)

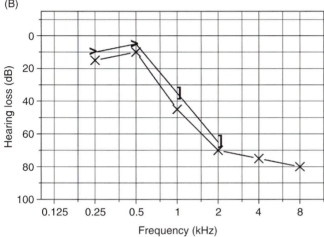

FIGURE 9.16 Audiograms of a welder exposed to shipyard noise for 30 years and who had conductive hearing loss in one ear (top audiogram). The bottom audiogram is from the ear without conductive hearing loss (data from Nilsson and Borg, 1983, with permission from Taylor & Francis).

as from ototoxic antibiotics, which affect the integrity of the cell bodies of hair cells.

Only exposure to extreme loud noise causes other structural damages besides the damage to hair cells. Thus exposure to sounds with levels in excess of 125 dB SPL seems to be necessary to cause mechanical damage to the cochlea of the guinea pig [308]. The level of noise exposure that causes structural damage varies between species and it may thus be different in humans from the values obtained in the guinea pig.

Hearing loss caused by injury to outer hair cells does not affect sensory transduction but rather the mechanical properties of the basilar membrane. Recall from Chapter 3 that the outer hair cells function as "motors" that increase the sensitivity and the frequency selectivity of the ear and that it is the inner hair cells that transduce the motion of the basilar membrane

and control the discharge pattern of auditory nerve fibers. Also, recall that the amplification caused by outer hair cells is most effective for sounds of low intensity and that it has little effect for sounds that are more than 50-60 dB above (normal) hearing threshold. This explains why hearing loss caused by impairment of the function of outer hair cells rarely exceeds 50 dB. It is also the reason why tests that employ high intensity sounds such as ABR and the acoustic middle ear reflex are largely normal in patients with hearing loss caused by malfunction of outer hair cells.

The most prominent physiological signs of noise induced hearing loss as revealed in animal studies are deterioration of the tuning of single auditory nerve fibers, loss of sensitivity at the fiber's CF and a downwards shift in frequency of the CF (Fig. 9.19) [51]. The widening of basilar membrane tuning after noise exposure is typical for loss of function of the active role of outer hair cells, that is to increase the sensitivity and frequency selectivity of the ear (cf. Chapter 3). The widening of the tuning of the basilar membrane broadens the "slices" of the spectrum of broad band sounds from which the cochlea provides information to the (temporal) analyzer in the central nervous system. This broadening may cause interference between different spectral components (impair "synchrony capture", see p. 109) and it may increase masking. The impairment of the cochlear amplifier from injury of the outer hair cells also impairs the amplitude compression that is prominent in the normal cochlea and that may be the reason why recruitment of loudness accompanies NIHL. The sensitivity of a single auditory nerve fiber for frequencies below a fiber's CF (in the tail region of the tuning curves) increases after noise exposure [179] and that may also contribute to the symptoms of NIHL.

While published reports of morphological changes of the cochlea as a result of noise exposure are abundant, few studies that concern the cause of these changes have been published. It is poorly understood how noise exposure causes the observed damage to the hair cells. It has been suggested that impairment of blood supply, or simple exhaustion of the metabolism could be the cause of the hair cell injury and destruction. These hypotheses have received little experimental support.

Oxygen free radicals have been implicated in causing injury to hair cells from noise exposure, aging and ototoxic antibiotics [94, 252]. It has been shown that the level of glutathione, an enzyme that defends cells against the toxic effects of reactive oxygen species, decreases with age and depend on the physiologic state of a person and on environmental challenges. It has been shown that oxygen free radical scavengers can reduce the effect of noise exposure on hearing. The best effect was obtained when a free radical scavenger

BOX 9.7

HAIR CELL LOSS, HEARING LOSS AND STEREOCILIA DAMAGE

Light microscopic studies of cochlear hair cells in animals that have been exposed to a moderately loud noise that causes hearing loss show loss of some hair cells, mainly outer hair cells (Fig. 9.17). Exposure to more intense sounds for longer periods causes more extensive damage and inner hair cells may be affected. An increment of only 5 dB in the intensity of the sound to which the animals were exposed caused a considerable increase in the injury of hair cells and in the PTS (Fig 9.17) [75]. Cell counts using surface preparation of the cochlea (cyto-cochleograms) reveal damage mainly to outer hair cells in the first row in an animal where the loss of sensitivity was moderate (30–40 dB) while high resolution light microscopy reveal abnormalities in stereocilia in both outer and inner hair cells (Fig. 9.17) [75]. An animal exposed to the same noise but studied at different times after noise exposure (right hand graphs in Fig. 9.18) showed much greater hearing loss and more extensive hair cell damage, including missing inner hair cells. There is a clear correlation between loss of hair cells and threshold shift at the characteristic frequency (CF) but there is considerable individual variation in the extent of the damage even in animals that are genetically similar and treated in similar ways.

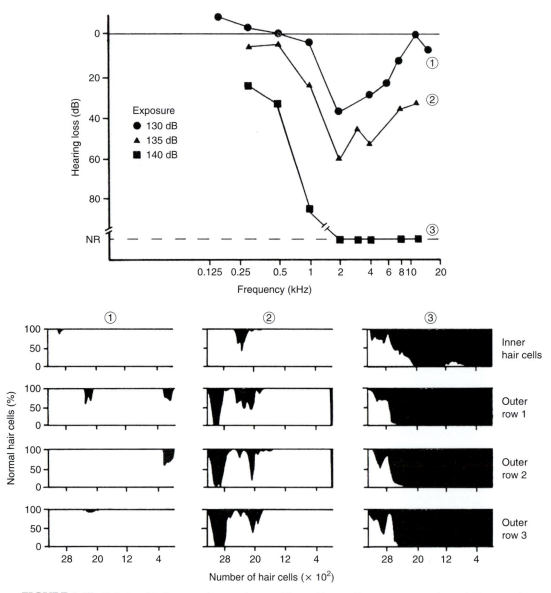

FIGURE 9.17 Relationship between hearing loss and loss of hair cells in cats exposed to 2 kHz tones for 1 h and three different intensities (reprinted from Dolan et al., 1975, with permission from Blackwell Publishing Ltd).

BOX 9.7   (*cont'd*)

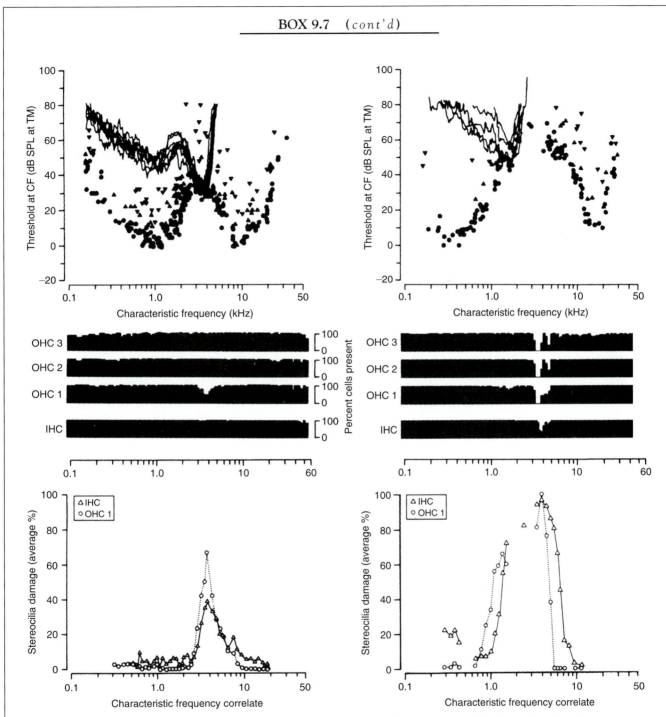

FIGURE 9.18   Results of recordings from single auditory nerve fibers and morphologic examination of the cochleae of two cats after exposure to 2 h of noise, 2 octaves wide, centered at 3 kHz and with an intensity of 115 dB SPL. The cats were examined, 620 (left panel) and 63 days (right panel) after noise the exposure. Upper graphs: sample tuning curves, centered at approximately 3.6 kHz of single auditory nerve fibers and threshold at CF. Middle graphs: cytocochleograms of the cochleae showing loss of hair cells. Bottom graphs: stereocilia damage in the first row of outer hair cells and inner hair cells as revealed by high resolution (Nomarsky) light microscopy with 100X objectives (reprinted from Liberman, 1987, with permission from Elsevier).

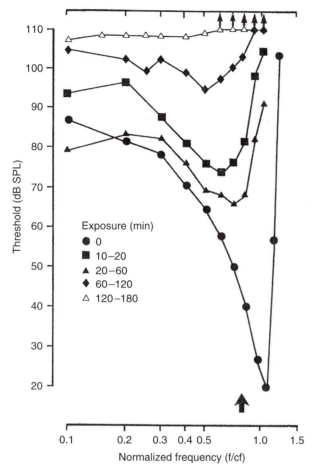

FIGURE 9.19 Deterioration in tuning and sensitivity of auditory nerve fibers as a result of exposure to pure tones. The data were pooled from many nerve fibers and the frequency scale is normalized. The arrows show the frequency of the exposure tones and the different curves represent different exposure times (reprinted from Cody and Johnstone, 1980, with permission from Elsevier).

function, or have a reduced function, without permanent injury occurring. That also explains the recovery of threshold shift after noise exposure of moderate degree (temporary threshold shift [TTS]). Only when the insult has reached a certain level does the recovery become incomplete and the result is permanent injury (PTS).

4.4. Implications of Hearing Loss on Central Auditory Processing

While NIHL is usually assumed to be caused only by the loss or injuries of outer hair cells it has been shown that NIHL is also associated with specific morphologic changes in the central nervous system [148, 205]. In addition to that, neural plasticity may result in functional changes in the nervous system because of the deprivation of input to specific groups of neurons that is caused by the injury to the cochlea [101]. This may alter the balance between inhibition and excitation, and that may cause hyperactivity (see Chapter 11).

Animal studies of evoked potentials recorded from the cerebral cortex showed enhancement of the responses after exposure to noise that caused hearing loss [318]. The authors concluded that their results indicate that the enhancement of the amplitude of the evoked potentials that are recorded from the auditory cortex is caused by changes in the processing of information in the central auditory nervous system. These changes are caused by expression of neural plasticity. Even exposure to sounds that do not cause hearing loss can cause changes in frequency tuning of neurons in the cerebral cortex of animals consisting of greater frequency selectivity and greater sensitivity to quiet sounds [88].

4.5. Modification of Noise Induced Hearing Loss

It has generally been assumed that exposure to loud sounds (noise) caused hearing loss only because it affected hair cells, either by mechanical stress or by changing the chemical composition inside or outside the hair cells. The finding that prior noise exposure can

was administered before the noise exposure but some effect was also achieved when it was administered after the noise exposure [252]. Oxygen free radicals are associated with activity of mitochondria, and the properties of mitochondria are inherited from mothers.

The finding that the cochlea can recover from noise induced hearing loss shows that hair cells can cease to

---

BOX 9.8

NOISE EXPOSURE CAUSES CHANGES IN THE COCHLEAR NUCLEUS

Animal experiments have shown morphological changes occur in the cochlear nucleus after noise exposure [204, 205]. Recordings made from the inferior colliculus shows signs of hyperactivity after noise exposure [320]. Several studies have shown that exposure to traumatizing noise alter frequency tuning of neurons in the auditory cortex.

affect the hearing loss from subsequent exposure to loud noise [42, 198, 302] brought a new and unexpected angle to the relations between the physical noise exposure and the acquired hearing loss. It became evident that the physiological mechanisms involved in noise induced hearing loss are more complex than earlier believed [171]. The finding that noise induced hearing loss is affected by prior stimulation and by simultaneous stimulation of the opposite ear may explain some of the individual variation in susceptibility to noise induced hearing loss.

It was shown by Miller et al. [198] in animal experiments that the TTS caused by noise exposure decreased gradually during repeated exposures. That was taken to indicate that the ear's susceptibility to noise exposure is affected by previous exposure. This "toughening" of the ear with regard to TTS from noise exposure has been extensively studied in a variety of animals and in humans by several investigators [42, 302] and it has been confirmed that it is also possible to reduce the effect of noise exposure on PTS by pre-exposure to noise. The exposure pattern of such "conditioning" is important for achieving this effect. Several studies have suggested this toughening of the cochlea against noise-induced injury is related to induced changes in the hair cells by the "conditioning" noise exposure.

The mechanism for such toughening is not completely understood but evidence has been presented from animal (guinea pig) experiments that both the medial and the lateral olivocochlear (efferent) system is involved [10, 171]. There is evidence that activity in the olivocochlear fibers can adjust the intracellular potential in the outer hair cells and thereby protect the hair cells from damage from noise exposure. Other possibilities that have been suggested involve intracellular pathways that can provide protection from noise-induced cellular damage in the cochlea. It has been suggested that pathways that regulate and react to levels of reactive oxygen species in the cochlea, stress pathways for the heat shock proteins, and neurotrophic factors may be involved [171]. However, none of these possibilities have been confirmed.

The concept of augmented acoustic environment has been pursued in other animal experiments [238], which showed that such "enriched acoustic environment," affects the tuning of neurons in the auditory cortex. Perhaps more surprising, it can alter the changes in the tuning that occur as a result of subsequent exposure to noise at levels that cause permanent damage to the ear. Such "enriched acoustic environment" also affect the development of changes in the cerebral cortex after exposure to traumatizing noise. Animal experiments have shown that NIHL can be reduced when the animals were placed in an enriched acoustic environment

after the noise exposure, thus, animals that were exposed to sounds at moderate levels had less hearing loss compared with similarly exposed animal that were placed for the same time in a quiet environment [238]. These authors also showed that the cortical re-organization that normally occurs after noise exposure was reduced in the animals that were exposed to such enriched acoustic environment after noise exposure.

These findings are important for two reasons: 1) Exposure to such enriched acoustic environment immediately after exposure to the traumatizing noise prevented the plastic tonotopic map changes in primary auditory cortex that normally occur after exposure to traumatizing noise; and 2) the hearing loss from the noise exposure was less than in animals that were placed in a quiet environment after the noise exposure. These studies also support the hypothesis that the nervous system is involved in noise induced hearing loss (see also tinnitus, p. 254).

4.6. Hearing Loss Caused by Ototoxic Agents (Drugs)

Many commonly used medications can cause hearing loss. Antibiotics of the aminoglycoside type can cause permanent hearing loss [94, 291]. Streptomycin (dihydrostreptomycin) was the first of this family of antibiotics found to cause hearing loss, but now commonly used antibiotics of the same family such as gentamycin, kanamycin, amikacin and tobramycin have also been found to be ototoxic but to a varying degree. Erythromycin and polypeptide antibiotics such as vancomycin have produce hearing loss but it is mostly reversible once the drugs are terminated. Commonly used agents in cancer therapy (chemotherapy) such as cisplatin and carboplatin are also ototoxic.

Aspirin (acetylsalicylic acid) can produce tinnitus and transient hearing loss but only at high dosages and the hearing loss normally resolves when the drug is terminated or the dosage reduced [234]. Administration of 5–10 g per day of acetylsalicylic acid (aspirin) can cause hearing loss and it can abolish the spontaneous otoacoustic emission [45]. Certain diuretic drugs such as furosemide and ethacrynic acid can produce transient hearing loss and tinnitus but they rarely cause permanent hearing loss. The same is the case for quinine. Hearing loss caused by these substances may be reversible when the drug treatment is terminated or it may be permanent.

These substances are used to treat diseases of different kinds and therefore they are almost always used in people with various kinds of illnesses that may increase the ototoxic effect. The experimental results upon which recommendations on the safe limits of such drugs are

BOX 9.9
_____

CARBOPLATIN AFFECTS INNER HAIR CELLS

While most ototoxic substances mainly affect outer hair cells, carboplatin causes injury mainly to inner hair cells in one animal species, the chinchilla, leaving outer hair cells intact. In the guinea pig, carboplatin injures outer hair cells, mainly in the basal region of the cochlea, thus similar to other ototoxic substances. This means that

the nature of the resulting hearing loss caused by carboplatin is different from that of other ototoxic substances because it affects neural transduction in the cochlea rather than the mechanical properties of the basilar membrane. Its effect in humans is unknown [70].

based were obtained in healthy individuals and these recommendations may not be applicable to humans with diseases for which these substances are administrated. The ototoxic effect of these substances varies widely among individuals and it is different in different animal species. There is evidence that older individuals have a higher susceptibility and individuals with diseases of various kinds may also be more susceptible.

Whether or not aminoglycoside antibiotics such as gentamycin and Kanamycin may cause hearing loss and vestibular disturbance depends on the way they are administered. Antibiotics that are ototoxic are often given in fixed dosages to treat life-threatening infections in individuals who are generally weakened and often have impaired kidney function. Many of these drugs are excreted through the kidneys and if kidney function is impaired, they are excreted more slowly than normal. The blood levels of the drug will therefore increase and become higher than anticipated if the excretion is slower than normal. The dosages used are designed to maintain a certain plasma level with normal excretion rates but in individuals with impaired kidney function such dosages may cause a pile up of the drug because it is excreted at a slower rate than it is administered. Since many of the ototoxic drugs are also nephrotoxic, a vicious circle may result from impaired kidney function that becomes aggravated by higher blood levels of an ototoxic drug. Monitoring of plasma levels of ototoxic antibiotics can reduce the risk of hearing loss considerably.

Antibiotics usually enter the cochlear fluid space from systemic administration but these substances may also enter the cochlear fluid space when administered in the middle-ear space, such as to treat infections. It is interesting that an ototoxic antibiotic, neomycin, that is not allowed for systemic administration because of its ototoxicity is approved for local administration including in the ear. Evidence has been presented that inflamed mucosa of the middle ear acts as a barrier for neomycin and prevents it from entering the

cochlea [305]. It is also possible that toxic substances generated by bacterial activity in the middle ear fluid can enter the cochlear fluid space through the membranes of the round and oval windows and cause injuries to hair cells [185, 305].

Ototoxic antibiotics may cause hearing loss by changing important biochemical processes leading to metabolic exhaustion of hair cells and that can eventually lead to cell death. It is generally assumed that oxygen free radicals are involved in causing injuries to the cochlea by ototoxic substances [94, 191]. Attempts have been made to prevent the ototoxic effect of drugs by administration of substances developed to protect against the effect of radioactivity but such drugs also reduce the ototoxic effect of these antibiotics [247, 274] and so far practical application of this method has not been demonstrated.

Most ototoxic drugs induce hearing loss by injuring outer hair cells and thus impairing the function of the cochlear amplifier, in a similar way as occurs in presbycusis and in NIHL. Inner hair cells are usually unaffected. However, the effect of toxic substances such as salicylate is different from that of noise in that it affects the cell bodies of the outer hair cells, while noise also causes a decoupling between the outer hair cell stereocilia and the tectorial membrane. Hearing loss caused by ototoxic drugs seldom exceeds 50–60 dB and it usually begins at high frequencies and extends gradually towards lower frequencies as it progresses. Most drugs cause the greatest damage to hair cells in the basal region of the cochlea and the greatest hearing loss thus occurs at high frequencies. High frequency audiometry (i.e., determination of the pure tone threshold at frequencies above 8 kHz) may therefore detect a beginning hearing loss before it reaches frequencies that affect speech discrimination.

While the effect on the cochlear hair cells from ototoxic substances have been studied extensively, little is known about the subsequent effect on the function of the central nervous system. Impairment of the function of outer hair cells affects tuning in auditory nerve fibers. As we have discussed earlier, impairment of cochlear function also affect the function of the auditory nervous system.

Studies in animals have shown that administration of ototoxic substances and metabolic insults to the cochlea can affect cochlear frequency tuning [90] (see Chapter 6). Tuning of single auditory nerve fibers in animals that were treated with Furosemide shows similar changes to those caused by anoxia. Treatment with Kanamycin also results in deterioration of tuning of auditory nerve fibers in a similar way to that caused by metabolic insult to the cochlea (see Chapter 6, Fig. 6.7).

4.7. Diseases that Affect the Function of the Cochlea

Several diseases may affect the function of the cochlea. The most common disease that causes hearing impairment is Ménière's disease. Hearing impairment may also result from hereditary causes. Infectious diseases such as meningitis and certain viral infections can also cause destruction to cochlear hair cells, causing hearing impairment.

Ménière's disease is a progressive disorder that is defined by a triad of symptoms, namely vertigo with nausea, fluctuating hearing loss and tinnitus [199]. It is one of a few kinds of sensorineural hearing loss that affects hearing initially at low frequencies. The incidence of Ménière's disease is different in different geographic locations. A study made in Rochester, Minnesota, showed an incidence of 15.3 per 100,000 people with a small preponderance for women (16.3 vs. 9.3 for men) [352]. A study in Italy [47] showed an incidence of 8.2, and a study in Sweden [311] arrived at an incidence of 46 per 100,000 people. A part of this variation is probably caused by differences in the definition of the disease.

In the early stages of the disease the patient experiences acute attacks of vertigo, often preceded by brief aural fullness in the affected ear, hearing loss and tinnitus that may last from several hours to 24 h. Longer lasting symptoms of vertigo are caused by other disorders of the inner ear. Typically, hearing loss in the early stages of Ménière's disease affects only low frequencies and it fluctuates and increases during an acute attack (Fig. 9.20). The hearing returns to normal after each attack in the beginning of the disease but as the disease progresses, residual hearing loss from each attack accumulates and the hearing loss spreads to higher frequencies. After years of disease, some patients may experience 'drop attacks," i.e., sudden severe vertigo that occurs without warning, and which causes the patient to fall to the ground. Over time, hearing loss progresses and extends to higher frequencies; but

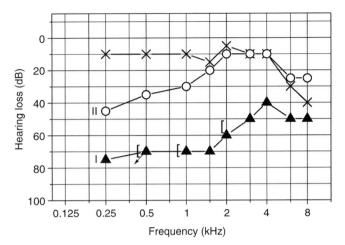

FIGURE 9.20 Typical audiogram from a person with Ménière's disease showing hearing loss during an attack (I) and between attacks (II) (data from Møller, M. B., 1994, with permission form Lippincott).

it rarely exceeds 50 dB. Speech discrimination is little affected in the early stages of the disease but may become affected in the advanced, late stage of the disease. The end stage of the disease, reached 10–15 years after its debut, is flat hearing loss of approximately 50 dB and speech discrimination scores of approximately 50%. The symptoms are initially unilateral but many patients experience bilateral symptoms after 10–15 years.

Ménière's disease can be diagnosed by the patient's history and standard audiological tests. Recently it has been hypothesized that the cochlear summating potential (SP) is abnormal in patients with Ménière's disease and recording of the SP is in common use for diagnosis of Ménière's disease and for monitoring treatment. The SP is the sum of the cochlear distortion products (Chapter 3) and its amplitude depends on several factors, one of which is endolymphatic pressure (or volume) and this is why it has been suggested as a way of detecting endolymphatic hydrops. The SP has been reported to be high in patients with Ménière's disease. SP varies considerably between individuals without signs of Ménière's disease and it may also vary from time to time in the same individual. The large variability of the SP in individuals without cochlear hydrops hampers the use of SP in diagnosis of disorders with hydrops including Ménière's disease. Some investigators have therefore expressed doubt about the significance of such findings and refer to the large individual variation in the SP. The value of recordings of SP as a diagnostic tool to diagnose Ménière's disease has therefore been set in question by some investigators [41, 83] whereas others [283] find that the SP anomaly and the ratio between the action potential (AP) and the SP are important signs of the disease.

BOX 9.10

SP/AP RATIO IS A MEASURE OF COCHLEAR HYDROPS

The ratio between the amplitude of the SP and the AP components are used as indication of cochlear hydrops. In response to clicks, the SP appears before the AP and the SP occurs at the same time as the cochlear microphonics (CM) making it difficult to distinguish the SP from the CM components of the electrocochleogram (ECoG). Some investigators [284] have made the SP appear more clearly by using clicks of high repetition rate as stimuli (Fig. 9.21), which reduces the amplitude of the AP component of the ECoG without affecting the SP. Subtracting the response to high stimulus rate from a response to low stimulus rate eliminates the SP component and thus shows a clean AP waveform. However, tone-bursts would be a more adequate stimulus than clicks for recording the SP. In the ECoG elicited by tone bursts, the SP appears as a plateau that occurs during and after the AP and it is thus easy to measure the amplitude of both AP and SP (Fig. 9.22). Note that the polarity of the SP is different in response to tones of different frequency.

It has been the ratio between the SP and the AP that has been used as an indicator of endolymphatic hydrops when Sass et al. [283] found a mean SP/AP ration of 0.26 in normal individuals with a standard deviation of 0.11. In patients with Ménière's disease, the mean SP/AP was 0.46 with a standard deviation of 0.15. The SP was significantly larger in Ménière's patients at 1 and 2 kHz but not at 4 and 8 kHz. Campbell et al. [41] used 6 kHz tones as stimuli and that may be the reason these investigators found little difference between the SP/AP ratio in patients with Ménière's disease compared with patients who did not have Ménière's disease. The sensitivity of transtympanic ECoG using the SP/AP ratio of the response to 1 kHz tone bursts was 82% and the specificity was reported to be 95% [283]. Thus the choice of stimuli affects the sensitivity if the SP/AP ratio as an indicator of endolymphatic hydrops, and that may be one of the reasons why different investigators have arrived at different values of sensitivity of this test.

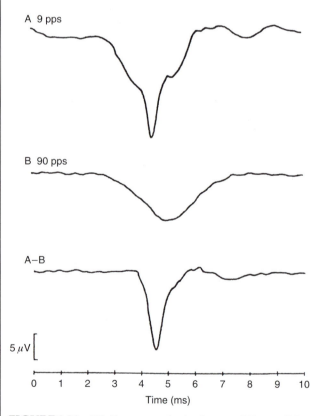

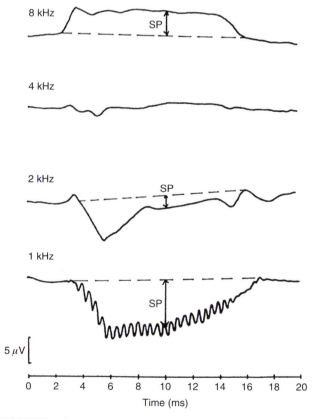

FIGURE 9.21 ECoG response obtained at two different click rates (A: 9 pps and B: 90 pps) in a patient with Ménière's disease, and the difference between the two responses (A-B) (reprinted from Sass et al., 1998, with permission from Blackwell Publishing Ltd).

FIGURE 9.22 ECoG in response to tone bursts of different frequencies obtained in a patient with Ménière's disease (reprinted from Sass et al., 1998, with permission from Blackwell Publishing Ltd).

Ménière's disease is probably not one disorder but rather a group of different disorders. Some variations of Ménière's disease have predominantly cochlear signs and some investigators have called these diseases cochlear Ménière's disease. Thus, while the classical definition of Ménière's disease is a triad of symptoms (fluctuating hearing loss, tinnitus, and vertigo), over time physicians have accepted patients with variations to that classical pattern and labeled these disorders Ménière's disease as well.

The fact that Ménière's disease has a distinct name while many disorders that have symptoms from the vestibular system have less distinct names or no names at all makes it attractive to use the name Ménière's disease for disorders that resembles Ménière's disease.

Recordings of the SP during operations for Ménière's disease may be of value because it involves comparison of the SP over a short time in the same individual and thus it is not subjected to the effect of individual variations. The SP is affected by vestibular nerve section [164] probably because of severance of the olivocochlear bundle that occurs in these operations. Neural activity in the olivocochlear bundle influences the function of hair cells, which contribute to the SP and that may explain the effect of severance of the olivocochlear bundle on the SP.

It is believed that the symptoms of Ménière's disease are caused by pressure (or rather volume) imbalance in the fluid compartments of the inner ear (endolymphatic hydrops). The hearing loss in Ménière's disease can be explained by a distension of the basilar membrane causing the largest distension where its stiffness is least, i.e., in the apical portion (Chapter 3). Permanent damage to hair cells does not seem to occur, at least not in the early stage of the disease. The fluctuations in hearing are assumed to be caused by varying degrees of endolymphatic hydrops in the cochlea. That is supported by studies by Kimura [150] who showed that blocking the endolymphatic duct in guinea pigs mimicked the signs of attacks of Ménière's disease.

Little is known about the effects of elevated perilymphatic pressure on the function of the ear and it is a matter of diverse opinion whether moderately abnormal pressure in the perilymphatic space causes any signs of pathology. It is, however, generally recognized that an increase in the volume of the endolymphatic space is associated with similar disturbances of hearing and balance as seen in Ménière's disease but it is not known whether the abnormal volume of inner-ear fluid compartments is a cause of the disorder or a result of the pathology of Ménière's disease.

Elevated endolymphatic volume causes Reissner's membrane to bulge and the basilar membrane to bow.

That is assumed to give rise to the low frequency hearing loss and perhaps tinnitus that are two of the triad of symptoms that defines Ménière's disease, at least in its early stage. If the volume of the endolymphatic space increases beyond a certain value Reissner's membrane may rupture resulting in fluids of widely different ionic composition mixing, which would have a dramatic effect on the function of the cochlea and the vestibular system [74]. It has been suggested that such an event increasing the concentration of potassium in the perilymphatic space was the cause of the most violent symptoms [74]. However, it is not known how the imbalance in pressure (or volume) comes about and why it only occurs at certain times.

That the pathophysiology of Ménière's disease is complex is evident from the finding that application of air puffs to the middle-ear cavity reduces the symptoms and normalizes electrophysiologic signs (SP) [68]. The applied air pressure affects the fluid pressure in the inner ear and thus presumably stimulates receptors in the labyrinth. The effect of that on the symptoms indicates that expression of neural plasticity probably is involved in generating the symptoms of Ménière's disease [213].

A few reports on single cases have found indications that the symptoms of Ménière's disease were related to vascular compression of the auditory-vestibular nerve roots [191]. However, these examples could have been misdiagnosed, disabling positional vertigo (DPV), which can be successfully treated by moving a blood vessel off the root of the vestibular nerve [231].

The pressure, or rather the volume, in the different fluid compartments of the cochlea is normally kept within narrow limits by mechanisms that are poorly understood [278]. It is known that imbalance of the volume in the endolymphatic and perilymphatic spaces causes malfunction of the cochlea and results in symptoms from both the auditory system and the vestibular system. Thus, proper balance in the pressure or rather the volume of the fluid in these compartments is essential to achieve optimal functioning of the cochlea.

It is not known what mechanisms keep the endolymphatic volume within its normal range but it seems reasonable to assume that pressure sensitive areas of membranes that limit the endolymphatic space may act as the sensors. Since the pressure in the perilymphatic space is closely coupled to that of the intracranial pressure (ICP) it seems unlikely that the pressure in the perilymphatic space can be regulated locally in the cochlea, at least in individuals in whom the cochlear aqueduct is patent. The role of the endolymphatic sac in pressure regulation in the inner ear is incompletely known but it is the target for some

BOX 9.11

NON-INVASIVE MEASUREMENT OF COCHLEAR FLUID PRESSURE

Measurement of pressure (or volume) in the cochlea has been done in animals for research purposes for many years, but it is only recently that it has become possible to measure intralabyrinthine pressure non-invasively. A method for measuring intralabyrinthine pressure that makes use of the effect of contractions of the middle ear muscles on the displacement of the tympanic membrane has been described [99, 187]. This method is based on the assumption that increased pressure in the perilymphatic space pushes the stapes footplate out of the oval window and that the displacement of the incus by contraction of the stapedius muscle will depend on how much the stapes is pushed out of the oval window. Displacement of the incus causes the tympanic membrane to displace and that can be measured as a small change in the air pressure in the sealed ear canal (Fig. 9.23) [336]. Normally, contraction of the stapedius muscle causes only a very small shift of the position of the incus as shown above for animals such as the cat or the rabbit (Chapter 2, Fig. 2.23). This is

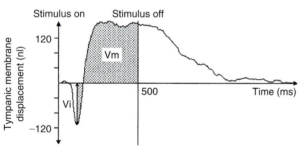

(A)   Normal pressure

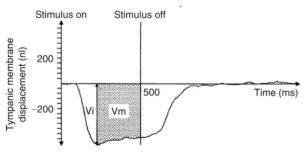

(B)   Hyperpressure

FIGURE 9.23   Illustration of how intracochlear (and intracranial) pressure can be measured by recording changes in the air pressure in the sealed ear canal (as a measure of the displacement of the tympanic membrane) during contraction of the stapedius muscle (reprinted from Wable et al., 1996, with permission from Springer-Verlag).

because contraction of the stapedius muscle normally causes the stapes to move perpendicular to the surface of the flat portion of the incudo-stapedial joint and that does not cause any movement of the incus and thus no displacement of the tympanic membrane. If the pressure in the perilymphatic space is abnormally elevated, the stapes tilts because the elasticity in the two ligaments of the stapes footplate is different. Contraction of the stapedius muscle then does not displace the stapes exactly perpendicular to the surface of the incudo-stapedial joint, but it will cause the incus to displace, and the tympanic membrane will move, and that results in a small change in the air pressure in the sealed ear canal. This test is used clinically to measure intralabyrinthine pressure non-invasively. The outcome of the test depends on fine details of the anatomy of the stapes and its suspension in the oval window, the incudo-stapedial joint and the orientation of its plane surface. This causes considerable individual variation in the displacement of the tympanic membrane from contraction of the stapedius muscle. The method is therefore best suited for measuring changes that occur over time in the same individual.

Measurement of the displacement of the tympanic membrane has also been proposed as a (non-invasive) method for measurement of intracranial pressure (ICP) or rather as an indicator of elevated ICP [89, 187]. The validity of this method for measuring ICP assumes that the perilymphatic space communicates with the intracranial space and that depends on the patentcy of the cochlear aqueduct (see Chapter 1).

Measurements of the change in the air pressure in the sealed ear canal from contraction of the stapedius muscle are technically difficult and the air pressure in the ear canal may change from other reasons such as pulsation of the blood. These unrelated changes act as background noise that interferes with measurements of the displacement of the tympanic membrane from contraction of the stapedius muscle. These difficulties may be overcome by using laser interferometry to measure the displacement of the tympanic membrane (mentioned in Chapter 2).

Since the method described above for detecting elevated intracochlear pressure or ICP relies on contraction of the stapedius muscle, the method is limited to individuals who have an acoustic middle ear reflex. Hearing loss of conductive type, lesions to the auditory nerve, presence of hypnotic drugs such as barbiturate, alcohol, anesthetics etc. are all factors that can affect or abolish the acoustic middle-ear reflex.

of the different treatments used in disorders that are believed to be caused by inner-ear hydrops for which Ménière's disease is one. The endolymphatic sac plays an important role in correcting imbalance of volume. The endolymphatic sac responds to endolymph volume disturbance and responds in opposition to volume increases and decreases. The endolymphatic sac can thereby correct volume disturbances caused by imbalance of the ion transport system in the labyrinth [278].

Treatment of Ménière's disease is mainly directed towards the vestibular symptoms. Vestibular nerve section was an early treatment used to relieve the vertigo in patients with Ménière's disease [95] and it is still often done [5, 299]. Other treatments aim at releasing the endolymphatic pressure and surgically establishing an artificial drainage of the endolymphatic sac (endolymphatic shunt [251, 292]). Modern treatments of Ménière's disease now include the use of infusion of Streptomycin or gentamycin (ototoxic antibiotics), into the middle-ear cavity [43] to destroy parts of the sensory epithelium. A method has been described to infuse gentamycin into the cochlea through a catheter that is passed through the tympanic membrane and the end of which is placed over the round window [298].

Medical treatment is successful in controlling the vestibular symptoms in 80% of patients with Ménière's disease but has little effect on hearing impairment and tinnitus. Acute treatment has consisted of intravenous Droperidol, atropine sulfate or diazepam (Valium). For long term treatment Valium, 2 mg, or alprazolam (Xanax), 0.25 mg B.I.D. (twice a day). Medrol dose-pak (a corticosteroid) has also been found useful for immune-mediated symptoms. Of the 20% of patients who do not respond satisfactorily to medical treatment, vestibular nerve section is effective in 90% of such patients [5].

The observation that changing the ambient pressure (using a pressure chamber) could influence the hearing threshold in patients with Ménière's disease [66] led to the development of a method for treatment of patients with Ménière's disease [67] that consists of applying pulses of air pressure to the inner ear through a device place in the sealed ear canal. Ventilation (PE) tubes in the tympanic membrane are a prerequisite for the use of this method. That such stimulation of the vestibular system has beneficial effect indicates as that expression of neural plasticity is involved in the development of the symptoms of the disease.

Thus, many different treatments are in use to treat Ménière's disease, but it is also a question of how much these treatments affect the normal course of the disease. The disease seems to be unpredictable in its short course but more predictable in the long term,

supporting the assumption that it is a complex disorder that is affected by many factors, probably including the autonomic nervous system and psychological factors. These assumptions are supported by the finding that some of the vestibular symptoms can be controlled by psychological counseling and controlled lifestyle with restricted diet [92]. Treatment of the vestibular symptoms of Ménière's disease has been summarized as avoiding caffeine, alcohol, tobacco and stress ("CATS") [5] and provides re-assurance to the patient.

4.8. Congenital Hearing Impairment

Congenital hearing disorders most often affect cochlear hair cells and result in hearing loss of a cochlear type. The hearing loss is usually bilateral and high frequencies are affected more often than low frequencies, but the audiograms may have widely different shapes. In some cases, the largest hearing loss in the mid-frequency range ("cookie bite" audiograms) (Fig. 9.24). The cause of most congenital hearing impairments is unknown, but conditions during pregnancy such as rubella or cytomegalovirus (CMV) infections can increase the risk of congenital hearing impairment. It has been shown that the gap junction protein connexin 26 is involved in many cases of congenital deafness [174].

Congenital hearing impairment may progress after birth and it may reach various degrees of severity. Hearing loss may accompany genetically related disorders. Rare congenital malformations include ear canal atresia and atresia of the internal auditory meatus [110]. Malformations of the internal auditory meatus are often accompanied by malformations of the inner ear. It is important to diagnose these malformations so

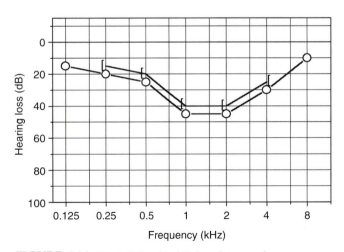

FIGURE 9.24 Typical "cookie bite" audiogram from a person with hereditary cochlear hearing loss (data from Lidén, 1985).

that the children with such causes of hearing loss do not receive cochlear implants. They should instead have auditory brainstem implants (ABI), where the auditory nerve is bypassed by directing the stimulation to the cochlear nucleus (see p. 277) [53].

Since the most common congenital hearing problems affect outer hair cells, newborns are now screened using recording of otoacoustic emission. Such screening will not find those with internal auditory meatus malformations, however.

Atresia of the ear canal can be detected by visual inspection whereas it is only recently that internal auditory meatus atresia has been recognized and diagnosed. Genetic factors account for at least half of all cases of profound congenital deafness [236]. Hearing loss occurs in malformation such as Mondini syndrome, Cogan's disease, Usher, Turners, Waardenburg and Pagett's disease.

4.9. Infectious Diseases

Infections diseases affect both the middle ear (see p. 207) and the cochlea. Bacterial meningitis was one of the most common causes of childhood hearing impairment before immunization came in common use. Several bacteria can cause meningitis and the hearing loss is a result of inflammation of the labyrinth that destroys hair cells and replaces the membranous labyrinth with fibrous tissue. The hearing loss is usually bilateral and permanent. Sometimes the cochlea fills with bone after meningitis, which makes it difficult to use cochlear implants to provide hearing. CMV infections can also cause congenital hearing impairment.

4.10. Perilymphatic Fistulae

Perilymphatic fistulae are small perforations that develop around the cochlear windows. They are most likely a result of slight weakening of the membranes that seal the cochlea fluid (perilymph). Such fistulae cause the perilymph of the cochlea to leak and the result is hearing loss and vestibular symptoms. Perilymphatic fistulae can appear spontaneously but they more often occur as a result of increased venous pressure from accidents, scuba diving, rapid descend by airplanes, extreme strain, etc., large or abrupt changes in middle ear pressure, as in barotraumas. Perilymphatic fistulae may present with similar symptoms and signs as Ménière's disease. The hearing loss is purely cochlear with normal or near normal acoustic middle-ear reflexes, and normal ABR.

Many children with hearing loss have hearing at birth but lose hearing at the time they begin to move around. It is possible that such hearing loss, often

called hereditary hearing loss, may be caused by perilymphatic fistulae that appear when the children begin to stand upright and thus experience fluctuations in the pressure of the inner-ear fluid. Perhaps the weakness cannot sustain normal fluctuations in the pressure of the inner-ear fluid that cause the hearing loss or deafness. It is possible to repair such leaks surgically although it is similar to repairing a leaking boat from the inside and therefore not always successful, at least not the first time.

Diagnosis of perilymphatic fistulae is a challenge and several tests have been designed to detect such leakage of cochlear fluid. Observation of vestibular responses (nystagmus) to a sudden change in air pressure in the ear canal is used to detect perilymphatic fistulae. The patient's eye movements are studied using either direct observation of eye movements or by using electrical recordings of eye movements (electronystagmography). ECoG recordings in connection with changes in posture have in animal experiments shown some promising results as indicators of perilymphatic fistulae. The changes in the ratio of the amplitudes of the summating potential and action potential elicited by click sounds or tone bursts (SP/AP) have been used as indicator of the presence of a fistula [40]. These methods are not precise indicators of fistulae and it is most important to take the patient's history into consideration.

4.11. Changes in Blood Flow in the Cochlea

Normal function of the cochlea depends on correct blood supply. The labyrinthine artery (see p. 16) is an end-artery and the inner ear has no collateral blood supply. Variation in perfusion may therefore give rise to abnormal function of the cochlea [235]. Thromboses or bleedings of the labyrinthine artery or surgical injury to the artery results in deafness on that ear.

4.12. Injuries to the Cochlea from Trauma

Injuries to the cochlea may be caused by trauma and skull fractures sometimes cause fractures of the cochlear bone causing total deafness. Auditory brainstem implants are now used to restore hearing in patients with bilateral traumatic cochlear injuries [54].

4.13. Sudden Hearing Loss

Sudden hearing loss (sudden deafness) [124, 263, 300] is characterized by sudden unexplained onset. The hearing loss is often total, fortunately almost always only in one ear. It can occur without any other symptoms, but often the patient has the feeling of a plugged ear or observes a "pop" in the ear before the hearing loss occurs. Tinnitus and imbalance or vertigo may accompany the loss of hearing at its onset. This is why the disorder is known as "idiopathic sudden sensorineural hearing loss" (SSNHL). It has been estimated that there are approximately 4,000 new cases of SSNHL in the USA every year [124].

SSNHL can occur during disorders such as myelogenous leukemia and other disorders where plasma viscosity is altered. It is often regarded to be caused by pathology of the cochlea although the exact anatomical location of the pathology is unknown. Perilymphatic fistulas may be one cause of sudden hearing loss and it has been suggested that SSNHL might also be a result of viral infections or interruption of the blood supply to the cochlea because it often results in total deafness in the affected ear. However, none of these causes have been proven and even suggestive evidence is rarely obtained.

There may in fact be several pathologies that can cause symptoms of sudden hearing loss, and the anatomical location of the pathology may not be the same in all patients with these symptoms. The symptoms are so characteristic that the disorders have often been treated as a single entity.

Many treatments have been tried [262, 263] but few have shown better results than no treatment. In approximately one-third of the patients hearing returns to near normal, one third will improve, and one third will remain deaf in the ear for life, with or without treatment. Treatment with antiviral agents and steroids [263] is common and studies have shown that such treatment is slightly better than no treatment in causing restoration of (some) hearing provided that it is done shortly after the hearing loss occurs [355]. Steroids injected directly into middle-ear cavity through the tympanic membrane have been reported to be more effective than systemic (intravenous) administration. Hyperbaric oxygen treatment [168] has also been tried. In general, treatment results are difficult to interpret because sudden hearing loss is a heterogeneous group of disorders, probably with different pathology. Adding to these difficulties is the fact that different investigators have selected their patients according to different criteria and treated the patients at different times after onset of the symptoms.

Recovery from SSNHL depends on many factors such as the patient's age and other symptoms that accompany the loss of hearing, such as vertigo, and it depends on the shape of the audiogram [167].

5. IMPLICATIONS OF HEARING LOSS ON CENTRAL AUDITORY PROCESSING

Morphological abnormalities of hair cells are often associated with changes in function of the auditory nervous system. The signs of such changes in function are deteriorated temporal resolution, (increased gap detection thresholds), and impaired speech discrimination [190]. The symptoms of presbycusis and other disorders that involve abnormalities of cochlear hair cells thus represent a combination of impaired function of the auditory periphery with altered function of the central auditory system. With a few exceptions, it has been difficult to detect morphological changes in the nervous system that could explain these functional changes. It has therefore been assumed that the changes in function may be caused by changes in synaptic efficacy and altered balance between inhibition and excitation. Expression of neural plasticity is most likely involved in causing these changes in function [213].

BOX 9.12

TEMPORAL BONES SHOW NO DETECTABLE ABNORMALITIES IN SSNHL

A study of temporal bones in 17 ears of individuals who were known to have had SSNHL [192] showed that there were no detectable histological abnormalities in two ears where hearing had recovered. Of the remaining 15 ears, 13 ears had loss of hair cells and supporting cells in the organ of Corti. One ear had loss of the tectorial membrane, supporting cells and stria vascularis. One ear had loss of auditory nerve fibers. Only one ear had signs of possible vascular cause of SSNHL. One ear of the 17 temporal bones that was acquired acutely during idiopathic SSNHL did not demonstrate any leukocytic invasion, hypervascularity, or hemorrhage within the labyrinth, as might be expected with a viral cochleitis. The authors of this study concluded that the most likely cause of SSNHL may be a pathologic activation of cellular stress pathways within the cochlea.

Studies in animals have shown that administration of ototoxic substances and metabolic insults to the cochlea can affect cochlear frequency tuning. Evans [90] in 1975 demonstrated that anoxia changes the frequency selectivity of single auditory nerve fibers. The changes in the tuning of auditory nerve fibers caused by anoxia consist of decreased sensitivity at a fiber's CF, broadening of the tuning and a shift of the CF towards lower frequencies (Fig. 6.7). These changes of the tuning of auditory nerve fibers have later been interpreted to be caused by impairment of the active function of outer hair cells (see Chapter 3). Tuning of single auditory nerve fibers in animals that were treated with Furosemide shows similar changes as those caused by anoxia. Treatment with Kanamycin also results in deterioration of tuning of auditory nerve fibers in a similar way as caused by metabolic insult to the cochlea (see Chapter 6, Fig. 6.7). Nerve fibers that did not respond to sound stimulation at all did respond to electrical stimulation of the cochlea, indicating that the nerve fibers were still excitable, thus showing the possibility of using electrical stimulation in cochlear prostheses in individuals who are deaf due to loss of hair cells.

It is not possible to record from single auditory nerve fibers in humans but estimates of the cochlear tuning in humans can be obtained by recording of the ECoG from the ear in connection with masking (two tone masking [59]). This method was used to study the effect of injuries to cochlear hair cells in humans and in animals (see Fig. 4.12) [114]. The results confirmed that cochlear tuning becomes broader when hair cells are injured by administration of Kanamycin. Comparison between the results obtained using electrophysiologic methods and psychoacoustic methods show good agreement, and the obtained tuning curves are similar to those obtained in recordings from single auditory

nerve fibers. Interestingly, simultaneous masking and forward masking gave different results in individuals with hearing loss, while in individuals with normal hearing the results of the two tests were similar.

Injuries to outer hair cells or other insults to the cochlea change the representation of sounds in the discharge pattern of auditory nerve fibers, thus changing the input to the central auditory nervous system. Changes in tuning of auditory nerve fibers from injuries to cochlear hair cells is the most apparent change but the altered balance between inhibitory and excitatory response areas of auditory nerve fibers also occurs because of insults to the cochlea, and these changes may have complex implications regarding the processing of sounds in the auditory nervous system.

5.1. Neural Components of Hearing Loss

Ototoxicity is normally associated with injury to the cochlea but some drugs affect neural processing of sounds. Thus the drugs salicylate and quinine that affect the cochlea (see Chapter 14) also change the function of the auditory nervous system. Neural discharges in neurons in the secondary auditory cortex (AII) that receive their input from the non-classical auditory system (see Chapter 5) are affected by administration of both salicylate and quinine [84]. The spontaneous activity of neurons in the AII area increased while administration of these drugs caused a decrease in the spontaneous activity of neurons in the primary auditory cortex (AI) and the anterior auditory field (AAF) that are parts of the classical auditory system (see Chapter 5). These drugs are known to cause tinnitus and these findings are therefore significant in understanding the pathophysiology of tinnitus and hyperacusis, which will be discussed in Chapter 10.

BOX 9.13

COCHLEAR TUNING DETERMINED USING MASKING

The use of masking to determine the tuning of the cochlea is based on the assumption that a weak tone activates only a few auditory nerve fibers. To obtain a tuning curve, the electrophysiologic response (AP, CAP from the auditory nerve or the ABR) to a weak tone (a few decibels above threshold) is recorded while a masking tone is applied. The intensity of the masking tone is adjusted so that the test tone evokes a reduced response (e.g., two-thirds

of the response without a test tone). The test tone and the masker are presented as short tone bursts, and the masker is usually applied immediately before the test tone (forward masking), but it can also be applied at the same time as the test tone (simultaneous masking). This procedure can be used in animals [59] as well as in humans [114]. A similar procedure can be used to obtain psychoacoustic tuning curves in humans [357].

5.2. Role of Expression of Neural Plasticity

The auditory nervous system possesses great abilities to change its function and there is increasing evidence that neural plasticity may be involved in such hyperactive hearing disorders as tinnitus and hyperacusis (Chapter 10). More recent studies indicate that neural plasticity may be involved in many more forms of hearing impairment than earlier believed. A variety of studies have shown that different brain functions such as processing of sensory information, pain and even motor functions can change more or less permanently as a result of altered input or lack of input. This means that the brain is plastic to a much greater extent than previously believed.

Earlier, it had been assumed that plastic changes could only occur early in life during ontogenetic development but it has become apparent that plastic changes can indeed occur in the adult nervous system, although to a lesser degree. The changes that occur are mainly a result of change in synaptic efficacy but outgrowth of new connections or degeneration of existing connections has also been demonstrated. The changes develop over time as a result of abnormal input such as overstimulation or deprivation of input but novel stimulation can also cause such changes. These changes may reverse spontaneously or become permanent after the processes that initially caused them have been eliminated.

The changes that occur over time in the function of the cochlea may result in deprivation of input to neurons that are tuned to high frequencies because of loss of hair cells in the basal portion of the cochlea.

Deprivation of input has been shown in many studies to promote expression of neural plasticity. Such expression may or may not include detectable morphological changes. The fact that mutant deaf mice have been shown to have fewer synapses and different synaptic organization in their auditory cortex show that deprivation can be associated with morphological abnormalities [243].

It is likely that slowly decreasing hearing such as occurs in presbycusis may have similar effects in reorganizing the central auditory system as NIHL and hearing loss from administration of ototoxic drugs.

Recent studies have provided evidence that other forms of hearing loss that have traditionally been associated with injuries to cochlear hair cells, such as presbycusis and drug induced hearing loss have neural components that are similar to the reorganization of the central nervous system discussed in connection with NIHL [281].

Animal studies have shown that deprivation of auditory input that results from hearing loss can affect auditory processing through expression of neural plasticity (see Chapter 10). The sound environment can alter maps in the cerebral auditory cortex [88, 146]. Evidence has been presented that exposure to some kinds of organized sounds such as music may be beneficial for the development of mental skills in the young individual and this has promoted the use of amplification in children with hearing loss and the use of cochlear implants in young children with severe hearing loss. A study in rats has shown that exposure in uteri to music improves the rats' ability to complete a maze test in a shorter time with fewer errors than animals exposed to white noise or silence [264].

Individuals with newly acquired cochlear injury have lower discrimination scores than individuals with similar hearing impairment of a congenital type when tested with distorted (low redundancy) speech [157]. The reason that people with congenital hearing loss have higher discrimination scores is probably an expression of neural plasticity, indicating that the auditory system can adapt to a poorly functioning cochlea.

Input to the central nervous system from the cochlea (via the auditory nerve) is not only excitatory but also inhibitory.

The changes in function of the central nervous system are assumed to be the result of changes in synaptic efficacy, change in the balance between inhibition and excitation, and degeneration of nerve fibers and nerve cells. Decrease in inhibition or increase in excitation may cause hypersensitivity and hyperactivity. Opening of dormant synapses[2] can result in rerouting of information. Such changes are known to cause certain types of pain [213] but only relatively recently has neural plasticity been implicated in disorders of the auditory system. Changes in the function of the cochlea can, however, also cause changes in the auditory nervous system that have morphologic correlates such as degeneration of axons. [148, 204, 205].

There is evidence that the gain of the central auditory pathways can be up- or down regulated to compensate for the amount of neural activity from the cochlea [281]. This is particularly pronounced for nerve cells that are tuned to high frequencies. Animal experiments have shown evidence of reduced inhibition after injury to cochlear hair cells (though noise exposure) [101] which over time may cause changes in the function of higher auditory centers such as the IC [320] (see Chapter 10). There are other causes for decrease in inhibition in the auditory nervous system. It has been shown that GABA in the central nucleus of the IC

---

[2]The term "dormant synapses" was coined in 1977 by Wall [339] as an explanation of certain forms of pain.

decreases with age, creating a deficit of an important inhibitory neurotransmitter [44]. This may cause an age-related shift in the balance between inhibition and excitation in the central nervous system. All these factors may contribute to the hearing impairment that is normally called presbycusis. These complex changes in the ear and auditory nervous system are also assumed to be involved in the development of tinnitus, which often accompanies presbycusis (see Chapter 10)

The old traditional view that the nervous system was hard wired has been replaced with a concept of dynamic connectivity and the ability to change function by changes in synaptic efficacy. Such changes can be brought about by novel stimulation of sensory systems, deprivation of stimulation, and by injuries of various kinds. The success in the use of prosthetic devices, such as cochlear implants and brainstem implants, has supported this view of flexibility of the function of the auditory system. That injury and loss of cochlear hair cells can cause profound changes in the structure and function of the central auditory system supports this hypothesis. Reorganization of frequency maps in the midbrain [113] and auditory cortex [146] and re-routing of information such as to non-classical auditory pathways [223] are other expressions of the ability of the nervous system to change its function. Such changes in function of nerve cells in the adult central auditory system are similar to the process of learning.

Many studies have shown that changes in the function of the auditory nervous system can be induced by deprivation of input [101], or overstimulation [119, 350], or by input that is abnormal in one way or another. Overstimulation may induce changes in the central auditory nervous system [148]. The changes reported in these early studies consist of increased sensitivity [100], altered temporal integration [101], broadening of tuning in cochlear nucleus units [119], or changes in the temporal pattern of responses from IC neurons [12, 341].

It has been known for many years that exposure to loud noise causes hearing loss. Until recently it has been assumed that such NIHL was the result of injury to cochlear hair cells. While a large part of NIHL can indeed be explained by injury to the cochlea [112] it is evident that exposure to loud noise can alter the function of parts of the auditory nervous system [148, 281].

Further, it has been shown that the amount of NIHL is affected by prior exposure to sound [42, 198, 302] and that this is likely to be caused by involvement of the central nervous system. It is thus evident that the hearing impairment from noise exposure is caused not only by alteration of the function of the cochlea. These changes in the auditory nervous system are only partly related to detectable morphological changes [205].

Babigian et al. [12] showed in 1975 that there is a central component to auditory fatigue and that the response from the inferior colliculus decreased more than the response from the ear and the auditory nerve during a period of temporary threshold shift caused by prior sound stimulation. Syka and co-workers [249] have shown that evoked potentials recorded from different places of the auditory nervous system are altered differently after exposure to loud noise [319]. The responses to sound stimulation were altered in different ways at three locations along the neural axis. Thus while the reduction in the amplitude of the evoked potentials was similar when recorded from the auditory nerve and the IC, the response from the cortex increased at a steeper rate as a function of sound intensity after noise exposure than before. Other investigators [279] found that noise exposure resulted in altered stimulus response functions of neurons in the IC. Mainly, the response increased at a steeper rate with increasing stimulus intensity after noise exposure, but these changes depended on the frequency that was tested and the spectrum of the noise to which the animals were exposed.

The amplitude compression in the cochlea might be impaired from noise exposure (see Chapter 3) but this should affect evoked potentials recorded from peripheral and central portions of the auditory nervous system equally. The paradoxical change in the evoked potentials recorded from the auditory cortex is likely to be a result of changes in synaptic efficacy somewhere in the ascending auditory pathways, brought about by expression of neural plasticity. Deprivation of input, perhaps due to neurons responding to high frequencies, or the changes in synaptic efficacy, could have been a result of the overstimulation during the noise exposure.

The changes in function in the central nervous system from noise exposure that can be demonstrated by electrophysiologic methods could be a result of change in synaptic efficacy or a change in the balance between inhibition and excitation. Some investigators have proposed that a disinhibition may occur in the auditory cortex after exposure to loud noise [317]. Morphologic studies by Morest and co-workers [204, 205, also 148] have shown that injuries to cochlear hair cells cause degeneration of not only auditory nerve fibers but also cells in the cochlear nucleus and that transneural degeneration of axonal endings occurs in the superior olivary complex and are thus signs of morphological changes.

Whatever the cause of these plastic changes are, the increased neural activity in the cortex may explain why some people with NIHL have an abnormal perception of the loudness of sounds and experience normal sounds to be unpleasantly loud and even perceive loud sounds as being unpleasant or painful (hyperacusis). The abnormal function of the auditory cortex that these changes reflect may also explain why some people with NIHL have lower than expected speech discrimination.

6. PATHOLOGIES FROM DAMAGE TO THE AUDITORY SYSTEM

Hearing impairments from disorders of the central nervous system are more difficult to assess than disorders affecting the conductive apparatus and the cochlea. Hearing impairments from disorders of the auditory nerve differs from hearing loss caused by cochlear impairments in the way that they affect the patient and in how such disorders alter the outcome of audiometric tests.

Symptoms from the auditory system from pathologies of central portions of the auditory nervous system manifest themselves with even more complex symptoms and signs than those caused by injury to the auditory nerve. Hearing impairments from disorders of the central nervous system are more difficult to assess than disorders affecting the conductive apparatus and the cochlea. The terms "psychogenic dysacusis," "functional deafness," or "non-organic deafness" have been used for disorders of the nervous system the (organic) cause of which could not be demonstrated by availably tests. That, however, does not mean that these disorders do not have an organic pathology and they are different from malingering where the patient knows that he/she can hear but pretends not to hear. If no pathology can be found with the methods of testing that are available, it does not mean that patients' complaints are false – it simply means that we are unable to find their cause with present knowledge and technology.

Diagnosis of lesions of the auditory nervous system requires more sophisticated audiological tests than diagnosis of lesions of the sound conducting apparatus and the cochlea. The patients' own description of his/her hearing loss is important for proper diagnosis of disorders of the auditory nervous system. Detailed knowledge about the anatomy and the function of the auditory nervous system is necessary in order to make an accurate diagnosis of central auditory disorders.

6.1. Auditory Nerve

Lesions to the auditory nerve are the most common cause of disorders of the auditory nervous system. Lesions to the auditory nerve may also affect the vestibular portion of the eighth cranial nerve and hearing deficits may thus be accompanied by symptoms from the vestibular (balance) system.

The most common disease process that affects the auditory nerve is vestibular Schwannoma, which is almost always associated with hearing impairment (and tinnitus, see p. 255). Other space occupying lesions in the cerebello-pontine angle are rare but any such lesion may cause symptoms and signs from the auditory nerve. Irritation or compression of the eighth cranial nerve from blood vessels may also cause symptoms such as tinnitus, hearing loss and vertigo. Surgical injury to the auditory nerve from operations in the cerebello pontine angle may cause hearing loss or deafness (together with tinnitus). Viral infections that affect the auditory nerve may cause hearing impairment.

It is assumed that normal speech discrimination depends on a high degree of temporal coherence. Normally, the conduction velocity of different auditory nerve fibers varies very little (see Chapter 5), ensuring a high degree of temporal coherence of nerve impulses that reach the cochlear nucleus. Mild injury to the auditory nerve makes nerve fibers conduct slower than normal and more severe injury can interrupt neural conduction in auditory nerve fibers. The reduced speech discrimination that is typical for injuries to the auditory nerve is assumed to be caused by impaired temporal coherence of nerve impulses that reach the cochlear nucleus. This occurs because the reduction in conduction velocity of auditory nerve fibers that occurs after injury is different for different nerve fibers.

The auditory nerve may be affected by disease processes such as vestibular Schwannoma and other space occupying lesions of the cerebello pontine angle. Viral infections may also affect neural conduction in the auditory nerve and close contact with a blood vessel can cause subtle changes in the function of the auditory nerve. The auditory nerve can be injured in operations in the cerebello pontine angle and head trauma may involve injuries to the auditory nerve. Close contact between the intracranial portion of the auditory nerve and a blood vessel (vascular compression) can cause symptoms from the auditory systems (mainly tinnitus, see p. 255).

Vestibular Schwannoma (earlier known as acoustic tumors) are benign tumors that grow (mainly) from the transition between peripheral (Schwann cell) myelin and central (oligodendrocyte) myelin (the Obersteiner-Redlich [OR] zone). Vestibular Schwannoma usually grow from the superior vestibular nerve (a portion of the eighth cranial nerve).

The earliest symptoms of vestibular Schwannoma are tinnitus and hearing loss in one ear, with a larger reduction in speech discrimination scores than occurs with a similar hearing loss of cochlear origin. It may be surprising that vestibular symptoms are not common.

BOX 9.14

VESTIBULAR SCHWANNOMA

Vestibular Schwannoma belong to a group of tumors known as skull base tumors as they occur in or near the base of the skull. Vestibular Schwannoma are benign tumors and they thus do not cause metastases. Vestibular Schwannoma make up 40% of all intracranial tumors. In most cases, the tumors originate from the superior vestibular nerve but they can also originate from the inferior vestibular nerve or the auditory nerve. The OR zone of the eighth cranial nerve is located inside the internal auditory meatus. Vestibular Schwannoma grow slowly [48]. The average growth rate of vestibular Schwannoma is 0.2 cm per year with a large individual variation. Some tumors may grow rapidly, or may decrease in size or even disappear.

In a study from Denmark, the incidence of vestibular Schwannoma was reported to be 0.78–0.94 per 100,000 [326]. These numbers are derived from diagnosed tumors and may thus be affected by the efficacy of diagnostic methods. There is also a geographical dependence, and since the incidence increases with age, it will depend of the longevity of a population.

The reason is that slowly decreasing vestibular function gives little or no symptoms because the other side's inner ear and other neural systems take over the function of the impaired vestibular system. This may be different when a tumor grows at a fast rate, especially if a person is in his or her sixties or older. This is because the loss of vestibular function cannot be compensated for by the remaining vestibular system in elderly individuals as well as it can in younger individuals. An elderly person who loses vestibular function on one side at that age typically has a persistent off-balance without vertigo.

The facial nerve travels in the internal auditory meatus together with the eighth cranial nerve but vestibular Schwannoma seldom cause noticeable signs from the facial nerve and impairment of facial function usually does not occur before the tumor is treated surgically. Injury to the facial nerve may occur as a result of surgical manipulations in connection with removal of a tumor but the risk of such damage can be reduced by the use of intraoperative neurophysiologic monitoring [212]. While surgical removal of these tumors is the most common treatment, gamma radiation therapy ("Gamma Knife") is also used to treat such tumors.

The presence of a vestibular Schwannoma must be ruled out in individuals who have asymmetric hearing loss. While the most common early sign of vestibular Schwannoma is tinnitus, only a few individuals with tinnitus have a vestibular Schwannoma and many people have asymmetric hearing loss without having a tumor. Recording of ABR is an effective test for vestibular Schwannoma because the tumor affects neural conduction in the auditory nerve. Recently it has been regarded to be more appropriate to use MRI scanning but the effectiveness of audiometric tests is equal to that of MRI. The combination of audiograms, acoustic middle-ear reflex test and ABR, using prolongation of the latency of peak V, is as good as MRI scans for diagnosis of vestibular Schwannoma.

When MRI scans are compared with other diagnostic methods, it should be noted that commonly made

---

BOX 9.15

AUDIOLOGICAL TESTS FOR DIAGNOSIS OF VESTIBULAR SCHWANNOMA

Selters and Brackmann [289] many years ago reported that ABR had a high sensitivity when compensations for age related changes in hearing threshold were made. More recently, Godey et al. [105] reported that the sensitivity of ABR alone is 92% and together with recordings of the acoustic middle ear reflex and caloric vestibular response, the sensitivity was 98% with all the false negative responses being in patients with tumors less than 1.8 cm in diameter [105]. These authors proposed ABR and the acoustic middle-ear reflex as first line screening tests.

estimates of the effectiveness of MRI scanning are subjected to misinterpretations. This is because MRIs are used both as a comparison between other methods and as the definitive proof of the presence of a tumor. MRI scans are thus used as the standard with which all other tests are compared. While negative MRI scans are interpreted to mean that the patient does not have a tumor, negative MRI scans cannot be confirmed unless the patient is operated on because there is no other way to find out if a patient in fact has a tumor. If a patient with a negative MRI scan has a tumor and other tests indicate the presence of a tumor, these results are normally judged to be false positive results. Only many years later, it may become known whether or not the MRI finding was a false negative result. Most positive MRI scans are verified because most patients with positive MRI scans for a vestibular Schwannoma are operated upon, although some are now treated by radiation ("Gamma knife"). Negative findings during operations are unlikely to be reported, however.

Thus, decisions to use MRI scans to rule out vestibular Schwannoma should be reconsidered and, instead, the use of ABR and acoustic reflex testing together with pure tone audiograms should be promoted as effective means to detect the presence of vestibular Schwannoma, the cost of which is much less than MRI scans. The use of ABR for diagnosis of vestibular Schwannoma requires interpretation of the ABR, and expertise for that is not always available.

The signs of hearing impairment from injury to the auditory nerve are more complex than those associated with cochlear injuries. Speech discrimination is reduced more than what it would have been from similar hearing loss caused by cochlear injury or conduction impairment. The audiogram often has irregular shapes with dips occurring at different frequencies (which appear clearly when the threshold is determined at half octave intervals) (Fig 9.25).

Anatomically, fibers from the apical portion of the auditory nerve are in the core of the nerve and fibers that are tuned to high frequencies are located superficially, at least in animals (the cat) [282] and the anatomical arrangement of the auditory nerve has been shown to be similar in humans [64]. The anatomical arrangement of the fibers in the auditory nerve may explain why injuries from compression of the auditory nerve such as occur in vestibular Schwannoma mostly affect hearing at high frequencies. The auditory nerve is longer in humans than in cats (2.5 cm [169] versus about 0.8 cm in the cat [97]) and it is twisted indicating that the anatomical arrangement of nerve fibers may be different in its intracranial course compared with its peripheral course (see p. 79). This is probably the reason why lesions to the auditory nerve

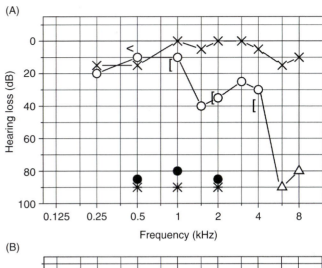

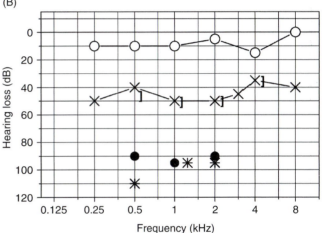

FIGURE 9.25 Examples of audiogram from patients with vascular compression of the auditory nerve. (A) Audiogram from a patient with vascular compression of the eighth cranial nerve on the right side near its entry into the brainstem showing a dip near 1.5 kHz (courtesy M.B. Møller). (B) Audiogram from a patient with hemifacial spasm on the left side showing a broad "cookie bite" dip around 2 kHz. Speech discrimination was 100% in both ears (data from Møller and Møller, 1985).

close to the brain stem, such as irritation from close contact with blood vessels or from surgical trauma causes hearing loss in the low to mid frequency range in humans (Fig. 9.26).

The threshold of the acoustic middle-ear reflex is elevated and the growth of the reflex response is impaired in individuals with hearing loss from auditory nerve injuries. The acoustic reflex may even be absent in patients with signs of injury to the auditory nerve. This is puzzling since mild injury to the auditory nerve is supposed to mainly affect the timing of the discharges and it might indicate that the acoustic middle-ear reflex depends on coherence of the nerve activity that reaches the cochlear nucleus. The growth of the amplitude of the reflex response is reduced in

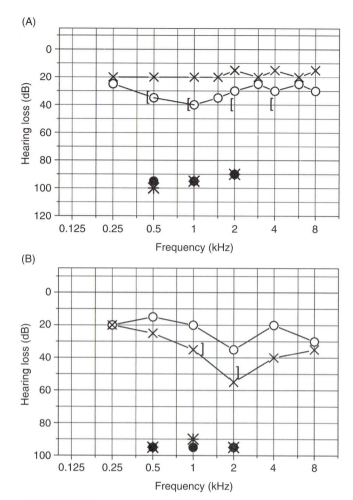

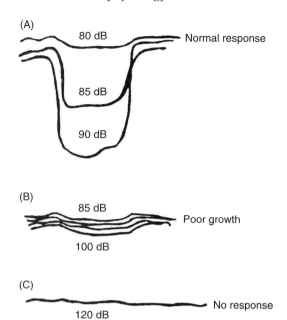

FIGURE 9.26 Examples of audiogram from patients with vascular compression of the auditory nerve. (A) Audiogram from a patient with vascular compression of the eighth cranial nerve near its entry into the brainstem showing a dip ("cookie bit") at 1 kHz (circles) (data from Møller, 1994). (B) Audiogram from a patient with hemifacial spasm (crosses) showing a dip around 2 kHz (data from Møller and Møller, 1985).

FIGURE 9.27 The growth of the acoustic middle ear reflex response in an individual with normal hearing (A), a patient with auditory nerve damage (B), and in a patient with an vestibular Schwannoma (C) (reprinted from Møller, 1994).

patients with lesions of the auditory nerve and the maximally obtainable amplitude of the response is much less than it is normally (Fig. 9.27).

Temporal integration[3] is affected by injuries to the auditory nerve. One way of obtaining an estimate of

temporal integration is to determine the threshold to tones of different duration e.g. 20 ms vs. 200 ms. The difference in threshold of tones of 20 and 200 ms duration is normally approximately 8 dB for tones of 1 and 4 kHz and it is less in individuals with vestibular Schwannoma where the auditory nerve is injured (approximately 5–6 dB). Cochlear injuries also result in a slightly reduced temporal integration with a difference between the threshold of tones of 20 ms duration and 200 ms duration of 6–7 dB [256].

Signs of decreased conduction velocity in the auditory nerve can be demonstrated by various kinds of electrophysiologic recordings. Of those methods, recordings of the ABR are the most commonly used for diagnosis of disorders caused by injury to the auditory nerve. Recordings made directly from the exposed eighth cranial nerve are used in surgical operations where the auditory nerve may be injured by surgical manipulations (see [212]. Such recordings show a prolonged latency of the compound action potential (CAP) recorded from the exposed CNVIII because of the decrease in conduction velocity (Fig. 9.28). The CAP also becomes broadened because different nerve fibers are affected differently.

Injury to the intracranial portion of the auditory nerve causes the latency of all peaks of the ABR to increase except peak I, which is generated by the most peripheral portion of the auditory nerve (see Chapter 7).

---

[3]The numerical value of a time constant for temporal integration is usually defined as the time it takes for the response to decay to an amplitude that is 1/e (e is the base of the natural logarithm, 2.718) of its original value. These definitions are taken from the engineering terminology which assumes an exponential decay or rise of a response such as an evoked response. In psychoacoustics, temporal integration causes the threshold to decrease with increasing duration of a response and it is often determined by obtaining the threshold to tones of different duration. The temporal integration is often expressed as the difference (in decibels) between the thresholds of tones of different durations such as 20 and 200 ms.

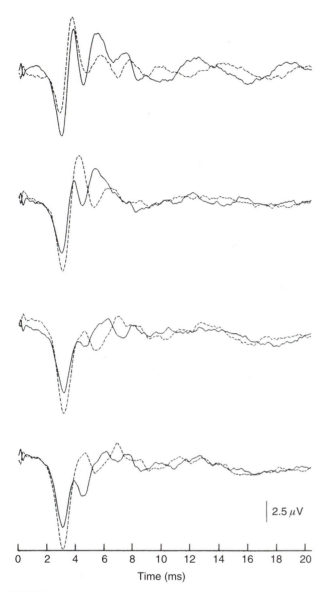

FIGURE 9.28 Typical CAPs recorded from the intracranial portion of the auditory nerve during an operation where the nerve was surgically manipulated (reprinted from Møller, 1995, with permission from Taylor & Francis).

Peaks II and III of the ABR may be obliterated and often only peak V and peak I are discernable in the ABR of individuals whose auditory nerve is injured. Subtle injuries to the auditory nerve such as from vascular compression typically cause less than 1 ms increase in latency. In patients with vestibular Schwannoma, it is not unusual that the latency of peak V is prolonged 3 ms or more. The normal conduction velocity of the auditory nerve is about 20 m/s [219]. The length of the auditory nerve is 2.5 cm, corresponding to a total conduction time of 1.25 ms. Prolongation of the conduction time by 3 ms implies a reduction of the conduction

velocity of the auditory nerve to about 8 m/s if the change was uniform along its entire length. If it is assumed that only the intracranial portion of the auditory nerve is affected, (length approximately 1 cm) the conduction velocity of that segment would have decreased from 20 m/s to approximately 3 m/s as a result of the insult from the vestibular Schwannoma.

6.2. Other Space-occupying Lesions

Other space-occupying lesions of the cerebellopontine angle (CPA) are benign tumors such as meningioma, cholesteatoma, neuroma of cranial nerves other than the eighth nerve and arachnoidal cysts. Such lesions are rare, but they may grow to large sizes without causing much hearing loss. Malignant tumors in that region of the brain are extremely rare and are usually metastases.

7. PATHOLOGIES OF THE CENTRAL AUDITORY NERVOUS SYSTEM

Disorders that are associated with morphologically detectable pathologies of nuclei and fiber tracts of the ascending auditory pathways of the brainstem and the auditory cortex are extremely rare. They are usually associated with multiple symptoms and signs from other brain systems. Lesions of the auditory cerebral cortex may cause auditory hallucinations consisting of hearing meaningful sounds such as speech and music. Similar signs have been reported in patients with lesions in the thalamic auditory nucleus (medial geniculate body [MGB]) [96].

Usually, only lesions to the nervous system that are associated with detectable morphologic changes have been considered. There is, however, another kind of change in the function of the auditory nervous system, which cannot be detected by the imaging techniques that are presently available. These changes in function are a result of neural plasticity and they result in various kinds of pathologic signs such as tinnitus and hyperacusis (discussed in Chapter 10). It is also possible that impaired speech discrimination can result from such plastic changes. These kinds of pathologies have earlier been largely disregarded mainly because they have no morphologic correlates. Without direct, visual, evidence that a pathology exists, disorders were often attributed to psychological causes and considered untreatable. There is increasing evidence, however, that such disorders have a true physiological basis and can be treated successfully by appropriate sound stimulation [132] (see Chapter 10).

7.1. Disorders of the Brainstem Auditory Pathways

Disorders that affect the brainstem nuclei of the auditory system and which are associated with detectable morphologic changes are extremely rare. Such disorders are tumors and vascular malformations and they usually give symptoms and signs from multiple systems. Rarely is hearing loss or other auditory signs present before signs occur from vital systems that are controlled from the brainstem. If a brainstem tumor manifests with auditory symptoms as the first sign, it usually does not take a long time before symptoms and signs from other systems will appear. Malformations such as the Arnold-Chiari malformation may be associated with hearing impairment [307]. The changes that occur in the pure tone audiograms from brainstem lesions are unpredictable. Changes in the ABR may be the most distinct and the anatomical location of a lesion can often be determined by considering the neural generators of the ABR. (The use of the ABR for topical diagnosis of brainstem lesions was discussed in more detail on, p. 240.)

Increased blood levels of bilirubin (hyperbilirubinemia) may occur when bilirubin, which is a breakdown product of the porphyrin ring of red blood cell hemoglobin, is not broken down further in the liver by the enzyme glucuronyl transferase. Hyperbilirubinemia (jaundice) often occurs in newborns because the enzyme that breaks down bilirubin is poorly developed. Bilirubin is neurotoxic and causes bilirubin encephalopathy [4]. Bilirubinemia has been studied extensively and an animal model of the disorder exists (Gunn rat model).

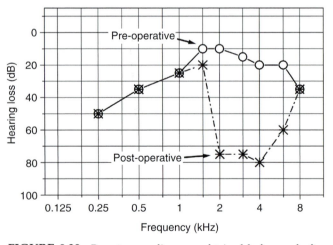

FIGURE 9.29 Pure tone audiograms obtained before and after an operation where the auditory nerve was injured by surgical manipulations. The speech discrimination was 96% before the operation and 0% after (obtained using recorded speech material).

The pathologic sign of hyperbilirubinemia is kernicterus, meaning yellow staining, that affects deep nuclei of the brain. Bilirubin encephalopathy affects the auditory nervous system specifically in addition to its general effect on the nervous system causing retardation, etc. The brainstem auditory nuclei and most noticeably the cochlear nuclei are most affected [78, 85, 294]. The effect on the auditory system is characterized by high frequency hearing loss that usually occurs bilaterally and symmetrically with severely decreased speech discrimination. The interpeak latency (IPL I-III) of the ABR is increased. The effect of hyperbilirubinemia is preventable, but it requires the condition to be detected early.

7.2. Auditory Cortices

Lesions that affect the auditory cortices are typically tumors of the temporal lobe but bleeding, strokes and trauma may also affect various parts of the auditory cortex. The symptoms are diffuse and sometimes consist of auditory hallucinations in the form of hearing meaningful sounds such as music or speech. This is unlike tinnitus, which involves hearing sounds that are not meaningful is regarded to originate in anatomically more peripheral levels of the ascending auditory nervous system (see Chapter 10).

Injury to the primary auditory cortex does not usually cause any noticeable degree of threshold elevation even for sounds presented to the ear that is contralateral to the side of the lesion. Speech discrimination is usually normal when assessed using standard tests. Lesions of the auditory cortex may be detected by low redundancy speech tests [24, 25], such as spectrally filtered speech, or speech that is presented at a very high rate (time compressed) or chopped speech. These three different ways to distort speech have a high degree of specificity to disorders of the central auditory nervous system [157]. Individuals with lesions of the temporal lobe affecting the auditory cortex have much lower discrimination scores for distorted speech when presented to the contralateral ear, while no abnormality can be detected when the speech is presented to the ear ipsilateral to the lesion (Fig. 9.30).

It is interesting to note that rather extensive lesions of the auditory cortex allow normal speech discrimination in contrast to lesions of the auditory nerve where subtle injury causes severe deterioration of speech even under ideal circumstances such as those usually used in testing the speech discrimination for clinical purposes.

Now, it is often regarded to be easier to use MRI scans for the diagnosis of cortical lesions but the high

(A)

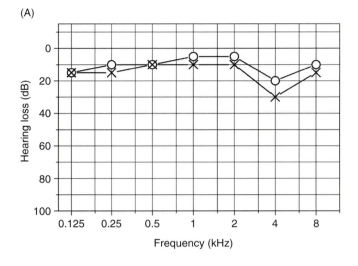

(B)

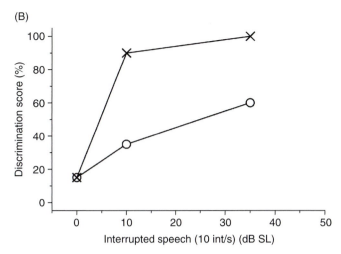

(C)

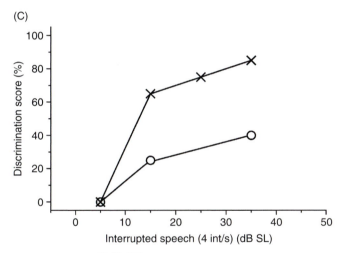

(D)

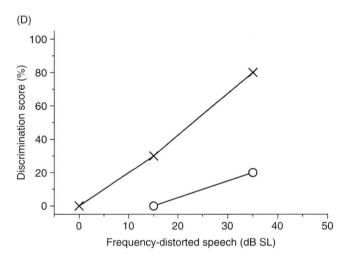

FIGURE 9.30   (A) Pure tone audiogram in a patient with a tumor (astrocytoma) in the left temporal lobe. (B) Speech discrimination in the same patient for interrupted speech (10 interruption per s) as a function of intensity with which the speech was presented (in decibels above the patient's threshold, SL). (C) Similar graph as in (B) showing discrimination of frequency filtered speech obtained in the same patient. (D) Similar graph as in (B) showing discrimination of time-compressed speech obtained in the same patient (data from Korsan-Bengtsen (aka Møller), 1973, with permission from Blackwell Publishing Ltd).

cost of MRI scans compared with speech tests makes low redundancy speech tests an attractive alternative for diagnosis of temporal lobe lesions that affect the auditory cerebral cortex. It may also be worth noting that not all disorders of the nervous system, including the auditory nervous system manifest themselves as detectable abnormalities in imaging studies, which are limited to detecting changes in structure. Plastic changes in the nervous system cannot be detected by imagining methods such as the MRI. Perhaps tests such as the low redundancy speech test should not have been abandoned totally.

BOX 9.16

LOW-REDUNDANCY SPEECH TESTS

Bocca and co-workers [25] were some of the first authors to publish studies showing that speech in which some of the redundancy of normal speech was removed could be used to detect auditory deficits in patients with central auditory lesions. The effect of such alterations of speech on its intelligibility has been studied in individuals with normal hearing [23, 39, 197] and it was found that individuals who did not have cortical lesions had normal speech discrimination when tested with the same low redundancy speech. Patients with lesions of the auditory cerebral cortex had normal speech discrimination when tested with undistorted speech while the same individuals had decreased speech discrimination scores when the test material was low-redundant speech (Fig. 9.30). Similar kinds of low redundancy speech were used to detect disorders that affect the auditory cortex, such as temporal lobe epilepsy and tumors of the temporal lobe, but also more general changes in the function of the nervous system such as that of aging [25, 158].

The efficiency of such low-redundancy speech in detecting central auditory lesions has been demonstrated [157]. Chopped and spectrally filtered speech have reduced redundancy because something has been removed from the speech, while time-compressed speech contains all components of the original speech, just presented in a shorter time than normal speech. Interruptions of 10/s using a ratio of speech to silence of 1:1 resulted in a marked reduction in speech discrimination in individuals with lesions to the central auditory nervous system, more so when the sound was presented to the contralateral ear than to the ipsilateral ear [157]. Normal hearing individuals could discriminate such interrupted speech at nearly 100% but a population of elderly individuals without specific disorders had reduced discrimination scores [227]. When speech is speeded up by a factor of two (from normally 110-140 words per minute to about 250 words per minute) discrimination is reduced even by individuals with normal hearing [157].

7.3. Efferent System

No specific disorders are known to affect the efferent auditory nervous system. However, due to its anatomical abundance it has been suggested that the efferent system is involved in several disorders, but little evidence has been presented to support such hypotheses. Because the cochlear efferent bundle travels together with the central portion of the vestibular nerve, it is severed in operations for vestibular neurectomy when done close to the brainstem. Vestibular neurectomy is often used to treat patients with vestibular disturbances such as Ménière's disease. Scharf and co-workers [286] found few measurable changes in hearing in patients who had their vestibular nerve sectioned intracranially to treat vestibular disorders and thus had a severed olivocochlear bundle.

7.4. Pathologies that Can Affect Binaural Hearing

Asymmetrical hearing loss of any kind may impair directional hearing and the perception of auditory space. Binaural hearing aids that are properly adjusted can restore some of these functions. The anterior commissure, a part of the corpus callosum, seems to be involved in perception of auditory space and there are some indications that this part of the brain may change with age so that it becomes more difficult to fuse the auditory input from the two ears into a single image. Little is known, however, about this aspect of impairments of auditory function. This has implications for fitting of hearing aids, and it may not always be to a patient's benefit to use binaural hearing aids [111, 138].

7.5. Viral Infections

Various forms of unspecific pathologies, such as viral infections, may affect the auditory nerve. Hearing loss can occur in connection with infections by the herpes virus such as the Ramsay Hunt syndrome [343], which often renders the patient deaf in one ear and may give symptoms from adjacent cranial nerves such as the facial nerve resulting in paralysis of face muscles. Viral infections of the auditory nerve may occur without any other signs than hearing loss. Establishing with any degree of certainty that a viral infection is the cause of hearing loss in an individual case is, however, difficult. Hearing loss is often extensive or total (deafness) and it often occurs within a short time (a few hours to a day) but it is not known how a viral infection can render all nerve fibers of a nerve non-conductive in such a short time. The auditory nerve and the vestibular nerve are usually not affected at the same time (viral infections affecting the vestibular nerve are known as vestibular neuronitis, causing violent vertigo).

Common childhood viral infections are suspected to cause hearing loss or deafness from inflammation of the auditory nerve (or the cochlea, with destruction of the organ of Corti, stria vascularis and tectorial membrane; see p. 234). The CMV is usually a harmless virus until some additional factors cause it to become activated, at which time it may cause sensorineural hearing loss. CMV infection is now regarded to be one of the more common causes of early cochlear hearing loss.

7.6. Ototoxic Drugs

Some ototoxic drugs not only affect the cochlea but may also affect the nervous system [84], as demonstrated in studies of the IC [135]. Drugs that do not seem to affect the function of the cochlea may affect neural processing. Thus it has been shown that administration of scopolamine reduces the discrimination score of low redundancy speech in young individuals with normal hearing [8].

7.7. Sudden Hearing Loss

Sudden hearing loss (SSNHL) was discussed above (p. 234) because it may be a result of injury to the cochlea. Other hypotheses about sudden hearing loss claim that it is a result of conduction block in the auditory nerve and it has been suggested that inflammation of the auditory nerve, perhaps as a result of viral infection, could cause sudden hearing loss. This would be similar to what is thought to occur to the facial nerve in Bell's Palsy. Rarely do vestibular nerve symptoms occur together with SSNHL and rarely is the loss of hearing associated with tinnitus.

8. ROLE OF NEURAL PLASTICITY IN DISORDERS OF THE CENTRAL AUDITORY NERVOUS SYSTEM

We discussed above how changes in the input to the central auditory nervous system from the cochlea in disorders that manifest with hearing impairment

could change the function of the central auditory nervous system. Pathologies of the auditory nervous system can also cause changes in the function of structures that are located central to the lesion. Such changes are caused by expression of neural plasticity.

Expression of neural plasticity is first and foremost involved in causing the symptoms and signs of disorders of hyperactivity such as severe subjective tinnitus, hyperacusis, and phonophobia, but even the symptoms and signs of other disorders of the auditory system such as some forms of hearing impairment may also have components that are caused by abnormal function of the auditory nervous system brought about by expression of neural plasticity. Even disorders of the conductive apparatus may cause changes in the function of the auditory nervous system.

When detectable morphological changes are present, such as in noise induced hearing loss, a component of the symptoms may be caused by changes in the function of the auditory nervous system. Determining the *cause* of disorders therefore becomes a matter of how one elects to define the word "cause." Intuitively, events such as diseases should have one well defined cause but in reality that is not always the case. In the examples given above where several factors are involved and no one of these factors alone can cause symptoms, the definition of cause becomes problematic. Expression of neural plasticity can also cause symptoms and signs of disease. Expression of neural plasticity can alleviate symptoms and signs of diseases such as tinnitus and pain.

8.1. What Is Neural Plasticity?

Neural plasticity is a broad term used to describe changes in the function of the CNS. Neural plasticity is an ability of the nerve cells to change their function. Altered excitability of specific neural structures, changes in synaptic efficacy or outgrowth of new connections (sprouting) are common changes attributed to neural plasticity. This can occur with or without morphological changes. Neural plasticity is necessary for the developing organism and it is beneficial to the mature organism because it can change the function of specific parts of the CNS to suit changing demands or compensate for the effect of injuries and diseases. Neural plasticity makes it possible to regain function after injuries such as trauma, strokes etc., which have destroyed neural tissue. Expression of neural plasticity can also cause symptoms and signs of disease.

It has been known for many years that the developing central nervous system is plastic but only relatively recently has it become evident that the mature nervous system is also plastic [125, 136, 154, 165, 181, 194, 259, 268, 277, 290, 316, 319, 320, 338, 339]. Apoptosis, formation of new connections and change of synaptic efficacy are all parts of normal development of the CNS, and that is controlled by internal factors, but many external factors can interfere with these normal processes. Failure to block synapses and inadequate pruning of the nervous system may play a role in developmental disorders of many different kinds. Ocular dominance and autism are examples of developmental disorders that may have similarities with disorders that are caused by expression of neural plasticity.

The ability of the central nervous system to change its function, including making new connections and eliminating old connections, is absolutely essential to the normal childhood development of functions that are controlled by the central nervous system, including cognitive functions. While expression of neural plasticity is most pronounced in childhood it is becoming more and evident that neural plasticity can be expressed not only in the developing individual but is also present in adults.

In adults, neural plasticity is important for shifting functions from impaired parts of the central nervous system to parts that are functioning normally. This occurs in recovery from, for instance, strokes and other insults to the brain. Treatment can facilitate recovery from such insults. The most common treatment is training but recently it has been shown that electrical stimulations of the cerebral cortex [32] can improve rehabilitation of stroke victims, probably by inducing neural plasticity. The use of neural plasticity in treatment of illnesses of the sensory system and the nervous system in general is increasing.

The changes that occur during childhood development as well as those that occur after injuries are promoted by input to the nervous system, or lack of input (deprivation). Activity related plasticity (Hebbian plasticity) has been described as "neurons that fire together wire together." This means that neural activity has control over morphological developments. Neural activity also affects synaptic efficacy. Expression of neural plasticity has many similarities with memory, and there is a short-term effect and a long-term effect of expression of neural plasticity. This means an understanding of neural plasticity is therefore important for understanding the pathology of many disorders and for treatment of disorders of the nervous system, including disorders of hearing.

Many forms of plasticity that cause signs and symptoms of disease occur without any known cause, but both external and internal events may cause expression of neural plasticity that causes symptoms of diseases.

Novel sensory stimulations, overstimulation, or, in particular, deprivation of sensory input are known to induce neural plasticity. The changes that occur as a result of neural plasticity can be transient or permanent, and may persist and even become aggravated after the events that caused the expression of neural plasticity have ceased or been reversed.

The absence of detectable morphologic abnormalities makes it difficult to diagnose disorders caused by neural plasticity. The anatomical location of the physiological abnormality that causes symptoms and signs of disease because of expression of neural plasticity may be different from that to which the patient refers the symptoms and it may be different from the anatomical location of an injury even when that injury was the initiator of the expression of neural plasticity.

Treatment of such disorders is hampered by lack of understanding of the pathophysiology of such disorders. Since plastic changes in function of the nervous system are not accompanied by tissue damage they are potentially reversible [13, 30, 132, 136, 348], often without side effects, which is rarely the case for pharmacological or surgical treatments. Changes in the function of the CNS to adapt to changing demands is important for the use of prostheses of limbs, or sensory prostheses such as cochlear implants.

Expression of neural plasticity triggered by hearing loss may increase the sensitivity of the ear and certain parts of the auditory nervous systems, but too much may cause tinnitus. Expression of neural plasticity is probably also involved in disorders affecting the balance systems (dizziness, vertigo).

Examples of effects of expression of neural plasticity that are of no advantage to the organism are hyperactivity, such as tinnitus, tremor and spasticity, and central neuropathic pain. What these signs have in common is that expression of neural plasticity may be regarded as misdirected attempts to compensate for neural deficits. In many cases, several conditions must be present at the same time to cause the expression of neural plasticity that is sufficient to cause noticeable symptoms and signs of disease. These expressions of neural plasticity are compensatory to loss of function through injury or other diseases, or they may be elicited by changes in demands.

Expression of neural plasticity is very important for the normal development of the nervous system and abnormal development is linked to deficits in the "re-wiring" of the nervous system that occurs during childhood development. Pruning of connection and competition of connections in the nervous system and competition of synaptic efficacy are also important parts of such development. Neural plasticity is also thought of as nature's means to replace or restore function after injury or diseases. There is, however, also another form or plasticity, which causes symptoms and signs of illness, and we may call that form of plasticity the "bad" plasticity. Neural plasticity of the central nervous system can cause widely different symptoms and signs, such as hyperactivity, hypersensitivity, different processing of sensory information and abnormal motor activity. Neuropathic pain and tinnitus are examples of hyperactive sensory disorders but hyperactive motor disorders such as spasticity, tremor, synkinesis, and ataxia may also be caused by functional changes that occur as a result of expression of neural plasticity.

Neural plasticity is induced when the normal homeostasis is upset and the plastic changes that occur can be regarded as attempts to restore the normal homeostasis, but when going wrong the induced plastic changes can cause symptoms and signs of diseases, not unlike many forms of therapies used in modern medicine. The plastic changes that occur in the mature nervous system are different from those that occur in the developing nervous system. The function of the nervous system normally changes during development and the abnormalities that occur are regarded as errors in the normal development. The mature nervous system has been regarded as being relatively stable, except for changes that are related to aging.

Neural plasticity consists of change in function of the nervous system and/or re-organization of the nervous system. The change in function can involve increase or decrease of excitability of neurons, change in processing of information, or re-routing of information. Expression of neural plasticity can cause change of synaptic efficacy, formation or elimination of synapses, and sprouting or elimination of dendrites and axons.

Expression of neural plasticity can cause functional changes in synaptic activity (altered synaptic efficacy or threshold), morphological changes such as outgrow of new connections and formation of new synapses, elimination of connections (axons and dendrites and synapses), apoptosis, or absence of normal apoptosis. These changes can cause altered processing of sensory information and altered motor control. Elimination of connections or establishment of new connections can cause re-routing of information in such a way that specific sensory information directed to other groups of neurons than normally process such sensory information or that other populations of neurons affect motor control than what is normally the case. Opening of dormant synapses is an example of neural plasticity that also can cause information to reach populations of nerve cells that normally do not receive such information. For example, neural plasticity may cause neural activity such as that elicited by sensory stimulation to reach brain regions that are not normally involved in such processing, or it may cause information to bypass certain populations of neurons. There are indications that re-routing of information may cause sensory information to reach the limbic system through a subcortical route, thus bypassing the processing that normally occurs in the cerebral cortices before it reaches limbic structures.

Expression of neural plasticity can be activity induced. Hebb's [118] principle states that neurons that are active together establish morphological connections. This has often been referred to as "Neurons that fire together, wire together." Neural plasticity can also be activated by inactivity, and that can be regarded as the reverse of Hebb's principle in that neurons that become inactive may lose established connections. That can be expressed by the statement "use it or lose it."

BOX 9.17

UNMASKING OF DORMANT SYNAPSES

One of the first studies that demonstrated one of the mechanisms for neural plasticity was published by Wall [339]. That study provided evidence that changes in synaptic efficacy of neurons in the dorsal horn of the spinal cord can be induced by deprivation of input accomplished by severing of a dorsal root. The deprivation of input that resulted changed the response of the target neurons of dorsal roots fibers and it caused cells in the dorsal horn of the spinal cord to respond to input from dermatomes from which they normally did not respond. This investigator coined the term "dormant synapses" to describe the event where synaptic connections are blocked because they have high synaptic thresholds or too low efficacy. Wall hypothesized that synapses that exist anatomically may not normally function but may become activated ("unmasked") by some abnormal event such as deprivation of input.

Re-organization means that the circuitry ("re-wiring") of the nervous system has changed. There are two different ways that this can occur. One way is by opening (unmasking) normally closed (dormant) synapses or closing normally open synapses. The other way is by forming new connections (sprouting of axons and formation of new synapses). Elimination of connections or of cells (apoptosis) are other ways in which the functional circuitry can change. Connections can be severed or created by sprouting of axons or severing of axons.

Connections can also be established by making non-functional synapses functional (unmasking of dormant synapses). Furthermore, neurons that are normally not activated by their input may become active by alterations in the input, such as increase of discharge rate may activate target neurons that are not activated by a lower rate. The excitatory post synaptic potentials (EPSP) in response to a low rate of incoming nerve impulses may not add up to produce membrane potentials that exceed the firing threshold of the neuron (see Fig. A1.1) because of insufficient temporal summation. Changes in synaptic efficacy or increased temporal integration may make it possible for an incoming train of nerve impulses to activate a target neuron. Changes in discharge pattern, for example from a regular pattern to burst pattern, may make it possible to exceeded the threshold of the target synapse which was not exceeded by the same average rate of discharges and thereby open new connections.

Reorganization may have different extents, and may change the wiring of local structures such as the cerebral cortex, or it may redirect information to population of neurons that have not normally received such input by opening dormant synapses. A third way that the function of the nervous system can change is by altering (enhancing) protein synthesis in target cells.

This means that change from sustained activity to burst activity in peripheral nerves, which is often seen in slightly injured nerves, may cause activation of target neurons that are normally not activated by sustained actively because the decay of the EPSP prevents temporal summation of input with large intervals to reach the threshold. The EPSP caused by impulses with short interval such as occur in burst activity may reach the threshold of some target neurons that are normally not activated by sustained activity. This would have the same effect as unmasking of the synapses in question.

Reduced inhibitory input to a central neuron may also lower its threshold and thereby unmask excitatory synapses.

Expression of neural plasticity is possible because of the existence of dormant connections which can be unmasked become functional. Connections that are functional can also be made non-functional. There are thus differences between the morphological circuitry of the nervous system and the functional circuitry which make it possible to change the function of the nervous system. Many of the morphological (intact) connections are normally not open because they make synaptic contacts that are ineffective.

8.2. What Can Initiate Expression of Neural Plasticity?

Neural plasticity can be evoked by many different kinds of events. One of the first demonstrations of neural plasticity was that of Goddard who showed that repeated of the amygdala nuclei in rats changed the function of these nuclei in such a way that the electrical stimulation began to evoke seizure activity after 4–6 weeks stimulation [104]. Goddard named this phenomenon "kindling." The kindling phenomenon has

---

BOX 9.18

COHERENT INPUT IS MOST EFFECTIVE IN UNMASKING DORMANT SYNAPSES

Wall and co-workers [339] showed that electrical stimulation was more efficient in activating dorsal horn neurons from distant dermatomes than natural stimulations. Electrical stimulation activates all fibers at the same time thus providing activations of the target neurons that are more coherent in time than what is the case for natural stimulation. These observations indicate that synapses on neurons in the dorsal horn that were normally dormant could be activated when stimulated coherently at a

high rate. Temporal and spatial integration may explain why coherent input at a high rate to these neurons could activate normally (unmask) dormant synapses. It is also in good agreement with the fact that high frequency stimulation is more effective in activating cells and it may activate cells that are unresponsive to low frequency stimulation. It is well known that bursts of activity can be more effective in activating the target neurons than continuous activity with the same average rate.

later been demonstrated in many other parts of the CNS [337] and even in motonuclei [290].

Plastic changes in nuclei of sensory systems can be induced by deprivation of input [100, 101, 136] by novel stimulation [290, 339] or by overstimulation [320]. Many animal studies have shown changes in the responses from cells in sensory cortices after stimulation or deprivation of input [147, 194].

Neural plasticity in the somatosensory system in response to deprivation of input was demonstrated by Patrick Wall [339] and Michael Merzenich [194], who in animal experiments showed changes in function of the neurons in the spinal cord and the primary somatosensory cortex respectively. These studies of the neural plasticity of the somatosensory system have been replicated and extended by many investigators.

Strengthening of synaptic efficacy is similar to long-term potentiation (LTP) and it may have similar functional signs as increased excitability of sensory receptors, decreased threshold of synaptic transmission in central neurons or that of decreased inhibition. Any one or more of such changes may be involved in generating the symptoms of hyperactivity and hypersensitivity that cause phantom sensations such as tinnitus, tingling and muscle spasm. Studies of LTP in slices of hippocampus in rats or guinea pigs show that LTP is best invoked by stimulation at a high rate. The effect may last from minutes to days, and glutamate and the NMDA receptor (N-methyl d-aspartate) have been implicated in LTP. Unmasking of ineffective synapses may occur because of increased synaptic efficacy or because of a decrease of inhibitory input that normally has blocked synaptic transmission [125, 126, 259].

Disorders where the symptoms and signs are caused by expression of neural plasticity are often labeled as "functional" because no morphological correlates can be detected. The label "functional" has often been used to describe psychiatric disorders, "Munchausen's" type of disorders and other disorders that do not exist except in the mind of the patient. *Stedman's Medical Dictionary* states the meaning of "functional" to be: "Not organic in origin; denoting a disorder with no known or detectable organic basis to explain the symptoms." This interpretation equates "not known" with "not detectable", which is interesting because something may indeed exist despite it not being detectable (with known methods). That means that many disorders have been erroneously labeled a "neurosis", which *Stedman's Medical Dictionary* defines as:

1. A psychological or behavioral disorder in which anxiety is the primary characteristic; defense mechanisms or any of the phobias are the adjustive techniques which an individual learns in order to cope with this underlying anxiety. In contrast to the psychoses, persons with a neurosis do not exhibit gross distortion of reality or disorganization of personality.
2. A functional nervous disease, or one for which there is no evident lesion.
3. A peculiar state of tension or irritability of the nervous system; any form of nervousness.

The fact that symptoms that arise from functional changes that are expressions of neural plasticity are not associated with detectable morphologic or chemical abnormalities is a major problem in treating disorders that are caused by neural plasticity because chemical testing and imaging techniques form the basis of diagnostic tools of modern medicine.

Knowledge about the physiology of neurological disorders can lead to adequate treatment of such disorders. Understanding of the pathophysiology of disorders that are caused by expression of neural plasticity can also reduce the number of patients who are diagnosed as "idiopathic" and instead directed as patients to effective treatment.

CHAPTER

10

Hyperactive Disorders of the Auditory System

1. ABSTRACT

1. Hyperactive disorders of the auditory system are subjective tinnitus, hyperacusis, and recruitment of loudness.
2. Tinnitus is of two kinds: objective and subjective tinnitus.
3. Objective tinnitus is caused by sound that is generated in the body and conducted to the cochlea.
4. Subjective tinnitus is perception of sound that is not originating from sound and can therefore only be heard by the person who suffers from the tinnitus.
5. Subjective tinnitus has many forms and its severity varies from person to person. It can be divided into mild, moderate and severe (disabling).
6. Severe subjective tinnitus is often accompanied by hyperacusis and phonophobia. Hyperacusis is a lowered threshold for discomfort from sound and phonophobia is fear of sound.
7. The anatomical location of the physiological abnormalities that cause tinnitus and hyperacusis is often the central nervous system.
8. Severe tinnitus is a phantom sensation that has many similarities with central neuropathic pain.
9. Tinnitus may be generated by neural activity in neurons other than those belonging to the classical auditory nervous system, thus a sign of re-organization of the nervous system.
10. Severe tinnitus is often accompanied by abnormal interaction between the auditory system and other sensory systems.
11. Hyperacusis and phonophobia are caused by reorganization of the central auditory nervous system.
12. Phonophobia may result from an abnormal activation of the limbic system through the non-classical auditory pathways, which are not normally activated by sound stimulation.
13. Expression of neural plasticity that is involved in the development of hyperactive conditions is often caused by overstimulation, or deprivation of stimulation.
14. Abnormal loudness perception (recruitment of loudness) is mainly associated with disorders of the cochlea.

2. INTRODUCTION

Hyperactive hearing disorders (subjective tinnitus and abnormal perception of sounds such as hyperacusis and phonophobia) are some of the most diverse and complex disorders of the auditory system and their causes are often obscure. Often it is not even possible to identify the anatomical location of the physiological abnormalities that cause these symptoms.

Tinnitus is the most common of the hyperactive disorders that affect the auditory system. Tinnitus is of two general types: 1) subjective tinnitus; and 2) objective tinnitus. Subjective tinnitus does not involve a physical sound and can only be heard by the individual who has the tinnitus. Objective tinnitus is not a hyperactive disorder. Objective tinnitus is caused by a physical sound generated within the body and conducted to the cochlea in a similar way as an external sound. An observer can often hear objective tinnitus and it is often caused by blood flow that passes a constriction in an artery causing the flow to become turbulent. This chapter will deal only with subjective tinnitus.

Since subjective tinnitus is perceived as a sound, the ear has often been assumed to be the location of the pathology. It is now evident that most forms of subjective tinnitus, hyperacusis (decreased tolerance of sound), phonophobia (fear of sound), and misophonia (dislike of certain sounds) are caused by changes in the function of the central auditory nervous system and these changes are not associated with any detectable morphological changes. The changes are often the result of expression of neural plasticity and the anomalies may develop because of decreased input from the ear or deprivation of sound stimulation and overstimulation or yet unknown factors. Tinnitus may be regarded as a phantom sensation [131]. Phantom sensations are referred to a different location on the body (usually the ear) than the anatomical location of the abnormality that causes the symptoms.

Altered perception of sounds often occurs together with severe tinnitus. Sounds may be perceived as distorted, or unpleasant (hyperacusis) or may be fearful (phonophobia). Such altered perception of sounds has received far less attention than tinnitus and yet, hyperacusis, and phonophobia, may be more annoying to the patient than their tinnitus.

Few effective treatment options are available for hyperactive disorders such as tinnitus and hyperacusis. Since most forms of severe tinnitus are caused by functional changes it should be possible to reverse the changes by proper sound treatment. This hypothesis has been supported by the experience that proper stimulation can alleviate tinnitus in some individuals [134] (p. 266). Medical treatment or surgical treatment such as microvascular decompression (MVD) operations can help some patients with tinnitus and hyperacusis.

While patients with severe tinnitus and hyperacusis or phonophobia are clearly miserable, it is not obvious which medical specialty is best suited for taking care of such individuals. It is, however, certain that whoever takes care of such patients must have the best possible knowledge and understanding of the changes in the function of the auditory system that can lead to tinnitus and hyperacusis in order to be able to help individuals with these disorders.

3. SUBJECTIVE TINNITUS

Subjective tinnitus is the perception of meaningless sounds without any sound reaching the ear from outside or inside the body. Tinnitus can be intermittent or continuous in nature and its intensity can range from a just noticeable hissing sound to a roaring noise that affects all aspects of life. Tinnitus may be a high frequency sound like that of crickets, a pure tone, or it may have the sensation of a sound with a broad spectrum. Some people hear intermittent noise; others hear continuous noise. Some hear their tinnitus as if it came from one ear; others hear their tinnitus as if it came from inside of the head, thus bilateral in nature. Tinnitus is often different from any known sound. Some people with tinnitus perceive their tinnitus as a slight bother while other people perceive their tinnitus as an unbearable annoyance that makes it impossible to sleep or to concentrate on intellectual tasks. Tinnitus is often accompanied by depression and tinnitus can cause people to commit suicide.

Subjective tinnitus is an enigmatic disease from which people suffer alone because they have no external signs of illness. Tinnitus thus has similarities with central neuropathic pain [213]. René Leriche, a French surgeon (1879–1955), has said about pain: "The only tolerable pain is someone else's pain", and that is true also for tinnitus.

Tinnitus is often the first sign of a vestibular Schwannoma and vestibular Schwannoma should always be ruled out in individuals who present with one-sided tinnitus with or without asymmetric hearing loss. This can be done by using suitable audiologic tests (see p. 239). However, very few individuals with tinnitus have a vestibular Schwannoma (the incidence of vestibular Schwannoma has been reported to be 0.78–0.94 per 100,000 [326]). The incidence of tinnitus is far greater although its prevalence is not known accurately.

3.1. Assessment of Tinnitus

Considerable efforts have been devoted to finding methods that can describe the character and intensity of an individual person's tinnitus objectively Attempts to match the intensity of an individual person's tinnitus to a (physical) sound have given the impression that the tinnitus is much weaker than the patient's perception of the tinnitus. Individuals who report that their tinnitus keeps them from sleeping or from concentrating on intellectual tasks often match their tinnitus to a physical sound of an intensity that is unbelievably low, often between 10–30 dB above threshold [330], thus sounds that would not be disturbing at all to a person without tinnitus.

Matching the character of a patient's tinnitus to that of an external sound has also been unsuccessful in confirming a patient's description of the character of his/her tinnitus. The results of having patients

---

BOX 10.1

ASSESSING TINNITUS SEVERITY WITH A VISUAL ANALOG SCALE

The individual whose tinnitus is to be evaluated marks the point on a line that he or she judges to correspond to the strength of the tinnitus. The line is divided in 10 equal segments (for example every other cm on a 20-cm long line) and a participant has to choose one of these segments as corresponding to the strength of the tinnitus. Extreme values such as 10 are regarded as being unusual reactions. Some investigators have used VAS with fewer categories (seven or even four). This way of evaluating tinnitus also includes the emotional value of "coping" with tinnitus, thus similar to evaluation of pain.

---

compare their tinnitus with a large variety of synthesized sounds to gain information about the frequency and temporal pattern of tinnitus have been equally disappointing. It is often difficult for a person with tinnitus to describe the sounds he or she hears because tinnitus often does not resemble any known physical sound. Only in a few individuals has it been possible to obtain a satisfactory match between the tinnitus and a real (synthesized) sound.

Because the results of the matching of the intensity of tinnitus to other sounds does not seem to correspond to the perceived intensity of tinnitus, other ways of evaluating the strength of tinnitus were sought. The visual analog scale (VAS) that is often used in evaluation of pain seems a better way of assessing tinnitus than loudness matching.

The best way to classify tinnitus may be to use the patient's own judgement about the severity of his/her tinnitus. Some investigators have used a classification in three broad groups of tinnitus: mild, moderate and severe tinnitus [230, 266]. Mild tinnitus does not interfere noticeably with everyday life; moderate tinnitus may cause some annoyance and it may be perceived as unpleasant; severe tinnitus affects a person's entire life in major ways, making it impossible to sleep and conduct intellectual work.

3.2 Disorders in which Tinnitus Is Frequent

Tinnitus is one of the three symptoms of Ménière's disease (the two other are attacks of vertigo and fluctuating hearing loss) (see p. 229). Tinnitus almost always occurs in patients with vestibular Schwannoma. Surgical injuries or other insults to the auditory nerve are often associated with tinnitus. Head injuries and strokes likewise may be accompanied by tinnitus.

Tinnitus is frequent in individuals who have noise induced hearing loss or other causes of impaired hearing but there is no direct correlation between the pure tone audiogram and the severity of the tinnitus. Some individuals with tinnitus have severe hearing loss and tinnitus can even occur in individuals who are deaf. Tinnitus may also occur together with moderate hearing loss or, in rare cases, normal hearing. Some patients with tinnitus have small dips in their audiogram that may be signs of vascular compression of the auditory nerve. Usually such small dips are only revealed when testing is done at half-octave frequencies.

3.3. Causes of Subjective Tinnitus and Other Hyperactive Symptoms

Tinnitus can have many different causes but it deserves to be mentioned that the cause of tinnitus is often unknown. As has been pointed out earlier in this book, there is rarely a disease with only a single cause and many disorders require multiple pathologies to become manifest. Tinnitus is not an exception to that and attempts to find the (single) cause of tinnitus are therefore often futile. For example, some forms of tinnitus can be cured by moving a blood vessel off the intracranial portion of the auditory nerve (microvascular decompression [MVD] operations) but similar close contact between the auditory nerve and a blood vessel is common [213, 214] and causes no symptoms.

Close contact between the auditory nerve and a blood vessel (vascular compression[1]) is associated with tinnitus in some patients (see Chapter 14) and probably also hearing loss with decreased speech discrimination in some individuals [230]. A blood vessel in close contact with the auditory nerve[2] can irritate the nerve and may give rise to abnormal neural activity and perhaps slight injury to the nerve. Over time such close contact

---

[1]Vascular contact with a cranial nerve is known as "vascular compression" but there is evidence that the pathology associated with close vascular contact between a cranial nerve and a blood vessel does not depend on a mechanical action (compression) but it is the mere contact that causes the pathology [208].

[2]Microvascular compression.

with a blood vessel may cause changes in more centrally located structures of the ascending auditory pathways and that is believed to be the cause of symptoms such as tinnitus, hyperacusis, and distortion of sounds. Many patients with vascular compression of the auditory nerve as a cause of these symptoms complain that sounds are distorted or sound "metallic." Tinnitus may be relieved by MVD operations, where the offending blood vessel is moved off the auditory nerve [130, 156, 230]. If such an operation is successful in alleviating tinnitus, it also often relieves the patient's hyperacusis, and distortion of sounds. The speech discrimination may improve. This indicates that at least some of the effects of vascular compression on neural conduction in the auditory nerve that are caused by vascular compression are reversible.

Small dips may be present in the audiogram of patients with tinnitus that can be alleviated by MVD operations of the auditory nerve (p. 241, Fig 9.26A) [226]. The audiograms of some patients with hemifacial spasm that is caused by vascular contact with the seventh cranial nerve had similar dips (Fig 9.26B) [229]. The reason for this is assumed to be irritation of the auditory nerve from the same vessel that was in contact with the facial nerve causing the patient's symptoms (HFS). These patients, however, did not have any symptoms from the auditory system and only the audiogram taken as a part of the preoperative testing done for patients to be operated for HFS revealed the involvement of the auditory nerve. The fact that these dips occurred in the mid-frequency range of hearing would indicate that nerve fibers originated from the middle portion of the basilar membrane are located superficially in the auditory nerve [64]. This would be different from what is seen in animals where high frequency fibers are located superficially on the nerve [282].

BOX 10.2

MICROVASCULAR COMPRESSION AS CAUSE OF DISORDERS

The reason that close contact between a cranial nerve and a blood vessel has been assumed to be the "cause" of diseases such as face pain (trigeminal neuralgia [TGN] or tic douleroux) and face spasm (hemifacial spasm [HFS]) is that these diseases can be effectively cured by moving a blood vessel off the respective nerve in an operation known as a MVD operation [18, 19, 208]. It has also been shown that close contact between a blood vessel and cranial nerves V or VII is rather common [314] and occurs in as much as approximately 50% of individuals who do not have any symptoms from these cranial nerves. However, the disorders that are associated with vascular contact with CNV and CNVII (TGN and HFS, respectively) are extremely rare with incidence of about 5 for TGN [144] and 0.8 per 100,000 for HFS [11]. Vascular contact with the eighth cranial nerve is also common although it is not known exactly how often that occurs. In fact, it is the experience from the author's observations of many operations in the cerebello pontine angle in patients undergoing MVD operations for TGN and HFS that close vascular contact with the eighth cranial nerve is common in such patients without any associated vestibular or hearing symptoms.

The reason that vascular contact with a cranial nerve only rarely gives symptoms and signs from the respective cranial nerve could be that vascular compression varies in severity but a more plausible reason is that vascular compression is only one of several factors all of which are necessary for causing symptoms [208]. The fact that vascular compression is common in asymptomatic individuals

means that vascular contact is not sufficient to give symptoms. The fact that MVD operations for TGN and HFS have a high success rate (80–85%) indicates that vascular compression is necessary to cause symptoms [18, 19]. Removal of the vascular contact with a cranial nerve can relieve symptoms despite the fact that the other factors are still present because vascular compression is necessary for producing the symptoms. Assuming that vascular compression is only one of the factors that are necessary to cause symptoms and signs of disease makes it understandable that vascular compression can exist without giving symptoms because other necessary factors are not present. Vascular contact with a cranial nerve alone can thus not cause symptoms and signs [208].

Subtle injuries to the auditory nerve or irritation from close contact with a blood vessel are thus present in a large number of individuals but only very few of such persons have any symptoms. Detecting the presence of a blood vessel is therefore not sufficient to diagnose these disorders. It has been attempted to use MRI scans for that purpose, but MRI scans are not effective in detecting the presence of close contact between vessels and cranial nerves. Recordings of ABR and the acoustic middle ear reflex response can detect the effect of vascular contact with the auditory nerve because it is associated with slower neural conduction in the auditory nerve. Prolongation of the latency of peak II in the ABR (see Chapter 11), and delays of all subsequent peaks are thus signs of slight injury to the auditory nerve.

This observation supports the findings discussed above that showed that vascular contact in itself does not cause symptoms and confirms that close contact between a blood vessel and the auditory nerve is only one of several factors that are necessary to cause symptoms such as tinnitus. This also means that tests that reveal contact between the auditory nerve and a blood vessel cannot alone provide the diagnosis of such disorders as tinnitus and hyperacusis and the case history must be taken into account to achieve a correct diagnosis of such disorders.

Surgical injury to the auditory nerve is a relatively recent cause of hearing loss, tinnitus, and hyperacusis, that began to appear when it became common to operate in the cerebellopontine angle for non-tumor causes (such as vascular compression of cranial nerves to treat pain and spasm of the face). Hearing loss from such operations is, however, less frequent now than earlier because of advances in operative technique, and the use of intraoperative monitoring of auditory evoked potentials [212, 222].

Surgical injuries can be caused either by compressing or by stretching the auditory nerve. Heat that spreads from the use of electrocoagulation to control bleeding can also injure the auditory nerve. Depending on the degree of compression, stretching or heating, the injuries may consist of slight decrease in conduction velocity, conduction block in some fibers or, in the more severe situation, conduction block in all auditory nerve fibers. The acute effect on neural conduction may recover completely with time or partially or not at all depending on the severity of the injury. Compression probably mostly affects fibers that are located superficially in the nerve whereas stretching is likely to affect all fibers. Surgically induced injuries to the auditory nerve caused by stretching of the nerve may affect all fibers of the auditory nerve [116], and this explains why hearing loss from surgically induced injury often affects both low and high frequencies. Surgically induced injury to the auditory nerve typically causes a moderate change in the pure tone audiogram and a marked impairment of speech discrimination (Fig. 9.29). In fact, moderate threshold elevation may be associated with total loss of speech discrimination. The effects of surgical injury to the auditory nerve at all degrees including total loss of hearing are almost always accompanied by tinnitus and hyperacusis.

Since many people have close contact between a blood vessel and their auditory nerve but no tinnitus, vascular contact is not sufficient to cause tinnitus. This means that vascular contact with a cranial nerve root is only one of several factors that are necessary to cause symptoms and signs. The fact that MVD operations can cure HFS and TGN and tinnitus in some patients

means that vascular contact with the respective cranial nerve root is a necessary factor for causing symptoms of these disorders. Removal of one factor, such as vascular compression, is an effective cure when that factor is necessary to cause the symptoms (although not sufficient). The other factor(s) that are necessary to cause symptoms are usually unknown and do not give symptoms [208].

Instead of attempting to find *the* cause of a certain form of tinnitus it may be more productive to try to identify the combination of factors that can cause tinnitus, each of which may not cause any symptoms when occurring alone. The inability to comprehend and deal with phenomena that depend on several causes may explain why it is common to find the diagnosis of "idiopathic tinnitus," which means "tinnitus of unknown origin."

The anatomical location of the abnormality that generates the neural activity that is perceived as a sound may be the ear, but it is more often the auditory nervous system. Since tinnitus presents as a sensation of sound it has often been assumed that tinnitus is generated in the ear and that it involves the same neural system as is normally activated by a sound that reaches the ear. More recently, evidence that plastic changes in the central auditory nervous system can cause symptoms such as tinnitus and hyperacusis has accumulated (see p. 247). The changes in the central auditory nervous system that cause such symptoms cannot be detected by the imaging techniques we now have available. Since the changes in the function of the central nervous system that are associated with tinnitus do not have any apparent morphologic abnormalities, these functional changes have for a long time escaped attention.

The finding that deaf people can have severe tinnitus and individuals with normal hearing without any signs of cochlear disorders can also have severe tinnitus shows clearly that tinnitus can be generated in other places of the auditory system than in the ear. Perhaps the strongest argument against the ear always being the location of the physiologic abnormalities that causes tinnitus is the fact that the auditory nerve can be severed surgically without alleviating tinnitus. Patients with vestibular Schwannoma almost always have tinnitus. That would indicate that the anatomical location of the physiological abnormality that generates the sensation of tinnitus would be the auditory nerve. However, the tinnitus often persists after removal of the tumor despite the fact that the auditory nerve has been severed during the operation [122], and that indicates a more central location of the generation of the tinnitus. The injury from the tumor to the auditory nerve may over time have caused changes in neural structures that are located more centrally, through expression of neural plasticity.

Auditory nerve section has, however, also been used to treat tinnitus [253, 254, 255], but not all patients were free of tinnitus after severing of the auditory nerve. That some individuals are relieved from their tinnitus by severing their auditory nerve, however, shows that in some individuals the cochlea is the anatomical location of the physiological abnormalities that generate the neural activity that is perceived as tinnitus [253], thus emphasizing the diversity of causes of tinnitus.

Other investigators have found evidence that the auditory cortex is re-organized in individuals with tinnitus [232]. The observations that some individuals with tinnitus get relief from tinnitus by transcranial magnetic stimulation [62] and by electrical stimulation of the auditory cortex by implanted electrodes [63] (see p. 265) are taken as further evidence that the cerebral auditory cortex is re-organized in some individuals with tinnitus.

It has been suggested that the olivocochlear efferent system may affect tinnitus. The fibers of the medial portion of the efferent bundle travel in the central portion of the inferior vestibular nerve, and join the cochlear nerve at the anastomosis of Oort. This bundle consisting of approximately 1,300 fibers is therefore severed in operations for vestibular nerve section eliminating efferent influence on the cochlea. If dysfunction of the efferent system were involved in tinnitus, vestibular nerve section would likely affect the tinnitus. However, a literature review reveals that it has little effect on tinnitus [15], and in fact, severing of the olivocochlear bundle has remarkably little effect on other aspects of hearing [286].

That individuals with tinnitus often have difficulties in selecting sounds that are perceived in the same way as their tinnitus indicates that neural circuits other than those normally activated by sound are involved in tinnitus. That many individuals with tinnitus who perceive their tinnitus to be unbearably strong but match their tinnitus to sounds that are only 10–30 dB above their hearing threshold [330] also indicates that tinnitus may be generated in parts of the central nervous system that do not normally process sounds.

Other studies have shown interaction between the somatosensory system and the auditory system in some patients with tinnitus [37, 223], indicating an abnormal involvement of the non-classical auditory pathways (see Chapter 5). Neurons in the non-classical pathways respond to more than one sensory modality [6, 216, 321], indicating that a cross-modal interaction occurs in the non-classical pathways between the auditory and the somatosensory pathways. Signs of

cross modal interaction in some individuals with tinnitus were therefore taken as a sign of involvement of the non-classical pathways in such individuals [223]. Such cross-modal interaction is a constant phenomenon in young children [225] but it occurs rarely in adults [223, 225]. This means that there are neural circuits that provide input from other sensory systems to the auditory system, but these neural pathways are not normally functional in adults, probably because of blockage of the synapses that provide connections from these other sensory systems to the auditory system. That stimulation of the somatosensory system may affect the perception of tinnitus in some patients indicates that these connections have been re-activated in some individuals with tinnitus [37, 223]. This re-activation may have occurred by unmasking of dormant synapses, as has been shown to occur in the somatosensory system after deprivation of input [339].

Other forms of abnormal interaction between the auditory and the somatosensory systems have been observed in patients with tinnitus. Touching the face, moving the head and changing gaze can change the tinnitus in some individuals with tinnitus [36, 37, 50]. Abnormal stimulation of the somatosensory system can occur from disease processes such as temporomandibular joint (TMJ) problems, which may also activate the non-classical auditory system, explaining why individuals with TMJ problems often have tinnitus [206]. Neck problems of various kinds are sometimes accompanied by tinnitus [176], thus another example of interaction with the auditory system from other systems. Some patients with tinnitus report that they hear sounds when touching the skin such as rubbing their back with a towel, thus a further indication that input from the somatosensory system can enter the auditory nervous system.

Neurons in the non-classical auditory pathways respond in a much less specific way than neurons in the classical (lemniscal) system and the neurons in the non-classical auditory system are broadly tuned (see Chapter 6), which may explain why many patients with hyperactive auditory disorders perceive sounds differently. The fact that neurons in the dorsal nuclei of the thalamus project to secondary auditory cortices (AII) [173, 216], thus bypassing the primary auditory cortex (AI), may explain why tinnitus is perceived differently from physical sounds that reach the ear in a normal way. Information that travels in the non-classical pathways reaches the AII and association cortices before information that travels in the classical pathways. Since information from the classical auditory pathway must pass the AI auditory cortex before it reaches the AII cortex, such information will arrive at the AII cortices later than the information from the non-classical pathways. That similar information arrives at the AII cortex at different times may contribute to difficulties in understanding speech that some patients with hyperactive auditory symptoms experience.

Functional imaging in individuals who can voluntarily alter their tinnitus [248] have supported the hypothesis that the neural activity that causes tinnitus is not generated in the ear. Other studies using the same technique have shown evidence that the neural activity in the cerebral cortex that is related to tinnitus is not generated in the same way as sound evoked activity and not generated in the ear [181]. These investigators found that tinnitus activated the auditory cortex on only one side whereas (physical) sounds activated the auditory cortex on both sides. These findings are in good agreement with the results of studies that show evidence that the non-classical auditory nervous system may be involved in tinnitus in some patients [223] and the hypothesis by Jastreboff [131] that tinnitus is a phantom sensation generated in the brain [35].

Neurons of the non-classical auditory system use the dorsal and medial thalamic nuclei and thus provide subcortical connections to the lateral nucleus of the amygdala [173, 213] and probably other structures of the limbic system.[3] This may explain why hyperactive disorders of the auditory system often are accompanied by symptoms of affective disorders such as phonophobia and depression (see p. 254).

Studies have shown indications of cross-modal interactions also may occur in the motor cortex in tinnitus patients resulting in increased intracortical facilitation [170].

3.4. Role of Expression of Neural Plasticity in Tinnitus

There is considerable evidence that expression of neural plasticity (see Chapter 9, p. 247) is involved in many forms of tinnitus. Deprivation of input to the central nervous system is a strong promoter of expression of neural plasticity but also overstimulation can promote reorganization of the nervous system that may result in symptoms of dysfunction of sensory and motor system [136, 195]. Studies in animals [340] have

shown alterations of tonotopic maps after exposure to loud sounds and deprivation of sounds has likewise been shown to alter tonotopic maps [281]. Recently it has been shown that patients with tinnitus have altered tonotopic maps in the auditory cortex [232].

Expression of neural plasticity may alter the balance between inhibition and excitation in the auditory nervous system. The dependence on gender of the incidence of tinnitus [57] may have to do with the fact that female reproductive hormones can modulate GABAergic transmission [86, 109]. The level of these hormones varies over the menstrual cycle of women in reproductive age and it is possible that the resulting (cyclic) variation in the potency of some GABA receptors can facilitate recovery from the changes in the central nervous system that cause tinnitus.

High frequency hearing loss is often accompanied with tinnitus. Such tinnitus may be caused by deprivation of input from the basal portion of the cochlea [100]. That hypothesis is supported by the efficacy of treating tinnitus in patients with high frequency hearing loss with electrical stimulation of the cochlea [273].

Some patients with otosclerosis have tinnitus, and 40% of such individuals obtain relief from successful stapedectomy [102, 122]. At a first glance these findings might be interpreted to show that the anatomical location of the pathology that generated the tinnitus is the conductive apparatus of the ear. However, it seems more likely that the cause of the tinnitus in such patients was changes in the function of the central nervous system brought about by sound deprivation due to the conductive hearing loss, and the observed reduction of tinnitus after restoring hearing may be explained by restoration of normal sound input to the cochlea and thereby to the CNS.

When the neural activity in many nerve fibers becomes phase locked to the same sound, the activity of each such fiber also becomes phase-locked to other's neural activity (spatial coherence). The central nervous system may use information about how many nerve fibers have neural activity that is phase locked to each other (temporal coherence) for detection of the presence of a sound and perhaps to determine the intensity of a sound [82, 215]. Spatial coherence of neural discharges may thus provide important information to higher centers of the auditory nervous system. In the absence of sound stimulation, any other cause of similar coherence of neural discharges in many nerve fibers may be interpreted as the presence of sounds. It has therefore been hypothesized that slight injury to the auditory nerve could facilitate abnormal cross talk between axons of the auditory nerve and cause phase-locking of neural activity in groups of nerve fibers [82, 215]. Such temporal coherence of

---

[3]The limbic system is a complex system of nuclei and connections consisting of structures such as the hippocampus, amygdala, and parts of the cingulate gyrus. These structures connect to other brain areas such as the septal area, the hypothalamus, and a part of the mesencephalic tegmentum. The limbic system also influences endocrine and autonomic motor systems and it affects motivational and mood states (see p. 19).

BOX 10.3

DEPRIVATION OF INPUT CHANGES TEMPORAL INTEGRATION

Gerken et al. [101] demonstrated in animal experiments that deprivation of input to the central auditory nervous system could change in the temporal integration in nuclei of the auditory systems. After impairment of hearing the threshold was lower both for electrical stimulation of the cochlear nucleus and the inferior colliculus, a sign of increased excitability. The threshold did not decrease when the number of stimulus impulses was increased, indicating that the temporal integration was reduced. Gerken et al. [101] concluded that the neural basis for temporal integration in the cochlear nucleus can be affected by deprivation of auditory input.

Hyperactivity in the cochlear nucleus after intense sound stimulation has been demonstrated by Kaltenbach [143]. Exposure to loud sounds causes increased amplitude of evoked responses from the inferior colliculus in animals [280, 315, 320]. Other animal studies have shown that similar noise exposure as that causing hyperactivity in the inferior colliculus [320] affects the function of the place cells[4] in the hippocampus [103]. This means that even the function of non-auditory systems of the brain may be altered in patients with tinnitus.

Other animal experiments have shown extensive changes in vital processes in nerve cells can occur after deprivation of input [271, 272, 297]. These investigators showed that severing the auditory nerve caused considerable

morphologic changes to develop in the cochlear nucleus The changes in cochlear nucleus cells were most prominent when the destruction of the cochlea was done in the developing animal. Protein synthesis in neurons of the cochlear nucleus is affected with very short delay after interruption of input (spontaneous or driven) [297]. Rapid changes in protein synthesis in cells, ribosomes and ribosomal RNA have been demonstrated in chick cochlear nucleus. Degeneration of dendrites can also occur rapidly [297]. This means that extensive changes in the function of nerve cells can occur with little delay in response to deprivation of input.

Studies have shown that removal of the cochlea to eliminate input to the cochlear nucleus caused a reduction in cell size of the cochlear nucleus neurons and a reduction in the size of the cochlear nucleus [327]. Changes in cells of nuclei in more centrally located structures of the ascending auditory pathways have also been demonstrated [340]. Keeping animals in a noise-free (sound-free) environment or reducing the sound input by occluding the ear canals causes similar changes in the nuclei of the ascending auditory pathway. Webster and Webster [345] showed in the newborn mouse that after sound deprivation, the cross-sectional areas of cells in the ventral cochlear nucleus and in the medial nucleus of the trapezoidal body were reduced.

[4]Cells that are involved in orientation in space.

---

discharges in many nerve fibers would mimic the response to sound stimulation and this might be interpreted by the central nervous system as a sound being present even in quiet conditions, thus tinnitus. Such pathologic cross-transmission (ephaptic transmission) between nerve fibers could occur when the myelin sheath becomes damaged and the normally occurring spontaneous activity in many nerve fibers could thereby become phase-locked to each other.

Sympathetic nerve fibers terminate close to the hair cells of the cochlea [69], and noradrenalin secreted from these adrenergic fibers may sensitize cochlear hair cells. It is conceivable that increased sympathetic activation can increase the sensitivity of cochlear hair cells to an extent that neural activity is generated even in the absence of sound. Stress activates the sympathetic nervous system and it is an indication of involvement of the sympathetic nervous system that

stress can aggravate tinnitus. Similar sensitization of receptors occurs in the somatosensory system, and that has been related to pain conditions (sympathetic maintained pain) (see [213]).

4. ABNORMAL PERCEPTION OF SOUNDS

Abnormal perception of sounds includes hyperacusis and recruitment of loudness. Hyperacusis is a lowered threshold for discomfort from sounds (lowered tolerance). Recruitment of loudness is a form of abnormal perception of loudness that is not associated with abnormal tolerance to sounds. Distortion of sounds is another anomaly that sometimes occurs, often together with tinnitus. Phonophobia, fear of sounds, may occur together with tinnitus but it can also occur together with other pathologies. Misophonia is an unpleasant perception of usually only a few, specific sounds.

4.1. Hyperacusis

The term hyperacusis [14] is used to describe a lowered threshold for discomfort from sounds that typical individuals do not find unpleasant. (Hyperacusis is also known as auditory hyperesthesia.) The decreased tolerance to sounds involves most sounds. Sounds above a certain level are normally perceived to be unpleasant but in patients with hyperacusis the sound level at which that occurs is lower than it is normally. When the sound level of discomfort is lowered the useable range of hearing is reduced. Hyperacusis can occur in individuals who have normal hearing threshold but it often occurs together with hearing loss and tinnitus.

Hyperacusis has many similarities with hyperpathia, which is a lowered tolerance to moderate pain stimulation [213, 217]. Hyperpathia often accompanies central neuropathic pain. The range between threshold of feeling of electrical stimulation of the skin and that which gives rise to pain sensation is narrower in some patients with neuropathic pain and the temporal integration of painful stimulation, that is evident in individuals without pain, is reduced or absent in some patients with central neuropathic pain [224].

Patients with what was earlier known as retrocochlear disorders (mostly disorders of the auditory nerve) have a higher threshold of discomfort than patients with cochlear types of disorders [121]. (In an attempt to adapt common neurological terminology to the auditory system, disorders of the auditory nerve are now known as auditory neuropathy [21, 312].) This difference in threshold of discomfort in cochlear injuries and in auditory neuropathy has been used to distinguish between disorders of the cochlea and disorders of the auditory nerve.

Hyperacusis often accompanies severe tinnitus, adding to the annoyance from tinnitus. Some patients judge hyperacusis to be worse than the tinnitus [145]. A few specific disorders are associated with hyperacusis. One is the Williams-Beuren syndrome (WBS) [29, 151, 177]. As many as 95% of individuals with WBS have hyperacusis and react adversely to sounds of moderate intensity [29, 151].

The fact that individuals with WBS have hyperacusis and higher than normal emotional reactions to sounds such as music and certain types of noise may indicate an abnormal activation of limbic structures [177]. It has been hypothesized that 5-HT (serotonin) may be involved in the disorder [188].

Lyme disease is another disorder that often is accompanied by hyperacusis (and tinnitus). Autism is also often associated with discomfort from loud sounds. Hyperacusis often occurs together with traumatic brain injuries and stroke and possibly also together with vestibular disorders such as those of superior canal dehiscence [17]. Different forms of intoxication can also cause hyperacusis. Tinnitus is one of the three symptoms that characterize Ménière's disease (see p. 229). Tinnitus (but not hyperacuris) almost always occurs in individuals with vestibular Schwannoma.

Expression of neural plasticity is assumed to be involved in causing hyperacusis. The abnormalities in processing of sound that cause hyperacusis may involve re-routing of information to parts of the nervous system that are normally not activated by sounds. Similar signs of re-direction of auditory information to non-classical pathways as has been shown to occur in severe tinnitus [223] has also been shown to occur in individuals with autism [221].

Increased arousal from sounds may contribute to the symptoms of hyperacusis. Sounds can cause arousal either through the reticular activating system, which receives input from ascending auditory pathways, or because of facilitation from the amygdala

---

BOX 10.4

INFANTILE HYPERCALCEMIA

Williams-Beuren syndrome (WBS), also known as infantile hypercalcemia, is characterized by high blood levels of calcium and is believed to be caused by hypersensitivity to vitamin D. Individuals with WBS have multiple congenital anomalies, such as cardiovascular disorders, prenatal and postnatal growth retardation, facial abnormalities and mental retardation including poor visuo-spatial skills but relatively preserved verbal skills, loquacity (talkativeness), motor hyperactivity and hyperacusis. Reports of the incidence of WBS differ between investigators from 1 in 20,000 live births [29] to 1 in 50,000 [9]. Individuals with WBS also have a high incidence of otitis media but their hyperacusis seems to be unrelated to that.

nuclei via the nucleus basalis. The amygdala may be activated either through the auditory cortex and association cortices, or through a subcortical route from the dorsal thalamus (see p. 90).

4.2. Phonophobia

Phonophobia is fear of sound [244] making it compatible with "photophobia," which is often experienced in connection with head injuries. Phonophobia is a sign that sensory stimuli evoke abnormal emotional reactions of fear. Phonophobia may occur together with severe tinnitus [214] and it may also occur in disorders such as multiple sclerosis [344].

Phonophobia is caused by changes in the function of the auditory pathways probably through expression of neural plasticity. Redirection of auditory information to limbic structures such as the amygdala is probably involved in the pathogenesis of phonophobia. (The amygdala is involved in fear, depression, anxiety, etc.) Establishment of subcortical connections from the auditory pathways to the lateral nucleus of the amygdala may be the cause of phonophobia. Auditory information can normally reach the lateral nucleus of the amygdala via the primary auditory cortex, secondary and association cortices (high route) (see Chapter 5, Fig. 5.13) [173]. The subcortical connections to the amygdala from the auditory system (the low route) involve the dorsal part of the thalamus, which is a part of the non-classical ascending auditory pathways.

4.3. Misophonia

Misophonia is a dislike of specific sounds. Unlike hyperacusis, misophonia is specific for certain sounds. Little is known about the anatomical location of the physiological abnormality that causes such symptoms but it is most likely high central nervous system structures.

4.4. Recruitment of Loudness

Recruitment of loudness is an abnormal (rapid) growth of loudness perception with increasing sound level. Recruitment of loudness involves impairment of the normal mechanisms for compression of the dynamic range of hearing (automatic gain control). Recruitment of loudness therefore causes a narrowing of the hearing range for loudness. (Abnormal perception of loudness of sounds has also been labeled dysacusis [244].) Hearing loss that is associated with cochlear injuries such as from noise exposure, ototoxic antibiotics, or age-related changes is often accompanied by various degrees of recruitment of loudness. Recruitment of loudness

may also be noted after paralysis of the stapedius muscle such as in Bell's Palsy (see Chapter 8), or after severance of the stapedius tendon that occurs after stapedectomy. Patients often adapt to recruitment of loudness, a sign that the brain can be retrained to process sounds normally.

Recruitment of loudness has sometimes incorrectly been included in the term hyperacusis but this form of abnormal perception of loudness is not directly associated with unpleasant perception of sounds as in hyperacusis.

The anatomical location of the physiological abnormality of recruitment of loudness is the ear, most often the cochlea, but absence of function of the acoustic middle-ear reflex can also cause an abnormal growth of the sensation of loudness above the normal threshold of the acoustic middle ear reflex (approximately 85 dB HL[5]) (see Chapter 8) [210].

The automatic gain control of the normal ear compresses the intensity range of sounds before they are coded in the discharge pattern of auditory nerve fibers. In the normal ear automatic gain control compresses the intensity range of sound before the sounds are coded in the discharge pattern of auditory nerve fibers. The automatic gain control depends on the function of the outer hair cells that act to amplify the motion of the basilar membrane at low intensities more than at high intensities. This dependence on the sound intensity of the action of outer hair cells results in amplitude compression (automatic gain control). Cochlear type of hearing loss is normally caused by impaired function of outer hair cells, and therefore impairment of the cochlear amplifier and impairment of the automatic gain control.

Recruitment of loudness frequently occurs together with noise induced hearing loss and other forms of hearing loss that affect outer hair cells such as that caused by administration of antibiotics and NIHL and in disorders such as Ménière's disease. When the acoustic middle ear reflex is impaired or absent, sounds of abnormally high intensities (above 85 dB HL) may reach the cochlea because the normal attenuation by the middle-ear reflex is absent. The most common cause of absence of the acoustic middle-ear reflex is facial nerve dysfunction such as occurs in Bell's Palsy. Severance of the stapedius tendon that occurs in stapedectomy operations eliminates the attenuation of sound that the acoustic middle ear reflex normally causes.

---

[5]Hearing level (HL): HL is the level in dB relative to the average hearing threshold of young individuals who do not have any disorders that are assumed to affect hearing.

BOX 10.5

RECRUITMENT OF LOUDNESS

It has earlier been assumed that recruitment of loudness is caused by an abnormal rapid growth of loudness. Recent studies have, however, shown that near the elevated threshold in individuals with cochlear hearing loss, loudness grows at a similar rate as in ears with normal hearing (with an exponent of 1.26 versus 1.31 in normal hearing ears [34]). Above threshold, loudness of sounds are perceived to be abnormally large and that is a better definition of recruitment of loudness than the classical definition of an abnormally rapid growth of loudness above an elevated threshold. The loudness at (elevated) thresholds has been shown to double for every 16 dB hearing loss. This, together with a larger exponent at 20 dB SL,[6] is in agreement with a near-normal loudness at high sound intensity in patients with hearing loss of a cochlear type. However, other studies [202] using loudness matching showed results that were inconsistent with this "softness imperception" hypothesis presented by Buus and Florentine [34]. These findings were synthesized in a model of loudness that is valid for normal as well as ears with cochlear injuries, and it also included the reduced loudness summation that is associated with recruitment of loudness [203].

[6]Sensation level (SL): SL specifies the level of a sound in terms of the person's own threshold. Regardless of whether the person has normal hearing or not, the person's threshold is defined as 0 dB. For example, a sound 30 dB more intense than the sound level at the person's threshold is described as 30 dB SL.

---

5. TREATMENT OF SUBJECTIVE TINNITUS

The fact that tinnitus is a complex disorder that has many forms and many different causes hampers finding effective treatments for the disorder. Treatments that have been used include medical treatment, sound treatment, and electrical stimulation of the ear and of the somatosensory system and, more recently, of the auditory cerebral cortex. Surgical treatments such as severance of the auditory nerve and MVD of the auditory nerve root are also used. Of the many different treatments that have been tried, beneficial effects have only been obtained in small groups of patients.

Like individuals with central neuropathic pain [213], tinnitus patients often invoke a suspicion of malingering, or having psychological disturbances or psychiatric disorders. Tinnitus is therefore not only a problem for the patient but also for the physician, who often does not know what to do to help the patient who is clearly miserable. It may be tempting for the person who treats a patient with tinnitus to state that "there is nothing wrong with you" because all test results are normal. We have to realize, however, that there are real disorders that are not associated with abnormal results of the tests we use at present. It would therefore be a more correct to state: "I do not know what is causing your tinnitus or how to treat it."

5.1. Medical Treatment

The fact that administration of the local anesthetic Lidocaine can totally abolish tinnitus in some individuals [99] has encouraged medical treatment, first done in patients with Ménière's disease. Lidocaine is not a practical treatment for tinnitus because it must be administrated intravenously. A similar drug to Lidocaine, Tocainide, which can be administrated orally, has considerable side effects [72, 73]. Some studies have found beneficial effects of local application of Lidocaine to the ear [73, 142].

Some medical treatments such as administration of benzodiazepines (Alprazalam, Clonazepam) [332] that are $GABA_A$ receptor agonists aim at restoring the balance between inhibition and excitation in the brain. A $GABA_B$ receptor agonist, baclofen, has also been tried but with little practical success. Carbamazepine, a sodium channel blocker [323] that is used in treatment of seizures and of pain, such as trigeminal neuralgia, has also been tried but with poor results. Also antidepressants have been tried.

In general, lack of controlled studies [72] together with the difficulties in making differential diagnosis of tinnitus have made the choice of drugs for medical treatment more an art than a science. It often happens that a drug that has shown promising effects in a pilot study or from experience by individual physicians fails when subjected to the rigor of standard evaluation such as double blind tests. Individuals with tinnitus are

---

BOX 10.6

EFFECT OF LIDOCAINE

Lidocaine is primarily thought of as a sodium channel blocker but it has many other effects and it has not been possible to determine which one of these effects is effective in treating tinnitus. It was originally thought that Lidocaine acts on cochlear hair cells but its effect may in fact be on the central nervous system. This hypothesis was supported by a recent study of patients who had

undergone translabyrinthine removal of vestibular Schwannoma [16], and thus had their auditory nerve severed. This study found a statistically significant beneficial effect on the tinnitus from Lidocaine compared with placebo [16]. The assessment used a visual analog scale for determining the intensity of the tinnitus.

---

not a homogeneous group regarding pathology and one drug may be effective in some individuals with tinnitus but not in others. This can have serious implications in testing of the efficacy of drugs using the double blind technique [72]. A drug that is effective in treating one kind of tinnitus may not reach significance in a group of individuals with different diseases. For example, the members of a group of patients with three different pathologies may benefit from (three) different treatments. If any one of these treatments is tested alone on such a heterogeneous group it may be impossible to obtain significant results, even in the situation where the treatment tested is effective in treating tinnitus with one particular kind of tinnitus. Unfortunately, the negative results of such double blind studies may discourage the use of treatments that are effective in some patients because of the great trust in double blind studies. A physician can try different treatments for an individual patient and thus achieve good results.

Combining two or more treatments that affect different "causes" of a disease may be the most effective therapy because the individual drugs may have an additive effect and could even have a synergistic effect. However, development of such combination treatments is hampered by difficulties in testing efficacy.

Anyhow, medical treatment of tinnitus with drugs is more an art than a science, and physicians often try different drugs, thus a trial and error approach, which by some patients may be interpreted as being used as "guinea pigs."

5.2. Electrical Stimulation

Electrical stimulation of the cochlea, the auditory nerve, the skin behind the ear or skin in other parts of the body, such as on fingers or peripheral nerves, have all been tried for alleviating tinnitus. More recently

electrical stimulation of the cerebral auditory cortex has been described for treatment of tinnitus. Some of the earliest attempts to apply electrical stimulation of the cochlea for tinnitus suppression used direct current (d.c.) while most subsequent attempts have used short impulses.

Electrical current (d.c.) that is passed through the cochlea can reduce tinnitus in some patients [46]. These investigators placed an electrode on the round window or the promontorium, and passed a positive current through the cochlea. Six of seven individuals with tinnitus obtained relief. It was assumed that the electrical current that passes through the cochlea affected the hair cells so that the spontaneous activity in auditory nerve fibers would decrease. However, the electrical current could also have affected the auditory nerve.

Stimulation with high frequency trains of electrical impulses applied to the cochlea seems to have a beneficial effect on certain forms of tinnitus where high frequency hearing loss is present [273]. Such stimulation probably restores the inhibitory influence of nerve fibers that are tuned to high frequencies and which have been reduced through the patient's hearing loss.

In deaf people with tinnitus the electrical stimulation provided by a cochlear implant can relieve tinnitus [200] because it stimulates the auditory nerve electrically. The electrical stimulation of the cochlea may also compensate for the deprivation of input in the high frequency range, which is a promoter of expression of neural plasticity (see p. 250).

Stimulation of the skin close to the ear has been used in attempts to stimulate the ear transcutaneously [288]. It is, however, unlikely that the electrical current from stimulation by electrodes placed behind the ear would reach the ear with sufficient strength to activate hair cells or auditory nerve fibers. It seems more likely that such stimulation might have had its effect by stimulating the trigeminal nerve fibers in the skin or somatosensory receptors that are innervated by the trigeminal nerve. Such cutaneous nerve stimulation seems to help a few (28%) individuals with tinnitus [331]. Other investigators who used electrical stimulation of peripheral nerves [223] or other forms of activation of the somatosensory system [141, 258] including the skin on fingers [87] also found that such stimulation could affect the perception of tinnitus. The explanation is likely to involve cross-modal interaction between the somatosensory system and the auditory system [37, 223] (see p. 86), through activation of the non-classical auditory pathways (see p. 85). Electrical stimulation of the somatosensory system never gained practical use in treatment of individuals with tinnitus.

Electrical stimulation of the auditory cortex has been done for treatment of tinnitus. For that purpose electrodes have been implanted near the auditory cerebral cortex. The electrical stimulation is generated by devices that are similar to those used in cardiac pacemakers. Transcranial magnetic stimulation that induces an electrical current in the cerebral cortex [63, 162] has been used as a test for patients with tinnitus to determine whether they would benefit from implants of stimulus electrode electrical stimulation of the auditory cortex. Electrical stimulation of the auditory cortex may reverse the re-organization of the cerebral cortex that is associated with tinnitus [232]. It is also possible that the effect of such electrical stimulation in fact does not have its beneficial effect by stimulation of the cerebral cortex but rather by affecting the thalamic auditory neurons through the abundant descending pathways (see Chapter 5, p. 89). It is possible that the neurons in the dorsal thalamus that are part of the non-classical pathways in that way become affected and the presumed hyperactivity becomes reversed.

5.3. Surgical Treatment

Surgical treatment of tinnitus has been mainly of three kinds, namely severance of the auditory nerve, MVD of the auditory nerve intracranially and sympathectomy.

Severing of the auditory nerve can alleviate tinnitus in many patients with Ménière's disease. As early as 1941, the neurosurgeon Dandy reported relief of tinnitus in approximately 50% of patients with Ménière's disease after sectioning of the eighth cranial nerve intracranially [60]. Labyrinthectomy and translabyrinthine section of the eighth nerve has been done in patients with vertigo and tinnitus. Pulec reported that auditory nerve section, medial to the spiral ganglion, provided relief of tinnitus in 101 of 151 patients that he treated in that way [253]. Other surgeons have reported success rates in the order of 40% [15, 127]. The beneficial effect on tinnitus from sectioning the eighth nerve is generally better in patients who have both vertigo and tinnitus [117, 122].

Microvascular decompression (MVD) of the auditory portion of the eighth cranial nerve [129, 130, 156] can alleviate tinnitus in some patients [230]. Microvascular decompression of cranial nerves is an established treatment for disorders such as HFS, TGN [18, 19, 218], and certain forms of vertigo (disabling positional vertigo [DPV]) [231]. The success rate of MVD for treatment of these three disorders has been reported to be approximately 85%. MVD operations for tinnitus have a much lower success rate, approximately 40% for total relief or much improved [230]. This is only about half of the success rate of microvascular decompression operations for TGN and HFS.

Sympathectomy or blockage of a cervical sympathetic ganglion (the stellate ganglion) has been done to treat tinnitus in patients with Ménière's disease [239]. Its effect may be explained by a reduction of secretion

---

BOX 10.7

SUCCESS RATE OF MVD OPERATIONS FOR TINNITUS

The success rate of microvascular decompression for tinnitus was different for men and women. Men had only 29.3% relief, while 54.8% of the women had relief of the tinnitus [230] while the success rate of MVD for TGN and HFS in men and women is similar [18, 19]. The success rate for MVD as a cure of tinnitus also depends on how long time a patient has had tinnitus. Patients who had total relief of their tinnitus or were markedly improved had only had their tinnitus for 2.9 and 2.7 years respectively, but patients who had only a slight improvement or no improvement at all had their tinnitus for an average of 5.2 and 7.9 years, thus a sign that the changes in the auditory system had become permanent [230]. The success rate for the MVD operation is higher in patients with unilateral tinnitus than with bilateral tinnitus [329].

BOX 10.8

STELLATE GANGLION BLOCK AS TREATMENT FOR TINNITUS

Stellate ganglion block is effective in treating the tinnitus in some patients with Ménière's disease [2, 239, 342] as demonstrated many years ago. Relief of tinnitus was obtained in 56% of patients with Ménière's disease but only 27% of patients with other causes of tinnitus benefited from that procedure [239, 342]. Other investigators [1] found that tinnitus in patients with Ménière's disease could either increase or decrease because of sympathectomy.

of noradrenalin near the hair cells and subsequent reduction of the sensitization of hair cells.

Otosclerosis often is associated with moderate tinnitus and stapedectomy for otosclerosis can significantly reduce tinnitus. In a study of 40 patients, 85% experienced reduced tinnitus after the operation [306], and in another study of 149 patients, 73% experienced complete relief after the operation [322].

5.4. Desensitization

Tinnitus retraining therapy (TRT) [134] is a behavioral treatment that has roots in the hypothesies that tinnitus is a phantom sensation caused by expression of neural plasticity [131, 132]. TRT consists of psychological treatment and exposure to sounds of moderate levels. The aim of the TRT method is to disconnect the patient psychologically from dependence on the tinnitus while subjecting the patient to moderate levels of sounds to reverse the effect of sound deprivation on the function of the central nervous system. The TRT method has shown some success in reducing tinnitus [134, 163].

The distinction between directive counseling and cognitive therapy points to the fact that this form of treatment – as so many other treatments of tinnitus – lacks the support of population studies using accepted methods for evaluating the efficacy of medical treatment [351].

6. TREATMENT OF HYPERACUSIS

Since hyperacusis is caused by abnormal function of the central nervous system and often involves expression of neural plasticity, treatment options consist of reversing these changes. Desensitization using exposure to sound of moderate level is an important part of treatment of hyperacusis. Hyperacusis in connection with tinnitus is relieved in some patients by TRT [133, 134]. Other methods that are successful in treating tinnitus normally also affect hyperacusis favorably.

11

Cochlear and Brainstem Implants

1. INTRODUCTION

Two kinds of auditory prostheses are in common use. One kind is known as cochlear implants and the other kind makes use of stimulation of the cochlear nucleus and is known as auditory brainstem implants (ABIs). Cochlear implants are devices that stimulate the endings of the auditory nerve in the cochlea with electrical impulses. ABIs stimulate cells in the cochlear nucleus. Modern cochlear implants use an array of 6–22 pairs of electrodes that are mounted on a plastic material and threaded into the cochlea through the oval window. ABIs are similar to the cochlear implants but the electrode array is placed on the surface of the cochlear nucleus. The electrical impulses that are applied to the cochlea or cochlear nuclei are derived from processing of sounds that are picked up by a microphone. While individuals who are deaf or have severe hearing loss caused by loss of cochlear hair cells use cochlear implants, ABIs are used in individuals whose auditory nerve does not function. This occurs most commonly in patients with neurofibromatosis type 2 (NF2) who have had bilateral vestibular Schwannoma removed with subsequent destruction of the auditory nerve. ABIs are also used in individuals who have had their auditory nerve transected through head trauma and in children with congenital auditory nerve disorders (auditory nerve aplasia). Cochlear implants are now the most successful of all prostheses of the nervous system and success of ABIs is rapidly improving with regard to providing good speech discrimination.

Electrical stimulation of the auditory nerve in the cochlea bypasses the complex function of the basilar membrane as a spectrum analyzer – a function that had been studied extensively and which has been believed to play a fundamental role in the function of the auditory system including the ability to understand speech. It was therefore met with great disbelief that such a simple device as cochlear implants could replace the function of the cochlea and it seemed unlikely that electrical stimulation of the auditory nerve could provide any useful hearing. While it was true that the early cochlear implants using only one electrode did not provide speech discrimination in the way we normally understand it, they did indeed provide valuable sound awareness to people who were deaf and provided an aid in lip-reading. Modern multi-electrode implants can provide good speech discrimination.

The success of cochlear and cochlear nucleus implants in providing useful hearing may still appear surprising because even multichannel cochlear implants cannot replicate the fine spectral analysis that normally occurs in the cochlea and some designs of cochlear implant processors do not use the temporal information in sound waves.

While the success of cochlear implants is a result of technological developments, the success would not have been achieved, at least not as rapidly, if brave individuals such as Dr House had not taken the bold step to try to provide some form of hearing sensations through electrical stimulation of auditory nerve fibers in individuals who were deaf because of loss of function of hair cells.

These auditory prosthetic devices have not only provided intelligibility of speech and recognition of many environmental sounds to individuals who are lacking hearing because of disorders of the cochlea and auditory nerve, but the introduction of these prostheses has initiated studies of the human auditory system and brought new perspectives on the importance of place and temporal coding of sounds.

In this chapter we will first discuss the development of cochlear implants and ABIs and then the physiological basis for these auditory prostheses.

2. COCHLEAR IMPLANTS

Cochlear implants are devices that stimulate auditory nerve endings in the cochlea with electrical impulses using an array of implanted electrodes. Electrical signals from a microphone that the user wears close to the ear are processed and applied to the cochlea through the electrodes that are placed along the basal part of the basilar membrane close to the endings of the auditory nerve. Early implants used only a single electrode and relatively simple processing strategies while contemporary devices use several electrodes and more sophisticated processing of sounds. These developments implied substantial improvement over the single electrode implants.

2.1. Development of Cochlear Implants

Cochlear implants were first introduced by Dr William House [123]. Pioneering work by Michaelson

regarding stimulation of the cochlea preceded the first clinical application of this technique [196].

Introduction of cochlear implants that used multiple implanted electrodes and improved processing of the signals from the microphone provided major improvements of speech discrimination. Using more than one electrode made it possible to stimulate different parts of the cochlea and thereby different populations of auditory nerve fibers with electrical signals derived from different frequency bands of sounds. When more sophisticated processing of the sound was added the results were clearly astonishing even to those individuals who had great expectations. Modern cochlear implants can provide speech discrimination under normal environmental conditions [77].

2.2. Function and Design of Cochlear Implants

All contemporary cochlear implants separate the sound spectrum using band pass filters so that the different electrodes are activated by different parts of the sound spectrum.

Three main kinds of processors are in use. One kind presents both spectral and temporal information and one kind presents only spectral information. A third kind of processor analyzes sounds and present information to the implanted electrodes about formant frequencies, etc.

BOX 11.1

HISTORY OF ELECTRICAL STIMULATION OF THE AUDITORY NERVOUS SYSTEM

While it had been shown that electrical current passed through an intact cochlea could give rise to sound sensation (electrophonic hearing), it was probably Djourno and Eyries [71] who first showed that electrical current passed through the auditory nerve in an individual with a deaf ear could cause sensation of sound, although only the noise of cricket-like sounds. Later, Simmons et al. [301] showed that stimulation of the intracranial portion of the auditory nerve using a bipolar stimulating electrode could produce a sensation of sound. The participant could discriminate the pitch of impulses below 1,000 pps with a difference limen of 5 pps. Above 1,000 pps the discrimination of pitch was absent but the test subject could distinguish between rising and falling pulse rates. Above 4000 pps no such discrimination could be done. This person had

normal cochlear function and it is naturally possible that what Simmons found was just another way of eliciting electrophonic hearing. However, the fact that the intracranial portion of the auditory nerve was stimulated using a bipolar electrode makes it unlikely that the stimulation could have activated hair cells in the cochlea.

An early study of electrical stimulation of the inferior colliculus did not provide any sensation of sound [301]. More recently, Colletti implanted electrodes in the inferior colliculus in a patient with bilateral auditory nerve section from removal of bilateral vestibular Schwannoma. The results showed that electrical stimulation of the inferior colliculus indeed can provide sound sensation and comprehension of speech (Colletti, personal communication, 2006).

Some processors provide a combination of these different principles of sound processing. Since the dynamic range of the electrical stimulation of the auditory nerve is much smaller than that of normal sounds, all cochlear implant processors compress the sounds before being applied to the electrode array and also the output of the band pass filters is compressed [182].

In its simplest version, the processing of the signals from the microphone consists of separating the sound spectrum into 4–8 frequency bands and then applying the output of these filters to the respective electrodes after gain control of various kinds (Fig. 11.1). This is known as the compressed-analog (CA) approach.

In the CA processors the signal is first compressed using an automatic gain control, and then filtered into four contiguous frequency bands, with center frequencies at 0.5, 1, 2, and 3.4 kHz (Fig. 11.1). The filtered waveforms pass through adjustable gain controls before being sent to four intracochlear electrodes. The electrodes are spaced 4 mm apart and operate in monopolar configuration through a percutaneous connection. An example of the four band-passed waveforms produced for the syllable "sa" using a simplified implementation of the CA approach is shown in Fig. 11.2. (The CA approach was originally used in the Ineraid device manufactured by Symbion, Inc., Utah [80]. The CA approach was also used in a UCSF/Storz device, which is now discontinued.)

Cochlear implants using the CA approach deliver continuous analog waveforms to four electrodes simultaneously. A major concern associated with simultaneous stimulation is the interaction between channels caused by the conduction of the electrical current between individual electrodes [347]. This causes stimuli from one electrode to interact with stimuli from other

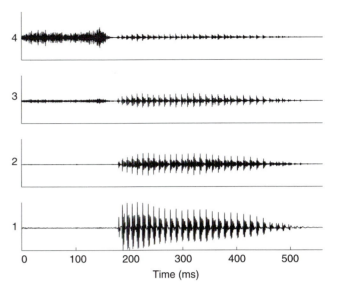

FIGURE 11.2   An example of the four band-passed waveforms produced for the syllable "sa" using a simplified implementation of the CA approach (reprinted from Loizou, 1998, with permission from the Institute of Electrical and Electronic Engineers).

electrodes thereby distorting the spectrum information. This problem was remedied by the introduction of continuous interleaved sampling (CIS) (Fig. 11.3) [347]. In the CIS configuration, the signals are delivered to the individual electrodes with a certain (small) delay. One manufacturer (Clarion) offers devices with processors that can be programmed with either the CA strategy or the CIS strategy.

Some cochlear implant devices have speech processors for enhancing the discrimination of speech. These sophisticated processors can extract information about formant frequencies and other speech features and then code these features in the pattern of impulses that are applied to the cochlea to stimulate the auditory nerve. Processors such as the Nucleus device that employ such feature-extraction were introduced in the 1980s. Automatic formant tracking was used for coding of vowel sounds. This was a fundamentally different approach from the CA or CIS principles of processing.

Signal processing principles that are based solely on extracting (power) spectral information are similar to that of the channel vocoder that was developed many years ago for serving in what was known as the analysis-synthesis telephony systems. In this type of cochlear implants the envelope of the output of band-pass filters controls the amplitude of electrical impulses that are applied to the electrode array that is implanted in the cochlea, usually using the CIS principle. Most cochlear implants are now based on the vocoder principle.

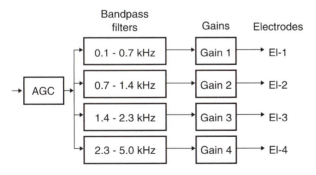

FIGURE 11.1   Four channel cochlear implant processor using the compressed analog (CA) principles. The signal is first compressed using an automatic gain control (AGC), and then filtered into four contiguous frequency bands, with center frequencies at 0.5, 1, 2, and 3.4 kHz. The filtered waveforms go through adjustable gain controls and then are sent directly through a percutaneous connection to four intracochlear electrodes (modified from Loizou, 1998).

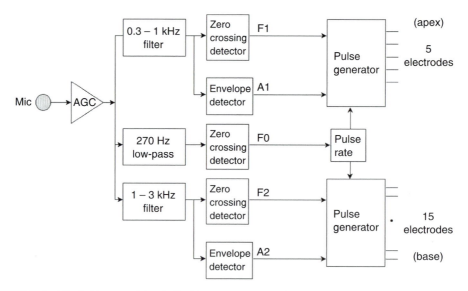

FIGURE 11.3   Block diagram of the F0/F1/F2 processor. Two electrodes are used for pulsatile stimulation, one corresponding to the F1 frequency and the other corresponding to the frequency of F2. The rate of the impulses is that of F0 for voiced sounds, and a quasi-random rate (average of 100 pps) for unvoiced segments (modified from Loizou, 1998).

BOX 11.2

DIFFERENT DESIGNS OF COCHLEAR IMPLANT PROCESSORS

One design of a processor for enhancing speech discrimination that was developed for the Nucleus device in the early 1980s (Fig. 11.3) uses a combination of temporal and spectral coding (F0/F1/F2 strategy). The fundamental (voice) frequency (F0) and the first and second formant (F1 and F2) are extracted from the speech signal using zero crossing detectors; F0 is extracted from the output of a 0.27 kHz low-pass filter, and F2 is extracted from the output of a 1–4 kHz band-pass filter (Fig. 11.3). The amplitude of F2 is estimated from the rectified and low-pass filtered (at 0.35 kHz) band-pass filtered signal. The output of these processors modulate impulses that are used to stimulate specific electrodes in the 20-electrode array that is implanted in the cochlea.

Another design, known as the MPEAK strategy, extracts the fundamental frequency (F0) and the formant frequencies (F1 and F2) using zero-crossing detectors of band pass filtered sounds. Band-pass filters followed by envelope detectors are used to determine the energy at high frequencies. This information is then coded in the pattern of the impulses that are applied to the implanted electrodes.

Extracting formant frequencies by processors of cochlear implants proved to be complex and did not work well in noisy environments [182]. It was therefore abandoned by some manufactures of cochlear implants and it was followed in the early 1990s by a signal processing strategy that did not require the extraction of any features of sound waves. This very simple strategy was based solely on the energy in a few frequency bands, thus the power spectrum of sounds (Fig 11.4).

This is why yet a another design, known as the Spectral Maxima Sound Processor (SMSP), has been developed. These processors do not extract speech features but treat all sounds equally. Spectral maxima are determined on the basis of the output of 16 band-pass filters. The output of the six band-pass filters with the largest amplitudes is coded in the impulses that are applied to the electrodes in the cochlea. The output modulates the amplitude of biphasic impulses with a constant rate of 250 pps. The Spectral Peak (SPEAK) strategy is similar to the SMSP strategy, but uses 20 filters instead of 16. (For details about these processing strategies, see Loizou, 1998 [182].) A modified CIS strategy, the enhanced CIS (EECIS), is used in cochlear implants manufactured by the Philips Corporation under the name of LAURA cochlear implants [242].

BOX 11.2   *(cont'd)*

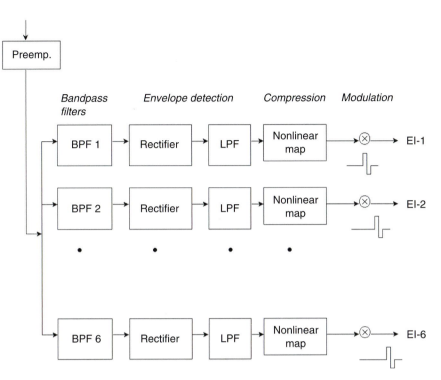

FIGURE 11.4 Block diagram of a processor using the continuous interleaved sampling (CIS) strategy in vocoder-type processor for cochlear implants. The signal is first passed through a network that changes the spectrum (pre-emphasis) and then filtered in six bands. The envelope of the output of these six filters is full-wave rectified and low pass filtered. The low pass-filters are typically set at 0.2 or 0.4 kHz cut-off frequency. The amplitude of the enveloped are compressed and then used to modulate the amplitude of biphasic impulses that are transmitted to the electrodes in an interleaved fashion. BPF = band-pass filter; and LPF = low pass filter (modified from Loizou, 1998).

BOX 11.3

HISTORY AND DESIGN OF THE CHANNEL VOCODER

The channel vocoder was developed in the 1950s–1960s for transmitting speech over long telephone lines. At that time, telephone lines consisted of copper wires and the capacity of these to transmitting speech signals was limited with regards to the frequencies (bandwidth) they could handle. In 1939, a research physicist, Homer Dudley, at Bell Laboratories, New Jersey, USA, described a device consisting of an analyzer at the transmitting end and a synthesizer at the receiving end that could reduce the bandwidth required to transmit speech sounds. This device

became known as the channel vocoder (Voice Operated reCorDER) [70, 287].

Dudley's vocoder converted information about speech in a series of control signals from which the speech could again be synthesized. Transmitting speech directly requires a bandwidth of approximately 3 kHz but the bandwidth required for transmitting these control signals was only a fraction (approximately 1/10) of that required to transmit the speech signal [287], thus allowing many more telephone calls to be transmitted on the same cable. This was

BOX 11.3   (cont'd)

important for transmitting more telephone calls on long distance cables. Such "analysis synthesis telephony" systems were expected to be economically feasible at the time when bandwidth was expensive because of the limitations of copper wires, especially in long telephone cables such as transoceanic cables.

The channel vocoder consists of band-pass filters that separate the frequency spectrum of speech sounds in a few frequency bands (Fig. 11.5). Information about the energy in each band was to be transmitted as slowly varying signals. At the receiving end these slowly varying signals were used to synthesize the speech sounds. The receiver consists of a similar bank of band-pass filters and the input to these filters was the fundamental frequency for voiced sounds and broadband noise for voiceless sounds (fricatives). The slowly varying signals that were received from the transmitting vocoder controlled the output amplitude of each band pass filter (Fig. 11.5). The sum of the output of these band-pass filters was then sent via normal telephone lines to the switching stations.

During the years 1960–1970 many other schemes in addition to the channel vocoder emerged for compression of speech with regard to the bandwidth necessary for

transmission of speech over long telephone lines [287]. One such proposed scheme analyzed the speech for the purpose of continuously determining the frequency of formants (formant tracking) [91]. Slowly varying signals that described the formant frequencies were then transmitted to the receiver where they controlled the formant frequencies of a speech synthesizer [91].

None of these techniques ever became fully developed before other less expensive possibilities were developed for transmission of many speech signals at the same time, first through the availability of satellites and later by fiber optic cables. Both of these techniques offered inexpensive bandwidth for transmission of speech sounds and even signals that require much larger bandwidth such as television signals and data of various kinds.

While channel vocoders or other analysis-synthesis systems never became used for the purpose for which they were developed, they did get some use in converting speech of deep sea divers that sounds like "Donald Duck" speech because the divers breathe a helium-oxygen mixture. Vocoder-type devices have been used in attempts to develop speech communication using the tactile sense [245].

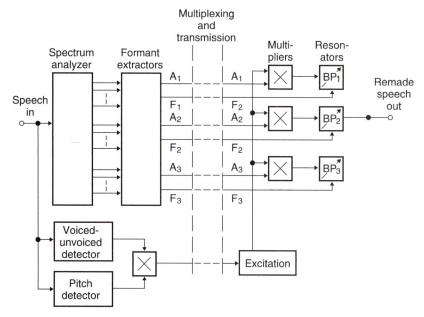

FIGURE 11.5 Schematic diagram of a vocoder that was developed in the early 1960s (reprinted from Schroeder, 1966, with permission from the Institute of Electrical and Electronic Engineers).

The results from the research on analysis-synthesis telephony systems are of value now because these same principles are applicable to the processors in cochlear implants. Studies in connection with the development of the vocoder that were done in the 1950s have shown that it is possible to obtain speech intelligibility on the basis of the energy in a few frequency bands. The channel vocoder did not preserve any temporal information about the speech except the voice frequency and the filters in the channel vocoder were much less frequency selective than the cochlear filters. These experimental vocoders also had much fewer filter bands than the cochlea and they could transmit almost all information that is necessary for synthesis of intelligible speech. (The frequency selectivity of the cochlea has been estimated to correspond to 28 independent filters [201].) Despite the success of cochlear implants that work as channel vocoders, it seems likely that including temporal coding in cochlear implants would improve performance, especially perhaps for non-speech sounds such as musical sounds.

2.3. Physiological Basis for Cochlear Implants

Cochlear implants bypass the frequency selectivity of the basilar membrane and replace it by a more coarse division of the audible spectrum than what the cochlea normally provides [93]. The success of cochlear implants in providing useful hearing may appear surprising because even multichannel cochlear implants cannot replicate the spectral analysis that occurs in the cochlea and do not include temporal coding of sounds. The fact that cochlear implants that use the vocoder principle are successful in providing good speech comprehension without the use of any temporal information sets the importance of the temporal code of frequency in question.

The success of cochlear implants in providing good speech comprehension, however, confirms earlier studies regarding development of the channel vocoder that showed that the auditory system could adequately discriminate speech sounds on the basis of information about the power in a few frequency bands [79, 287].

Three main reasons why cochlear implants are successful in providing speech intelligibility may be identified:

1. Much of the natural speech signal is redundant, which may explain why cochlear implants only need to transmit a small fraction of the information that is contained in speech sounds to achieve good speech intelligently. This was recognized as early as 1928 when Dudley conceived the "vocoder" for transmitting speech over telephone lines [79] and the observation has been confirmed in many later studies.

2. Much of the processing capabilities of the ear and the auditory nervous system are redundant. Individuals with normal hearing can understand speech solely on the basis of temporal information [293], and studies of the vocoder principle have shown equally convincingly that speech can be understood solely on spectral (place) information as well [79, 287]. This means that frequency discrimination can rely on either the place or the temporal hypothesis. The importance of the frequency analysis that takes place in the normal cochlea is different from what was believed for many years. The analysis that occurs in the auditory nervous system is far more important for discrimination of sounds than generally recognized.

3. The central nervous system has an enormous ability to adapt ("re-wire") to changing demands through expression of neural plasticity.

The finding that good speech comprehension can be achieved on the basis of only the spectral distribution of sounds seems to contradict the results of animal studies of coding of the frequency of sounds in the auditory nerve [276, 354]. Such studies have shown that temporal coding of sounds in the auditory system is more robust than spectral coding. On these grounds it has been concluded that temporal coding is important for frequency discrimination (see Chapter 6). These and other studies have provided evidence that the place principle of coding of frequency is not preserved over a large range of sound intensities [276] and that it is not robust [209]. On the basis of these findings it was concluded that the place principle is of less importance for frequency discrimination than temporal information [358]. It was always assumed that frequency discrimination according to the place principle would require narrow filters and many filters covering the audible frequency range but studies in connection with development of channel vocoders, and more recently in connection with cochlear implants, showed clearly that speech comprehension could be achieved using much broader and much fewer filters.

Channel vocoders and cochlear implants that use the vocoder principle have similarities with trichromatic color vision in humans. The trichromatic system uses only three channels and provides the basis for discrimination of small nuances of color, thus small differences in the wavelength (or spectrum) of light. Color discrimination is accomplished by combining the information about the intensity in three broad spectral bands of the visual spectrum. The three kinds of photo pigment that are present in the cones of the retina in the human eye act as spectral filters (see [216]). The relationship between energy in these three bands of the visual spectrum is sufficient to provide detailed information about the spectrum of light and thus many nuances of colors. The output of these receptors also depends on the intensity of the light but that affects all three types equally and therefore does not affect the relationship between the output of the three receptors. Any color (wavelength of light) will therefore result in a unique relationship between the output of these three overlapping spectral filters (Fig. 11.6) and that is the basis for color discrimination in humans and in the animals that have trichromatic color vision.

The central nervous system of vision is capable of discriminating light with different wavelength (thus the color) on the basis of the response from these three kinds of receptors and it is not necessary to have receptors that are sensitive to each wavelength of light that can be discriminated. That trichromatic color vision is the basis for our color discrimination means that light with different spectra (or wavelength), thus nuances of colors, generates a unique relationship between the output of these three types of receptors. That is the basis for discrimination of light of many different wavelengths.

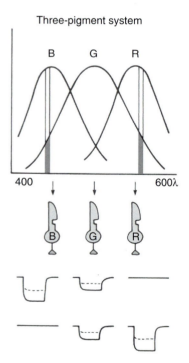

FIGURE 11.6 Illustration of how a three-pigment system can distinguish colors (wavelength of light) independently of the intensity of the light, provided that the intensity is sufficient to elicit a response from at least two of the three kinds of receptors (adapted from Shepherd, 1994).

The similarity between the basis for trichromatic color discrimination and the vocoder is thus obvious. To illustrate how frequency discrimination in the auditory system can be achieved by using a few (three) filters, assume that the task is to determine the frequency of a single spectral component, such as a pure tone. When the bands of frequencies covered by each filter overlap (because the slope of the attenuation of the filters is finite) a tone, the frequency of which is within the range covered by the filter bank, will cause output of more than one of the individual filters. The relationship between the output of the different filters is unique for any frequency of the tone and therefore, like in the visual system, the relationship between the output of three or more band pass filters will provide unique information about the frequency of a pure tone. It is essential that filters overlap so that the tone produces an output of more than one filter. It is probably also important that the filters have a rounded passband rather than having a flat top as is often preferred in man-made spectral filters.

In the same way, the relationship between the outputs of three or more filters can provide the basis for discrimination sounds that have broad spectra as that of speech sounds.

One of the strongest arguments against the place hypothesis for frequency discrimination has been that the frequency to which a certain point on the basilar membrane shifts when the sound intensity is changed (see p. 44). This lack of robustness of cochlear spectral analysis has been regarded an obstacle to the place hypothesis for frequency discrimination [209, 358]. Since the band pass filters in cochlear implants do not change with sound intensity (see p. 271) the vocoder-type cochlear implants may actually have an advantage over the cochlea as a "place" frequency analyzer. The spectral acuity of the cochlea also changes with sound intensity, which is not the case for the filters used in cochlear implants.

For practical reasons it is important to consider how many band-pass filters are necessary in cochlear implant processors to ensure good speech discrimination. Development of the channel vocoder revealed that speech recognition does not require that fine spectral details are preserved [79, 287] and a total of 15 frequency bands was found to be sufficient to synthesize speech and achieve satisfactory intelligibly for telephone communication (Fig 11.5). The frequency selectivity of the cochlea is regarded to be equivalent to approximately 28 independent filters in the frequency range of speech [7, 201]. More recently, studies regarding design of cochlear implant processors have shown that a dramatic improvement in intelligibility occurs when the number of channels (thus the number of band-pass filters) is increased from one to four. The increase in speech intelligibility, however, increases only slightly when the number of filters is increased above eight [93]. Other studies in individuals with normal hearing have shown that 4–5 channels are sufficient for a high degree of speech discrimination (90%) [183].

2.4. Coding of Sound Intensity

The function of cochlear implants that use the vocoder principle depends on proper coding of sound intensity in a wide range of sound intensities. Sound intensity is coded in auditory nerve fibers by the discharge rate but only a few auditory nerve fibers seem to code sound intensity over the physiological range of sound intensities. The discharge rate of most nerve fibers reaches saturation only 20–30 dB above hearing threshold [233] (see Chapter 6). Most nerve fibers, however, seem to code changes in sound intensity over a much larger range of sound intensities [58].

The number of channels that are required depends on the resolution of coding of the intensity of sounds in these frequency bands [184]. If the resolution of the coding of intensity is reduced, more channels are needed. Using six channels, the speech discrimination was reduced significantly when the intensity coding had only eight steps and the number of channels had to be increased to 16 to obtain good speech discrimination (92%) with that resolution. Limited spectral resolution and interaction between electrodes have been regarded to be a cause of susceptibility to background noise [20].

Cochlear implants code the intensity of sounds (the energy in respective frequency bands) by the amplitude of the electrical signals that are used to stimulate the auditory nerve. In the normal cochlea, increasing stimulus strength of a sound causes an increasing number of nerve fibers to become activated because of the widening of the segment of the basilar membrane that cause activation of nerve fibers (see Fig. 6.2) and, in addition, the discharge rate of at least some nerve fibers increases with increasing stimulus intensity (see Fig. 6.21). In cochlear implants increasing sound intensity causes more nerve fibers to be activated but the discharge rate is independent of the sound intensity.

2.5. Functions that Are Not Covered by Modern Cochlear Implants

Cochlear implants generally do not convey information about the fine temporal pattern of sounds.

It is assumed that the normal auditory system can discriminate fine frequency information because small changes in frequency of sounds are preserved in the temporal patterns of discharges in the auditory nerve fibers, much like that of the formant frequencies [354]. The envelope of the output of the different filters of the vocoder type cochlear implant processors preserve modulation information up to 0.2 or 0.4 kHz but the temporal fine-structure (frequency modulation) is thrown away. It may be advantageous to preserve the fine temporal structure of sounds because of its importance in coding fine frequency information such as in music sounds. Cochlear implants perform poorly for perception of music (melody recognition).

The response areas of auditory nerve fibers are surrounded by inhibitory bands [275], known as two-tone suppression (see Chapter 6, Fig. 6.5). It is believed that these areas of suppression are important for the normal function of the auditory system but that function is not covered by cochlear implants. Two-tone inhibition resembles lateral inhibition that is best known from the visual system where it has been shown to enhance contrast [261]. Two-tone suppression may enhance responses to sounds such as sounds with rapidly varying frequency [81, 207].

Cochlear implants cause synchronous (coherent) activation of many nerve fibers, which is not the case for the normal activation of the auditory nerve. The importance of that is not known but some hypotheses suggest that temporal coherence of activity in the auditory nerve is important for detection of sounds and for discrimination of sound intensity (loudness) (see Chapter 6).

The electrodes in cochlear implants can only be placed in the basal portion of the cochlea, which means that low frequency sounds activate auditory nerve fibers that normally respond to high frequency sounds. Cochlear implants therefore do not stimulate auditory nerve fibers according to the frequencies to which they are normally tuned. Thus, the tonotopic map that is normally based on the separation of sounds by the basilar membrane of the cochlea will be different in cochlear implant users than it is in individuals with normal hearing. This tonotopic (or cochleotopic) organization exists throughout the auditory nervous system, including the cerebral auditory cortex (see Chapter 6), but functional importance of this anatomical organization is unknown.

Another feature of the cochlea that is not included in modern cochlear implants is the difference in the travel times of the motion of the basilar membrane for sounds of different frequencies, but its importance is unknown. Since the waves on the basilar membrane travel relatively slowly from the basal portion towards the apical portion of the basilar membrane, low frequency components will normally activate nerve fibers later than high frequency components. It could be studied in individuals with normal hearing using synthetic speech where the timing of the different spectral components can be manipulated.

2.6. Success of Cochlear Implants

Dorman [76] has shown the relationship between the success in different principles of cochlear implants, manufactured by different companies and having a different number of channels (filters). This investigator compared the speech discrimination (word correct) in adult cochlear implant wearers with speech discrimination scores obtained in individuals with normal hearing as a function of the number of channel (filters).

Monosyllabic words were used as test material. It is seen that the results from cochlear implant wearers have a large individual spread, with median values from 30 to 50%, the lowest being in cochlear implants that use the SPEAK processing algorithms and 22 channels. The best (average) results are from implants that use eight channels (clarion) and the CIS or CA principles. The number of correct words showed large individual variation in all the cochlear implants that were included in this study, ranging from near 100 to 0% (Fig. 11.7) [76].

That of providing meaningful input to the developing brain is important for normal development of the

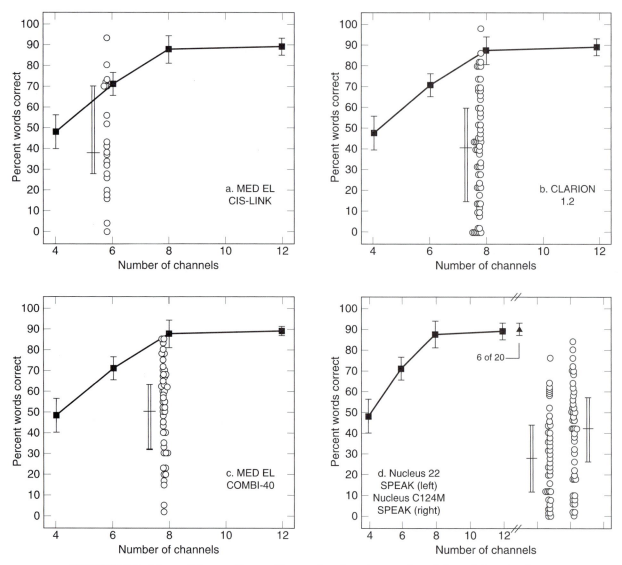

FIGURE 11.7   Monosyllabic word recognition as a function of the number of channels in a signal processor for normal-hearing listeners (filled squares and solid lines). Performance of cochlear implant wearer is shown by open circles. The open rectangles indicate the interquartile range of performance. Horizontal bars indicate median scores (data from Dorman, 2000).

brain is well established [153, 160, 161]. The benefit that children acquire from cochlear implants includes an advantage in language development compared with children with the same hearing loss who did not have cochlear implants. This advantage is enormous [267] and some advantage may even be noticeable from correcting lesser degrees of hearing loss that occur in the critical period for development of the auditory nervous system.

2.7. Selection Criteria for Cochlear Implant Candidates

When cochlear implants first became available they were given only to individuals who were essentially deaf (profound sensorineural hearing loss). More recently a broader indication is accepted [52, 257] because it has become evident that individuals with severe hearing loss can benefit from cochlear implants. Bilateral implantation is now accepted.

Providing young children with cochlear implants was slow to become accepted but it is now regarded to be essential to provide cochlear implants as early as possible [159, 295]. Cochlear implants should naturally not be performed in individuals whose hearing loss is caused by auditory nerve problems such as occur in children with narrow internal auditory canal (IAC). Such children should instead have ABIs. It is therefore important that candidates for cochlear implants have an MRI scan to show the structure of the IAC and not only the anatomy of the middle and inner ear [108].

3. COCHLEAR NUCLEUS IMPLANTS

Cochlear nucleus implants, known as auditory brainstem implants (ABIs), were introduced later than cochlear implants [31, 250] and much less is known about the success of ABIs compared with that of cochlear implants. ABIs stimulate the cochlear nucleus with electrical impulses by an array of electrodes placed on the surface of the cochlear nucleus. The stimuli are generated by processors that are similar to those used in cochlear implants.

When first introduced, ABIs were almost exclusively used in patients with neurofibromatosis type 2 (NF2) who had bilateral vestibular Schwannoma removed [31, 250]. More recently, auditory brainstem implants have been used in patients with traumatic injuries to the auditory nerve bilaterally [53, 55] and in children with malfunction of the auditory nerve such as may occur from internal auditory meatus malformation (atresia) causing auditory nerve aplasia [53].

3.1. Function and Design of Auditory Brainstem Implants

ABIs stimulate cells in the cochlear nucleus by signals derived from the sound that is picked up by a microphone placed near the wearer's ear in a similar way as cochlear implants. The processors that are used separate sounds according to their spectrum and some processors perform similar sophisticated processing as described for cochlear implants.

Electrodes for stimulating the cochlear nucleus are placed in the lateral recess of the fourth ventricle through the foramen of Luschka [166] in a similar way as electrodes that have been used for recording evoked potentials from the cochlear nucleus in neurosurgical operations [166, 211, 220]. Placement of electrodes for ABIs is technically more demanding than placements of cochlear implants. Not only is it more difficult to maintain a stable electrode placement in the brain than in the cochlea, but it is also more difficult to place the electrode array so that an optimal population of nerve cells is stimulated.

3.2. Physiological Basis for Auditory Brainstem Implants

Cochlear nucleus implants stimulate cells in the cochlear nucleus and thereby bypass not only the processing of sounds that occurs in the cochlea and the neural transduction in the hair cells but ABIs also bypass the auditory nerve and probably some neural processing that normally occurs in the cochlear nucleus.

At first it may sound surprising that stimulation of the cochlear nucleus can provide good speech comprehension. However, the main difference between cochlear implants and cochlear nucleus implants is that the latter also bypass the auditory nerve, which does not provide any processing of information but merely acts as a connection that conveys information from the cochlea to the cochlear nucleus. Provided that proper placement of the stimulating electrode can be arranged, ABIs can therefore be expected to perform as well as cochlear implants. Which neurons of the cochlear nucleus are being stimulated is difficult to control and less than optimal placement of the stimulating electrode array most likely accounts for some of the problems that have been experienced with ABIs in getting a high degree of speech discrimination. When these technical problems regarding the implantation of electrodes are solved, the success of ABIs most likely will be improved.

If the cochlear nucleus stimulation is arranged so that it stimulates primary-like nerve cells the stimulation may act in a similar way as cochlear implants because

---

BOX 11.4

ANATOMY OF THE COCHLEAR NUCLEUS

The cochlear nucleus has three main divisions: the dorsal cochlear nucleus; the anterior ventral cochlear nucleus; and the posterior ventral cochlear nucleus (see Chapter 6, Fig. 6.12). The surface of the ventral cochlear nucleus and that of the dorsal cochlear nucleus share the floor of the lateral recess of the fourth ventricle. The anterior ventral nucleus occupies the most rostral part of the cochlear nucleus [166]. The cochlear nucleus includes a complex network of many different types of cells that are interconnected and which have excitatory and inhibitory influence on each other.

---

such cells only receive a few auditory nerve fibers. Electrical stimulation may therefore be expected to activate these cells in a similar way as they are activated normally by auditory nerve fibers.

Each auditory nerve fiber innervates cells in all three main divisions of the cochlear nucleus (for details, see Chapter 6). This is the beginning of parallel processing that is prominent in the ascending auditory pathways. Cochlear nucleus implants are likely to activate only one of these three divisions of the cochlear nucleus and, therefore, only one of the parallel pathways to higher nervous centers is activated by a cochlear nucleus implant. The implications of that are unknown.

While electrical stimulation of the auditory nerve in the cochlea activates nerve fibers in a similar way, ABIs can stimulate different types of cells as well as nerve fibers that terminate on cells of the cochlear nucleus. The excitability of nerve cells depends on the size of the cells, their membrane potentials and threshold, all of which varies considerably among cells. A specific type of stimulation may therefore stimulate a specific population of cochlear nucleus cells whereas other types of stimulation may activate different types of cells and fibers. The duration of the electrical impulses that are applied to the cochlear nucleus is important in that aspect.

The cochlear nucleus is tonotopically organized (see Chapter 6, Fig. 6.12) [269], but it is not known if it is important to stimulate the cochlear nucleus cells according to this tonotopic organization. Since the orientation of the tonotopic maps of the cochlear nucleus is insufficiently known it is not possible to orient the electrode array so that activity in different frequency bands stimulates cells that are normally activated by the same spectrum of sounds.

The limitation of cochlear implants to stimulate auditory nerve fibers that normally respond to high frequency sounds does not exist in connection with cochlear nucleus implants and with correctly placed electrode arrays it should be possible to stimulate all neurons that normally respond to sounds within the entire audible hearing range. This means that ABIs have the potential of providing better hearing than cochlear implants.

3.3. Success of Auditory Brainstem Implants

ABIs have been less effective in providing useful hearing than cochlear implants [175]. ABIs are less effective in restoring functional hearing in patients with NF2 [175] than cochlear implants in patients with cochlear injuries. ABIs in NF2 patients provide assistance in lip-reading but no real speech discrimination. However, recent experience shows that ABIs can be equally efficient in providing speech comprehension as cochlear implants where used in patients with traumatic injuries to the auditory nerve [54] bilaterally and in patients with auditory nerve aplasia from internal auditory meatus atresia [53, 55, 56].

It is not known why ABIs cannot provide useable speech discrimination in NF2 patients. It has been indicated that a separate auditory pathway process amplitude modulation of sounds and that pathway is important for speech discrimination and it has been hypothesized that the difference between the success of ABIs in patients with NF2 and in patients with other causes of auditory nerve dysfunction is that this modulation pathway is damaged in NF2 patients [56]. Severance of the auditory nerve as often occurs in operations for large vestibular Schwannoma causes degeneration of the nerve fibers that terminate on cells in the cochlear nucleus, resulting in changes in these cells [65, 297]. Other forms of lesions to the auditory nerve may not have the same effect on cells of the cochlear nucleus.

3.4. Patient Selection for Auditory Brainstem Implants

Patients for ABIs were originally limited to patients who were bilaterally deaf because they had bilateral vestibular Schwannoma removed. Almost all of these patients had neurofibromatosis type 2 (NF2) [31, 250]. Later it was found that some congenital deaf children had internal auditory canal (IAC) atresia strangling the auditory nerve, causing their deafness. These patients together with patients with traumatic destruction of their auditory nerves bilaterally were found to be excellent candidates for ABIs [53, 55, 56].

4. ROLE OF NEURAL PLASTICITY

Cochlear implants and cochlear nucleus implants (ABIs) do not accurately replace all the normal functions of the ear. These devices activate the auditory nervous system in a way that is different from what occurs in the normal ear and they do not activate all the parts of the auditory nervous system that are normally activated by sound. This requires the nervous system to "learn" a new code. It has been known for a long time that expression of neural plasticity helps to regain function after trauma or insults, such as from strokes (see [213]). Expression of neural plasticity that enables the auditory nervous system to adapt to changing demands plays an important role in the success of cochlear and cochlear nucleus implants. Training is a powerful method for activating neural plasticity and training is a part of all cochlear and cochlear nucleus implant programs.

The ability of the nervous system to change its function is greatest in a short period after birth [159], which makes it easier to adapt cochlear implants to young individuals than adults [159, 160, 295]. Proper training can also improve the success of cochlear implants in adults.

The nervous system in animals that are born deaf has a rudimentary tonotopic organization [115, 172], and that organization is refined through sound stimulation [172, 304]. Studies in congenital deaf animals (cats and guinea pigs) have shown that electrical stimulation of the cochlea can modify the cochleotopic organization that exists even in animals that never have had any auditory input [153, 160, 162]. Studies in animals that have hearing have shown that tonotopic maps of the auditory cortex changes after sound stimulation [146]. In humans such rudimentary tonotopic organization that exists at birth before sound stimulation is normally refined by the sound that the child experiences. Expression of neural plasticity makes it possible for cochlear and cochlear nucleus implants to impose a new tonotopic organization of the auditory nervous system. The tonotopic organization that exists in individuals who have had hearing and became deaf can also be modified by the input from cochlear and cochlear nucleus implants. That input from cochlear implants can change the function of the auditory nervous system has been demonstrated by recording of auditory evoked potentials (event related potentials [ERP]) [295].

Studies on cochlear implant patients indicate that the central auditory neural processor not only identifies the neurons of the ascending auditory pathway by their anatomical connections to hair cells at different locations along the basilar membrane but also by their "signature" firing pattern. Neurons may be "tagged" by the properties (frequency etc.) of the sounds that activate the neurons.

SECTION III REFERENCES

1. Adams DA and Wilmot TJ. Longterm results of sympathectomy. *J Laryngology and Otology* 92: 705–710, 1982.

2. Adlington P and Warrick J. Stellate ganglion block in the management of tinnitus. *J Laryngol Otol* 85: 159–168, 1971.

3. Aggarwal R and Saeed SR. The genetics of hearing loss. *Hosp Med* 66: 32–36, 2005.

4. AhdabBarmada M and Moossy J. The neuropathology of kernicterus in the premature neonate: diagnostic problems. *J Neuropatholog Exp Neurol* 43: 45–56, 1984.

5. Ahrenberg IK. Ménière's disease: diagnosis and management of vertigo and endolymphatic hydrops. In: *Dizziness and balance disorders*, edited by Ahrenberg IK. Amsterdam: Kugler Publications, 1993, p. 503–509.

6. Aitkin LM. *The auditory midbrain, structure and function in the central auditory pathway.* Clifton, NJ: Humana Press, 1986.

7. ANSI. *Methods for the calculation of the speech intelligibility index.* New York: American Standards Institute, 1997.

8. Antonelli AR and Calearo C. Drug effects on the auditory speech discrimination mechanisms. *Acta Otolaryng (Stockh)* 58: 105, 1964.

9. Arnold R, Yule W, and Martin N. The psychological characteristics of infantile hypercalccaemia: a preliminary investigation. *Developmenetal Medicine & Child Neurology* 27: 49–59, 1985.

10. Attanasio G, Barbara M, Buongiorno G, Cordier A, Mafera B, Piccoli F, Nostro G, and Filipo R. Protective effect of the cochlear efferent system during noise exposure. *Ann N Y Acad Sci* 884: 361–367, 1999.

11. Auger RG and Whisnant JP. Hemifacial spasm in Rochester and Olmsted County, Minnesota, 1960 to 1984. *Arch Neurol*: 1233–1234, 1990.

12. Babighian G, Moushegian G, and Rupert AL. Central auditory fatigue. *Audiology* 14: 72–83, 1975.

13. Bach S, Noreng MF, and Tjellden NU. Phantom limb pain in amputees during the first 12 months following limb amputation, after preoperative lumbar epidural blockade. *Pain* 33: 297–301, 1988.

14. Baguley DM. Hyperacusis. *J R Soc Med* 96: 582–585, 2003.

15. Baguley DM, Axon P, Winter IM, and Moffat DA. The effect of vestibular nerve section upon tinnitus. *Clin Otolaryngol Allied Sci* 27: 219–226, 2002.

16. Baguley DM, Jones S, Wilkins I, Axon PR, and Moffat DA. The inhibitory effect of intravenous lidocaine infusion on tinnitus after translabyrinthine removal of vestibular schwannoma: a double-blind, placebo-controlled, crossover study. *Otol Neurotol* 26: 169–176, 2005.

17. Banerjee A, Whyte A, and Atlas MD. Superior canal dehiscence: review of a new condition. *Clin Otolaryngol Allied Sci* 30: 9–15, 2005.

18. Barker FG, Jannetta PJ, Bissonette DJ, Larkins MV, and Jho HD. The long-term outcome of microvascular decompression for trigeminal neuralgia. *N Eng J Med* 334: 1077–1083, 1996.

19. Barker FG, Jannetta PJ, Bissonette DJ, Shields PT, and Larkins MV. Microvascular decompression for hemifacial spasm. *J Neurosurg* 82: 201–210, 1995.

20. Berger EH and Royster LH. In search of meaningful measures of hearing protector effectiveness. *Spectrum* Suppl. 1, 13: 29, 1996.

21. Berlin CI, Morlet T, and Hood LJ. Auditory neuropathy/dyssynchrony: its diagnosis and management. *Pediatric Clinics of North America* 50: 331–340, 2003.

22. Bernstein JM. Middle ear mucosa: histological, histochemical, immunochemical, and immunological aspects. In: *Physiology of the ear*, edited by Jahn AF and Santos-Sacchi J. New York: Raven Press, 1988, p. 59–80.

23. Bocca E. Clinical aspects of cortical deafness. *Laryngoscope* 68: 301, 1958.

24. Bocca E. Distorted speech tests. In: *Sensory-neural hearing processes and disorders*, edited by Graham BA. Boston: Little, Brown & Co, 1965.

25. Bocca E, Calearo C, and Cassinari V. A new method for testing hearing in temporal lobe tumours. *Acta Otolaryngol (Stockh)* 44: 219, 1954.

26. Borg E. Noise induced hearing loss in normotensive and spontaneously hypertensive rats. *Hear Res* 8: 117–130, 1982.

27. Borg E, Canlon B, and Engstrom B. Noise-induced hearing loss. Literature review and experiments in rabbits. *Scand Audiol* 24 (Suppl 40): 1–147, 1995.

28. Borg E and Møller AR. Noise and blood pressure: effects on lifelong exposure in the rat. *Acta Physiol Scand* 103: 340–342, 1978.

29. Borsel van J, Curfs LMG, and Fryns JP. Hyperacusis in Williams syndrome: a sample survey study. *Genetic Counseling* 8: 121–126, 1997.

30. Brach JS, Van Swearingen JM, Lenert J, and Johnson PC. Facial neuromuscular retraining for oral synkinesis. *Plastic and Reconstructive Surgery* 99: 1922–1931, 1997.

31. Brackmann DE, Hitselberger WE, Nelson RA, Moore J, Waring MD, Portillo F, Shannon RV, and Telischi FF. Auditory brainstem implant: 1. Issues in surgical implantation. *Otolaryngol Head Neck Surg* 108: 624–633, 1993.

32. Brown JA, Lutsep HL, Cramer SC, and Weinand M. Motor cortex stimulation for enhancement of recovery after stroke: case report. *Neurol Res* 25: 815–818, 2003.

33. Burns W and Robinson DW. *Hearing and noise in industry.* London: HMSO, 1970.

34. Buus S and Florentine M. Growth of loudness in listeners with cochlear hearing losses: recruitment reconsidered. *J Assoc Res Otolaryngol* 3: 120–139, 2002.

35. Cacace AT. Expanding the biological basis of tinnitus: cross-modal origins and the role of neuroplasticity. *Hear Res* 175: 112–132, 2003.

36. Cacace AT, Cousins JP, Parnes SM, McFarland DJ, Semenoff D, Holmes T, Davenport C, Stegbauer K, and Lovely TJ. Cutaneous-evoked tinnitus. II: review of neuroanatomical, physiological and functional imaging studies. *Audiol Neurotol* 4: 258–268, 1999.

37. Cacace AT, Lovely TJ, McFarland DJ, Parnes SM, and Winter DF. Anomalous cross-modal plasticity following posterior fossa surgery: some speculations on gaze-evoked tinnitus. *Hear Res* 81: 22–32, 1994.

38. Caiazzo AJ and Tonndorf J. Ear canal resonance and temporary threshold shift. *Otolaryngology* 86: 820, 1978.

39. Calearo C and Antonelli AR. "Cortical" hearing tests and cerebral dominance. *Acta Otolaryngol (Stockh)* 56: 17, 1963.

40. Campbell KCM and Abbas PJ. Electrocochleography with postural changes in perilymphatic fistula. *Ann Otol Rhin Laryng* 103: 474–482, 1994.

41. Campbell KCM, Harker LA, and Abbas PJ. Interpretation of electrocochleography in Ménière's disease and normal subjects. *Ann Otol Rhin Laryngol* 101: 496–500, 1992.

42. Canlon B, Borg E, and Flock A. Protection against noise trauma by pre-exposure to a low level acoustic stimulus. *Hear Res* 34: 197–200, 1988.

43. Carey J. Intratympanic gentamicin for the treatment of Meniere's disease and other forms of peripheral vertigo. *Otolaryngol Clin North Am* 37: 1075–1090, 2004.

44. Caspary DM, Raza A, Lawhorn, Armour BA, Pippin J, and Arneric SP. Immunocytochemical and neurochemical evidence for age-related loss of GABA in the inferior colliculus: implications for neural presbycusis. *J Neurosci* 10: 2363–2372, 1990.

45. Cazals Y. Auditory sensori-neural alterations induced by salicylate. *Prog Neurobiol* 62: 583–631, 2000.

46. Cazals Y, Negrevergne M, and Aran JM. Electrical stimulation of the cochlea in man: hearing induction and tinnitus suppression. *J Am Audiol Soc* 3: 209–213, 1978.

47. Celestino D and Ralli G. Incidence of Ménière's disease in Italy. *Am J Otol* 12: 135–138, 1991.

48. Charabi S, Thomsen J, Tos M, Charabi B, Mantoni M, and Børgesen SE. Acoustic neuroma/vestibular schwannoma growth: past, present and future. *Acta Otolaryngol (Stockh)* 118: 327–332, 1998.

49. Chmiel R, Jerger J, Murphy E, Pirozzolo F, and Tooley-Young C. Unsuccessful use of binaural amplification by an elderly person. *J Am Acad Audiol* 8: 1–10, 1997.

50. Coad ML, Lockwood A, Salvi R, and Burkard R. Characteristics of patients with gaze-evoked tinnitus. *Otol Neurotol* 22: 650–654, 2001.

51. Cody AR and Johnstone BM. Single auditory neuron response during acute acoustic trauma. *Hear Res* 3: 3–16, 1980.

52. Cohen NL. Cochlear implant candidacy and surgical considerations. *Audiol Neurootol* 9: 197–202, 2004.

53. Colletti V, Carner M, Fiorino F, Sacchetto L, Morelli V, Orsi A, Cilurzo F, and Pacini L. Hearing restoration with auditory brainstem implant in three children with cochlear nerve aplasia. *Otol Neurotol* 23: 682–693, 2002.

54. Colletti V, Carner M, Miorelli V, Colletti L, Guida M, and F. F. Auditory brainstem implant in posttraumatic cochlear nerve avulsion. *Audiol Neurootol* 9: 247–255, 2004.

55. Colletti V, Fiorino FG, Sacchetto L, Miorelli V, and Carner M. Hearing habilitation with auditory brainstem implantation in two children with cochlear nerve aplasia. *Int J Pediatric Otorhinolaryngol* 60: 99–111, 2001.

56. Colletti V and Shannon RV. Open set of speech perception with auditory brainstem implant? *Laryngoscope* 115: 1974–1978, 2005.

57. Cooper Jr. JC. Health and nutrition examination survey of 1971–75: Part II. Tinnitus, subjective hearing loss, and well-being. *J Am Acad Audiol* 5: 37–43, 1994.

58. Cooper NP, Robertson D, and Yates GK. Cochlear nerve fiber responses to amplitude-modulated stimuli: variations with spontaneous rate and other response characteristics. *J Neurophysiol* 70: 370–386, 1993.

59. Dallos P and Cheatham MA. Compound action potential tuning curves. *J Acoust Soc Am* 59: 591–597, 1976.

60. Dandy WE. Surgical treatment of Ménière's disease. *Surg Gynecol Obstet* 72: 421–425, 1941.

61. Davis RR, Kozel P, and Erway LC. Genetic influences in individual susceptibility to noise: a review. *Noise Health* 5: 19–28, 2003.

62. De Ridder D, De Mulder G, Walsh V, Muggleton N, Sunaert S, Verlooy J, Van de Heyning P, and Møller AR. Transcranial magnetic stimulation for tinnitus: a clincial and pathophysiological approach: influence of tinnitus duration on stimulation parameter choice and maximal tinnitus suppression. *Otol Neurotol* 147: 495–501, 2005.

63. De Ridder D, De Mulder G, Walsh V, Muggleton N, Sunaert S, and Møller A. Magnetic and electrical stimulation of the auditory cortex for intractable tinnitus. *J Neurosurg* 100: 560–564, 2004.

64. De Ridder D, Ryu H, Møller AR, Nowe V, Van de Heyning P, and Verlooy J. Functional anatomy of the human cochlear nerve and its role in microvascular decompressions for tinnitus. *Neurosurgery* 54: 381–388, 2004.

65. Deitch JS and Rubel EW. Rapid changes in ultrastructure during deafferentiation-induced dendritic atrophy. *J Comp Neurol* 281: 234–258, 1989.

66. Densert B and Densert O. Overpressure in treatment of Ménière's disease. *Laryngoscope* 92: 1285–1292, 1982.

67. Densert B and Sass K. Control of symptoms in patients with Ménière's disease using middle ear pressure applications: two years follow-up. *Acta Otolaryngol* 21: 616–621, 2001.

68. Densert B, Sass K, and Arlinger S. Short term effects of induced middle ear pressure changes on the electrocochleogram in Ménière's disease. *Acta Otolaryngol* 115: 732–737, 1995

69. Densert O. Adrenergic innervation in the rabbit cochlea. *Acta Otolaryngol (Stockh)* 78: 345–356, 1974.

70. Ding DL, Wang J, Salvi R, Henderson D, Hu BH, McFadden SL, and Mueller M. Selective loss of inner hair cells and type-I ganglion neurons in carboplatin-treated chinchillas. Mechanisms of damage and protection. *Ann N Y Acad Sci* 884: 152–170, 1999.

71. Djourno A and Eyries C. Prothese auditive par excitation electrique a distance du nerf sensoriel a l'aide d'un bobinage inclus a demeure. *Presse Med* 35: 1417, 1957.

72. Dobie RA. A review of randomized clinical trials in tinnitus. *Laryngoscope* 109: 1202–1211, 1999.

73. Dodson KM and Sismanis A. Intratympanic perfusion for the treatment of tinnitus. *Otolaryngol Clin North Am* 37: 991–1000, 2004.

74. Dohlman GF. Mechanism of the Meniere attack. *ORL J Otorhinolaryngol Relat Spec* 42: 10–19, 1980.

75. Dolan TR, Ades HW, Bredberg G, and Neff WD. Inner ear damage and hearing loss after exposure to tones of high intensity. *Acta Otolaryngol (Stockolm)* 80: 343–352, 1975.

76. Dorman MF. Speech perception by adults. In: *Cochlear implants*, edited by Walzman SB and Cohen NL. New York: Thieme, 2000.

77. Dorman MF, Loizou PC, Kemp LL, and Kirk KI. Word recognition by children listening to speech processed into a small number of channels: data from normal-hearing children and children with cochlear implants. *Ear & Hearing* 21: 590–596, 2000.

78. Dublin WB. The cochlear nuclei pathology. *Otolaryngol Head & Neck Surgery* 93: 448–463, 1985.

79. Dudley H. Remaking speech. *J Acoust Soc Am* 11: 169–177, 1939.

80. Eddington D. Speech discrimination in deaf subjects with cochlear implants. *J Acoust Soc Am* 68: 885–891, 1980.

81. Eggermont JJ. Between sound and perception: reviewing the search for a neural code. *Hear Res* 157: 1–42, 2001.

82. Eggermont JJ. On the pathophysiology of tinnitus: a review and a peripheral model. *Hear Res* 48: 111–124, 1990.

83. Eggermont JJ. Summating potentials in Ménière's disease. *Arch Otorhinolaryngol* 222: 63–75, 1979.

84. Eggermont JJ and M. K. Salicylate and quinine selectively increase spontaneous firing rates in secondary auditory cortex. *Hear Res* 117: 149–160, 1998.

85. El Barbary A. Auditory nerve of the normal and jaundiced rat. II. Frequency selectivity and twotone rate suppression. *Hear Res* 54: 91–104, 1991.

86. Elkind-Hirsch KE, Stoner WR, Stach BA, and Jerger JF. Estrogen influences auditory brainstem responses during the normal menstrual cycle. *Hear Res* 60: 143–148, 1992.

87. Engelberg M and Bauer W. Transcutaneous electrical stimulation for tinnitus. *Laryngoscope* 95: 1167–1173, 1985.

88. Engineer ND, Percaccio CR, Pandya PK, Moucha R, Rathbun DL, and Kilgard MP. Environmental enrichment improves response strength, threshold, selectivity, and latency of auditory cortex neurons. *J Neurophys* 92: 73–82, 2004.

89. Ernst A, Snik AFM, Mylanus IAM, and Cremers CWRJ. Noninvasive assessment of the intralabyrinthine pressure. *Arch Otolaryngol Head & Neck Surg* 121: 926–929, 1995.

90. Evans EF. Normal and abnormal functioning of the cochlear nerve. *Symp Zool Soc Lond* 37: 133–165, 1975.

91. Fant G. Acoustic analysis and synthesis of speech with applications to Swedish. *Ericsson Technics* 1:1–106, 1959.

92. Filipo R and Barbara M. Natural history of Meniere's disease: staging the patients or their symptoms? *Acta Otolaryngol (Stockh)* Suppl. 526: 10–13, 1997.

93. Fishman KE, Shannon RV, and Slattery WH. Speech recognition as a function of the number of electrodes used in the SPEAK cochlear implant speech processor. *J Speech Lang Hear Res* 40: 1201–1215, 1997.

94. Forge A and Schacht J. Aminoglycoside antibiotics. *Audiol Neurotol* 5: 3–22, 2000.

95. Frazier CH. Intracranial division of the auditory nerve for persistent aural vertigo. *Surg Gynecol Obstet* 15: 524–529, 1912.

96. Fukutake T and Hattori T. Auditory illusions caused by a small lesion in the right geniculate body. *American Acad Neurol* 51: 1469–1471, 1998.

97. Fullerton BC, Levine RA, Hosford Dunn HL, and Kiang NYS. Comparison of cat and human brain stem auditory evoked potentials. *Hear Res* 66: 547–570, 1987.

98. Gates GA, Couropmitree NN, and Myers RH. Genetic associations in age-related hearing thresholds. *Arch Otolaryngol Head &Neck Surg* 125: 654–659, 1999.

99. Gejrot T. Intravenous xylocaine in the treatment of attacks of Ménière's disease. *Acta Otolaryngol (Stockh)* Suppl 188: 190–195, 1963.

100. Gerken GM, Saunders SS, and Paul RE. Hypersensitivity to electrical stimulation of auditory nuclei follows hearing loss in cats. *Hear Res* 13: 249–260, 1984.

101. Gerken GM, Solecki JM, and Boettcher FA. Temporal integration of electrical stimulation of auditory nuclei in normal hearing and hearing-impaired cat. *Hear Res* 53: 101–112, 1991.

102. Glasgold A and Altman F. The effect of stapes surgery on tinnitus in otosclerosis. *Laryngoscope* 76: 1524–1532, 1966.

103. Goble TJ, Farmer GE, Frank J, Møller AR, and Thompson LT. Acute noise exposure alters hippocampal place-fields: evidence for extralemniscal sensory pathway plasticity. *Society for Neuroscience* 2004.

104. Goddard GV. Amygdaloid stimulation and learning in the rat. *J Comp Physiol Psychol* 58: 23–30, 1964.

105. Godey B, Morandi X, Beust L, Brassier G, and Bourdiniere J. Sensitivity of auditory brainstem response in acoustic neuroma screening. *Acta Otolaryngol (Stockh)* 118: 501–504, 1998.

106. Gordon MA. The genetics of otosclerosis: a review. *Am J Otol* 10: 426–438, 1989.

107. Gottesman II and Hanson DR. Human development: biological and genetic processes. *Annu Rev Psychol* 56: 263–286, 2005.

108. Gray RF, Ray J, Baguley DM, Vanat Z, Begg J, and Phelps PD. Cochlear implant failure due to unexpected absence of the eighth nerve–a cautionary tale. *J Laryngol Otol* 112: 646–649, 1998.

109. Gruber CJ and Huber JC. Differential effects of progestins on the brain. *Maturitas* 46: S71–75, 2003.

110. Guirado CR. Malformations of the inner auditory canal. *Rev Laryngol Otol Rhinol (Bord)* 113: 419–421, 1992.

111. Hallgren M, Larsby B, Lyxel B, and Arlinger S. Cognitive effects in dichotic speech testing in elderly persons. *Ear Hear* 22: 120–129, 2001.

112. Harding GW and Bohne BA. Noise-induced hair-cell loss and total exposure energy: analysis of a large data set. *J Acoust Soc Am* 115: 2207–2220, 2004.

113. Harrison JM, Ibrahim D, and Mount RJ. Plasticity of tonotopic maps in auditory midbrain following partial cochlear damage in the developing chinchilla. *Exp Brain Res* 123: 449–460, 1998.

114. Harrison RV, Aran J-M, and Erre JP. AP tuning curves in normal and pathological human and guinea pig cochlea. *J Acoust Soc Am* 69: 1374–1385, 1981.

115. Hartmann R, Shepherd RK, Heid S, and Klinke R. Response of the primary auditory cortex to electrical stimulation of the auditory nerve in the congenitally deaf white cat. *Hear Res* 112: 115–133, 1997.

116. Hatayama T and Møller AR. Correlation between latency and amplitude of peak V in brainstem auditory evoked potentials: intraoperative recordings in microvascular decompression operations. *Acta Neurochir (Wien)* 140: 681–687, 1998.

117. Hazell JWP. Tinnitus. In: *Scott-Brown's diseases of the ear, nose and throat, 4th edition: the ear*, edited by Ballantyne J and Groves J. London: Butterworth, 1979, p. 81–91.

118. Hebb DO. *The organization of behavior*. New York: Wiley, 1949.

119. Henderson D and Møller AR. Effect of asymptotic threshold shift in neural firing patterns of the rat cochlear nucleus. *J Acoust Soc Am* 57: 53, 1975.

120. Henry KR. Age-related changes in sensitivity of the postpubertal ear to acoustic trauma. *Hear Res* 8: 285–294, 1982.

121. Hood JD and Poole JP. Tolerable level of loudness. *J Acoust Soc Am* 40: 47–53, 1966.

122. House JW and Brackmann DE. Tinnitus: surgical treatment. In: *Tinnitus (Ciba foundation symposium 85)*. London: Pitman Books Ltd, 1981.

123. House WH. Cochlear implants. *Ann Otol Rhinol Laryngol* 85 (Suppl. 27): 3–91, 1976.

124. Hughes GB, Freedman MA, Haberkamp TJ, and Guay ME. Sudden sensorineural hearing loss. *Otolaryngol Clin North Am* 29: 393–405, 1996.

125. Irvine DR and Rajan R. Injury- and use-related plasticity in the primary sensory cortex of adult mammals: possible relationship to perceptual learning. *Clin Exp Pharmacol Physiol* 23: 939–947, 1996.

126. Irvine DR and Rajan R. Injury-induced reorganization of frequency maps in adult auditory cortex: the role of unmasking of normally-inhibited inputs. *Acta Otolaryng (Stockh)* 532: 39–45, 1997.

127. Jackson P. A comparison of the effects of eighth nerve section with lidocaine on tinnitus. *J Laryngol Otol* 99: 663–666, 1985.

128. Jacobson M, Kim S, Romney J, Zhu X, and Frisina RD. Contralateral suppression of distortion-product otoacoustic emissions declines with age: a comparison of findings in CBA mice with human listeners. *Laryngoscope* 13: 1707–1713, 2003.

129. Jannetta PJ. Neurovascular cross compression in patients with hyperactive dysfunction symptoms of the eighth cranial nerve. *Surg Forum* 26: 467–469, 1975.

130. Jannetta PJ. Observations on the etiology of trigeminal neuralgia, hemifacial spasm, acoustic nerve dysfunction and glossopharyngeal neuralgia. Definitive microsurgical treatment and results in 117 patients. *Neurochirurgia (Stuttg)*: 145–154, 1977.

131. Jastreboff PJ. Phantom auditory perception (tinnitus): mechanisms of generation and perception. *Neurosci Res* 8: 221–254, 1990.

132. Jastreboff PJ. Tinnitus as a phantom perception: theories and clinical implications. In: *Mechanisms of tinnitus*, edited by Vernon JA and Møller AR. Boston, MA: Allyn & Bacon, 1995, p. 73–93.

133. Jastreboff PJ and Hazell JWP. A neurophysiological approach to tinnitus: clinical implications. *Brit J Audiol* 27: 7–17, 1993.

134. Jastreboff PJ and Jastreboff MM. Tinnitus retraining therapy (TRT) as a method for treatment of tinnitus and hyperacusis patients. *J Am Acad Audiol* 11: 162–177, 2000.

135. Jastreboff PJ and Sasaki CT. Salicylate-induced changes in spontaneous activity of single units in the inferior colliculus of the guinea pig. *J Acoust Soc Amer* 80: 1384–1391, 1986.

136. Jenkins WM, Merzenich MM, Ochs MT, Allard T, and Guic–Robles E. Functional reorganization of primary somatosensory cortex in adult owl monkeys after behaviorally controlled tactile stimulation. *J Neurophysiol* 63: 82–104, 1990.

137. Jerger J, Alford B, Lew H, Rivera V, and Chmiel R. Dichotic listening, event-related potentials, and interhemispheric transfer in the elderly. *Ear Hear* 16: 482–498, 1995.

138. Jerger J, Moncrieff D, Greenwald R, Wambacq I, and Seipel A. Effect of age on interaural asymmetry of event-related potentials in dichotic listening task. *J Am Acad Audiol* 11: 383–389, 2000.

139. Johnsson LG and Hawkins HL. Sensory and neural degeneration with aging, as seen in microdissections of the human inner ear. *Ann Otol Rhinol Laryngol* 81: 179–193, 1972.

140. Jones PA and Takai D. The role of DNA methylation in mammalian epigenetics. *Science* 293: 1068–1070, 2001.

141. Kaada B, Hognestad S, and Havstad J. Transcutaneous nerve stimulation (TNS) in tinnitus. *Scand Audiol (Stockh)* 18: 211–217, 1989.

142. Kalcioglu M, T., Bayindir T, Erdem T, and Ozturan O. Objective evaluation of the effects of intravenous lidocaine on tinnitus. *Hear Res* 199: 81–88, 2005.

143. Kaltenbach JA and Afman CE. Hyperactivity in the dorsal cochlear nucleus after intense sound exposure and its resemblance to tone-evoked activity: a physiological model for tinnitus. *Hear Res* 140: 165–172, 2000.

144. Katusic S, Beard C, Bergstralh E, and Kurland L. Incidence and clinical features of trigeminal neuralgia, Rochester, Minnesota 1945–1984. *Ann Neurol*: 89–95, 1990.

145. Katzenell U and Segal S. Hyperacusis: review and clinical guidelines. *Otol Neurotol* 22: 321–326, 2001.

146. Kilgard MP and Merzenich MM. Cortical map reorganization enabled by nucleus basalis activity. *Science* 279: 1714–1718, 1998.

147. Kilgard MP and Merzenich MM. Plasticity of temporal information processing in the primary auditory cortex. *Nature Neurosci* 1: 727–731, 1998.

148. Kim J, Morest DK, and Bohne BA. Degeneration of axons in the brainstem of the chinchilla after auditory overstimulation. *Hear Res* 103: 169–191, 1997.

149. Kim S, Frisina DR, and Frisina RD. Effects of age on contralateral suppression of distortion product otoacoustic emissions in human listeners with normal hearing. *Audiol Neurootol* 7: 348–357, 2002.

150. Kimura R. Experimental blockage of the endolymphatic duct and sac and its effect on the inner ear of the guinea pig. *Ann Otol Rhinol Laryngol* 76: 664–687, 1967.

151. Klein AJ, Armstrong BL, Greer MK, and Brown FR. Hyperacusis and otitis media in individuals with Williams syndrome. *J Speech Hear Dis* 55: 339–344, 1990.

152. Kleinjung T, Eichhammer P, Langguth B, Jacob P, Marienhagen J, Hajak G, Wolf SR, and Strutz J. Long-term effects of repetitive transcranial magnetic stimulation (rTMS) in patients with chronic tinnitus. *Otolaryngol Head Neck Surg* 132: 566–569, 2005.

153. Klinke R, Hartmann R, Heid S, Tillein J, and Kral A. Plastic changes in the auditory cortex of congenitally deaf cats following cochlear implantation. *Audiol Neurootol* 6: 203–206, 2001.

154. Kohama I, Ishikawa K, and Kocsis JD. Synaptic reorganization in the substantia gelatinosa after peripheral nerve neuroma formation: aberrant innervation of lamina II neurons by beta afferents. *J Neurosci* 20: 1538–1549, 2000.

155. Kokko E. Chronic secretory otitis media in children. *Acta Otolaryngol (Stockh)* Suppl. 327: 7–44, 1974.

156. Kondo A, Ishikawa J, Yamasaki T, and Konishi T. Microvascular decompression of cranial nerves, particularly of the seventh cranial nerve. *Neurol Med Chir (Tokyo)* 20: 739–751, 1980.

157. Korsan-Bengtsen M, (aka MB Møller). Distorted speech audiometry. *Acta Otolaryng (Stockholm)* Suppl. 310, 1973.

158. Korsan-Bengtsen M, (aka MB Møller). The diagnosis of hearing loss in old people. In: *Geriatric audiology*, edited by Liden G. Stockholm: Almqvist & Wiksell, 1968, p. 24–36.

159. Kral A, Hartmann R, Tillein J, Heid S, and Klinke R. Delayed maturation and sensitive periods in the auditory cortex. *Audiol Neurootol* 6: 346–362, 2001.

160. Kral A, Hartmann R, Tillein J, Heid S, and Klinke R. Hearing after congenital deafness: central auditory plasticity and sensory deprivation. *Cereb Cortex* 12: 797–807, 2002.

161. Kral A, Hartmann R, Tillrin J, Heid S, and Klinke R. Congenital auditory deprivation reduces synaptic activity within the auditory cortex in layer specific manner. *Cerebral Cortex* 10: 714–726, 2000.

162. Kral A, Tillein J, Heid S, Hartmann R, and Klinke R. Postnatal cortical development in congenital auditory deprivation. *Cereb Cortex* 15: 552–562, 2005.

163. Kroener-Herwig B, Biesinger E, Gerhards F, Goebel G, Verena Greimel K, and Hiller W. Retraining therapy for chronic tinnitus. A critical analysis of its status. *Scand Audiol* 29: 67–78, 2000.

164. Krueger WW and Storper IS. Electrocochleography in retrosigmoid vestibular nerve section for intractable vertigo caused by Meniere's disease. *Otolaryngol Head Neck Surg* 116: 593–596, 1997.

165. Kuroki A and Møller AR. Facial nerve demyelination and vascular compression are both needed to induce facial hyperactivity: A study in rats. *Acta Neurochir (Wien)* 126: 149–157, 1994.

166. Kuroki A and Møller AR. Microsurgical anatomy around the foramen of Luschka with reference to intraoperative recording of auditory evoked potentials from the cochlear nuclei. *J Neurosurg*: 933–939, 1995.

167. Laird N and Wilson WR. Predicting recovery from idiopathic sudden hearing loss. *Am J Otolaryngol* 4: 161–164, 1983.

168. Lamm K, Lamm H, and Arnold W. Effect of hyperbaric oxygen therapy in comparison to conventional or placebo therapy or no treatment in idiopathic sudden hearing loss, acoustic trauma, noise-induced hearing loss and tinnitus. A literature survey. *Adv Otorhinolaryngol* 54: 86–99, 1998.

169. Lang JJ, Ohmachi N, and Lang JS. Anatomical landmarks of the rhomboid fossa (floor of the 4th ventricle), its length and its width. *Acta Neurochir (Wien)* 113: 84–90, 1991.

170. Langguth B, Eichhammer P, Zowe M, Kleinjung T, Jacob P, Binder H, Sand P, and Hajak G. Altered motor cortex excitability in tinnitus patients: a hint at crossmodal plasticity. *Neurosci Lett* 380: 326–329, 2005.

171. Le Prell CG, Dolan D, Schacht J, Miller JM, Lomax MI, and Altschuler RA. Pathways for protection from noise induced hearing loss. *Noise Health* 5: 1–17, 2003.

172. Leake PA, Snyder RL, Rebscher SJ, Moore CM, and Vollmer M. Plasticity in central representation in the inferior colliculus induced by chronic single- vs. two-channel electrical stimulation by cochlear implant after neonatal deafness. *Hear Res* 147: 221–241, 2000.

173. LeDoux JE. Brain mechanisms of emotion and emotional learning. *Curr Opin Neurobiol* 2: 191–197, 1992.

174. Lefebvre PP and Van De Water TR. Connexins, hearing and deafness: clinical aspects of mutations in the connexin 26 gene. *Brain Res Brain Res Rev* 32: 159–162, 2000.

175. Lenarz M, Matthies C, Lesinski-Schiedat A, Frohne C, Rost U, Illg A, Battmer RD, Samii M, and Lenarz T. Auditory brainstem implant part II: subjective assessment of functional outcome. *Otol Neurotol* 23: 694–697, 2002.

176. Levine RA. Somatic (craniocervical) tinnitus and the dorsal cochlear nucleus hypothesis. *Am J Otolaryngol* 20: 351–362, 1999.

177. Levitin DJ, Menon V, Schmitt JE, Eliez S, White CD, Glover GH, Kadis J, Korenberg JR, Bellugi U, and Reiss AL. Neural correlates of auditory perception in Williams syndrome: an fMRI study. *Neuroimage* 18: 74–82, 2003.

178. Liberman MC. Chronic changes in acoustic trauma: serial-section reconstruction of stereocilia and cuticular plates. *Hear Res* 26: 65–88, 1987.

179. Liberman MC and Mulroy MJ. Acute and chronic effects of acoustic trauma: cochlear pathology and auditory nerve pathology. In: *New perspectives in noise–induced hearing loss*, edited by Hamernik RP, Henderson D, and Salvi R. New York: Raven Press, 1982.

180. Lidén G. *Audiology*. Stockholm: Almquist & Wiksell, 1985.

181. Lockwood A, Salvi R, Coad M, Towsley M, Wack D, and Murphy B. The functional neuroanatomy of tinnitus. Evidence for limbic system links and neural plasticity. *Neurology* 50: 114–120, 1998.

182. Loizou PC. Introduction to cochlear implants. *IEEE Signal Processing Magazine* September: 101–130, 1998.

183. Loizou PC. On the number of channels needed to understand speech. *J Acoust Soc Am* 106: 2097–2103, 1999.

184. Loizou PC, Dorman M, and Fitzke J. The effect of reduced dynamic range on speech understanding: implications for patients with cochlear implant. *Ear Hear* 21: 25–31, 2000.

185. Lundman L, Juhn SK, Bagger-Sjöbäck D, and Svanborg C. Permeability of the normal round window membrane to Haemophilus influenzae type b endotoxin. *Acta Otolaryngol* 112: 524–529, 1992.

186. Maison SF and Liberman MC. Predicting vulnerability to acoustic injury with a non-invasive assay of olivocochlear reflex strength. *J Neurosci* 20: 4701–4707, 2000.

187. Marchbanks RJ. Hydromechanical interactions of the intracranial and intralabyrinthine fluids. In: *Intracranial and intralabyrinthine fluids*, edited by Ernst A, Marchbanks R and Samii M. Berlin: Springer-Verlag, 1996.

188. Marriage J and Barnes NM. Is central hyperacusis a symptom of 5–hydroxytryptamine (5–HT) dysfunction? *J Laryngol Otol* 109: 915–921, 1995.

189. Matsushima T, Inoue T, and Fukui M. Arteries in contact with the cisternal portion of the facial nerve in autopsy cases: microsurgical anatomy for neurovascular decompression surgery of hemifacial spasm. *Surg Neurol* 34: 87–93, 1990.

190. Mazelova J, Popelar J, and Syka J. Auditory function in presbycusis: peripheral vs. central changes. *Exp Gerontol* 38.: 87–94, 2003.

191. McCabe BF and Harker LA. Vascular loop as a cause of vertigo. *Ann Otol Rhinol Laryngol* 92: 542–543, 1983.

192. Merchant SN, Adams JC, and Nadol JB. Pathology and pathophysiology of idiopathic sudden sensorineural hearing loss. *Otol Neurotol* 26: 151–160, 2005.

193. Merchant SN, Rosowski JJ, and Ravicz ME. Middle ear mechanics of type IV and type V tympanoplasty: II. Clinical analysis and surgical implications. *Am J Otol* 16: 565–575, 1995.

194. Merzenich MM, Kaas JH, Wall J, Nelson RJ, Sur M, and Felleman D. Topographic reorganization of somatosensory cortical areas 3b and 1 in adult monkeys following restricted deafferentation. *Neuroscience* 8: 3–55, 1983.

195. Merzenich MM, Nelson RJ, Stryker MP, Cynader MS, Schoppmann A, and Zook JM. Somatosensory cortical map changes following digit amputation in adult monkeys. *J Comp Neurol* 224: 591–605, 1984.

196. Michaelson RP. Stimulation of the human cochlea. *Arch Otolaryngol* 93: 317–323, 1971.

197. Miller G and Licklider JCR. The intelligibility of interrupted speech. *J Acoust Soc Am* 22: 167, 1950.

198. Miller JM, Watson CS, and Covell WP. Deafening effects of noise on the cat. *Acta Oto Laryng Suppl* 176: 1–91, 1963.

199. Minor LB, Schessel DA, and Carey JP. Ménière's disease. *Curr Opin Neurol* 17: 9–16, 2004.

200. Miyamoto RT and Bichey BG. Cochlear implantation for tinnitus suppression. *Otolaryngol Clin North Am* 36: 345–352, 2003.

201. Moore BC. Coding of sounds in the auditory system and its relevance to signal processing and coding in cochlear implants. *Otol Neurotol* 24: 243–254, 2003.

202. Moore BC. Testing the concept of softness imperception: loudness near threshold for hearing-impaired ears. *J Acoust Soc Am* 115: 3103–3111, 2004.

203. Moore BC and Glasberg BR. A revised model of loudness perception applied to cochlear hearing loss. *Hear Res* 188: 70–88, 2004.

204. Morest DK, Ard MD, and Yurgelun-Todd D. Degeneration in the central auditory pathways after acoustic deprivation or over-stimulation in the cat. *Anat Rec* 193: 750, 1979.

205. Morest DK and Bohne BA. Noise-induced degeneration in the brain and representation of inner and outer hair cells. *Hear Res* 9: 145–152, 1983.

206. Morgan DH. Tinnitus of TMJ origin. *J Craniomandibular Practice* 10: 124–129, 1992.

207. Møller AR. Coding of sounds with rapidly varying spectrum in the cochlear nucleus. *J Acoust Soc Am* 55: 631–640, 1974.

208. Møller AR. Cranial nerve dysfunction syndromes: pathophysiology of microvascular compression. In: *Neurosurgical topics book 13, "surgery of cranial nerves of the posterior fossa," chapter 2*, edited by Barrow DL. Park Ridge, IL: American Association of Neurological Surgeons, 1993, p. 105–129.

209. Møller AR. Frequency selectivity of single auditory nerve fibers in response to broadband noise stimuli. *J Acoust Soc Am* 62: 135–142, 1977.

210. Møller AR. *Hearing: its physiology and pathophysiology*. San Diego, CA: Academic Press, 2000.

211. Møller AR. *Intraoperative neurophysiologic monitoring*. Luxembourg: Harwood Academic Publishers, 1995.

212. Møller AR. *Intraoperative neurophysiologic monitoring, 2nd edition*. Totowa, NJ: Humana Press Inc., 2006.

213. Møller AR. *Neural plasticity and disorders of the nervous system*. Cambridge: Cambridge University Press, 2006.

214. Møller AR. Pathophysiology of tinnitus. In: *Otolaryngologic clinics of north america*, edited by Sismanis A. Amsterdam: W.B. Saunders, 2003, p. 249–266.

215. Møller AR. Pathophysiology of tinnitus. *Ann Otol Rhinol Laryngol* 93: 39–44, 1984.

216. Møller AR. *Sensory systems: anatomy and physiology*. Amsterdam: Academic Press, 2003.

217. Møller AR. Similarities between severe tinnitus and chronic pain. *J Amer Acad Audiol* 11: 115–124, 2000.

218. Møller AR. The cranial nerve vascular compression syndrome: I. A review of treatment. *Acta Neurochir (Wien)* 113: 18–23, 1991.

219. Møller AR, Colletti V, and Fiorino FG. Neural conduction velocity of the human auditory nerve: bipolar recordings from the exposed intracranial portion of the eighth nerve during vestibular nerve section. *Electroenceph Clin Neurophysiol* 92: 316–320, 1994.

220. Møller AR and Jannetta PJ. Auditory evoked potentials recorded from the cochlear nucleus and its vicinity in man. *J Neurosurg* 59: 1013–1018, 1983.

221. Møller AR, Kern JK, and Grannemann B. Are the non-classical auditory pathways involved in autism and PDD? *Neurol Res* 27: 625–629, 2005.

222. Møller AR and Møller MB. Does intraoperative monitoring of auditory evoked potentials reduce incidence of hearing loss as a complication of microvascular decompression of cranial nerves? *Neurosurgery* 24: 257–263, 1989.

223. Møller AR, Møller MB, and Yokota M. Some forms of tinnitus may involve the extralemniscal auditory pathway. *Laryngoscope* 102: 1165–1171, 1992.

224. Møller AR and Pinkerton T. Temporal integration of pain from electrical stimulation of the skin. *Neurol Res* 19: 481–488, 1997.

225. Møller AR and Rollins P. The non-classical auditory system is active in children but not in adults. *Neurosci Lett* 319: 41–44, 2002.

226. Møller MB. Audiological evaluation. *J Clin Neurophysiol* 11: 309–318, 1994.

227. Møller MB. Changes in hearing measures with increasing age. In: *Hearing and balance in the elderly*, edited by Hinchcliffe R. Edinburgh: Churchill Livingstone, 1983, p. 97–122.

228. Møller MB. Hearing in 70 and 75 year-old people. Results from a cross-sectional and longitudinal population study. *Am J Otolaryngol* 2: 22–29, 1981.

229. Møller MB and Møller AR. Audiometric abnormalities in hemifacial spasm. *Audiology* 24: 396–405, 1985.

230. Møller MB, Møller AR, Jannetta PJ, and Jho HD. Vascular decompression surgery for severe tinnitus: selection criteria and results. *Laryngoscope* 103: 421–427, 1993.

231. Møller MB, Møller AR, Jannetta PJ, Jho HD, and Sekhar LN. Microvascular decompression of the eighth nerve in patients with disabling positional vertigo: selection criteria and operative results in 207 patients. *Acta Neurochir (Wien)* 125: 75–82, 1993.

232. Mühlnickel W, Taub E, and Flor H. Reorganization of auditory cortex in tinnitus. *Proc Nat Acad Sci USA* 95: 10340–10343, 1998.

233. Müller M, Robertson D, and Yates GK. Rate-versus-level functions of primary auditory nerve fibres: evidence of square law behavior of all fibre categories in the guinea pig. *Hear Res* 55: 50–56, 1991.

234. Myers EN and Bernstein JM. Salicylate ototoxicity. *Arch Otolaryngol* 82: 483–493, 1965.

235. Nakashima T, Naganawa S, Sone M, Tominaga M, Hayashi H, Yamamoto H, Liu X, and Nuttall AL. Disorders of cochlear blood flow. *Brain Res Brain Res Rev* 43: 17–28, 2003.

236. Nance WE. The genetics of deafness. *Ment Retard Dev Disabil Res Rev* 9: 109–119, 2003.

237. Nilsson R and Borg E. Noise-induced hearing loss in shipyard workers with unilateral conductive hearing loss. *Scand Audiol* 12: 135, 1983.

238. Norena AJ and Eggermont JJ. Enriched acoustic environment after noise trauma reduces hearing loss and prevents cortical map reorganization. *J Neurosci* 25: 699–705, 2005.

239. Passe EG. Sympathectomy in relation to Ménière's disease, nerve deafness and tinnitus. A report of 110 cases. *Proc Roy Soc Med* 44: 760–772, 1951.

240. Payne MC and Gikla F.J. Effects of perforations of the tympanic membrane on cochlear potentials. *Arch Otolaryngol* 54: 666–674, 1951.

241. Peake WT, Rosowski JJ, and Lynch TJI. Middleear transmission: acoustic versus ossicular coupling in cat and human. *Hear Res*: 245–268, 1992.

242. Peeters S, Offeciers FE, Kinsbergen J, Van Durme M, Van Enis P, Dykmans P, and Bouchataoui I. A digital speech processor and various encoding strategies for cochlear implants. *Prog Brain Res* 97, 283–291, 1993.

243. Perier O, Alegria J, Buyse M, D'Alimonte G, Gilson D, and Serniclaes W. Consequences of auditory deprivation in animals and humans. *Acta Otolaryngol* Suppl. 411: 60–70, 1984.

244. Phillips DP and Carr MM. Disturbances of loudness perception. *J Am Acad Audiol* 9: 371–379, 1998.

245. Pickett JM. Advances in sensory aids for the hearing-impaired: visual and vibrotactile aids. *Ann Otol Rhinol Laryngol* 89: 74–78, 1980.

246. Pierson LL, Gerhardt KJ, Rodriguez GP, and Yanke RB. Relationship between outer ear resonance and permanent noise-induced hearing loss. *Am J Otolaryngol* 15: 37–40, 1994.

247. Pierson MG and Møller AR. Prophylaxis of kanamycin-induced ototoxitity by a radioprotectant. *Hear Res* 4: 79–87, 1981.

248. Pinchoff RJ, Burkard RF, Salvi RJ, Coad ML, and Lockwood AH. Modulation of tinnitus by voluntary jaw movements. *Am J Otol* 19: 785–789, 1998.

249. Popelar J, Syka J, and Berndt H. Effect of noise on auditory evoked responses in awake guinea pigs. *Hear Res* 26: 239–248, 1987.

250. Portillo F, Nelson RA, Brackmann DE, Hitselberger.W.E, Shannon.R.V, Waring MD, and Moore JK. Auditory brain stem implant: electrical stimulation of the human cochlear nucleus. *Adv Oto-Rhino-Laryngol* 48: 248–252, 1993.

251. Portmann G. The saccus endolymphaticus and an operation for draining the same for the relief of vertigo. *J Laryng Otol* 42: 809, 1927.

252. Priuska EM and Schacht J. Formation of free radical by gentamycin and iron and evidence for an iron/gentamycin complex. *Biochem Pharmacol* 50: 1749–1752, 1995.

253. Pulec JL. Cochlear nerve section for intractable tinnitus. *ENT Journal* 74: 469–476, 1995.

254. Pulec JL. Tinnitus: surgical therapy. *Am J Otol* 5: 479–480, 1984.

255. Pulec JL, Hodell SF, and Anthony PFT. Tinnitus: diagnosis and treatment. *Ann Otol Rhin & Laryngol* 87: 821–833, 1978.

256. Quaranta A, Sallustio V, and Scaringi A. Cochlear function in ears with vestibular schwannomas. In: *Third international conference on acoustic neurinoma and other CPA tumors*, edited by Sanna M, Taibah A, Russo A, and Mancini F. Rome: Monduzzi Editore, 1999, p. 43–50.

257. Quaranta N, Bartoli R, and Quaranta A. Cochlear implants: indications in groups of patients with borderline indications. A review. *Acta Otolaryngol (Stockh)* Suppl 552: 68–73, 2004.

258. Rahko T and Kotti V. Tinnitus treatment by transcutaneous nerve stimulation (TNS). *Acta Otolaryngol (Stockh)* Suppl 529: 88–89, 1997.

259. Rajan R and Irvine DR. Neuronal responses across cortical field AI in plasticity induced by peripheral auditory organ damage. *Audiol Neurootol* 3: 123–144, 1998.

260. Raser JM and O'Shea EK. Noise in gene expression: origins, consequences, and control. *Science* 309: 2010–2013, 2005.

261. Ratliff F. *Mach bands. Quantitative studies on neural networks in the retina.* San Francisco, CA: Holden-Day, Inc., 1965.

262. Rauch SD. Intratympanic steroids for sensorineural hearing loss. *Otolaryngol Clin North Am* 37: 1061–1074, 2004.

263. Rauch SD, Chen CY, and Halpin CF. Our experience in diagnostics and treatment of sudden sensorineural hearing loss. *Otol Neurotol* 26: 317, 2005.

264. Raucher FH, Robinson KD, and Jens JJ. Improved maze learning through early music exposure in rate. *Neurol Res* 20: 427–432, 1998.

265. Ravicz ME, Rosowski JJ, and Merchant SN. Mechanisms of hearing loss resulting from middle-ear fluid. *Hear Res* 195: 103–130, 2004.

266. Reed GF. An audiometric study of 200 cases of subjective tinnitus. *Arch Otolaryngol* 71: 94–104, 1960.

267. Robbins AM. Language developement. In: *Cochlear implants*, edited by Waltzman SB and Cohen NL. New York: Thieme Medical Publishers, Inc., 2000.

268. Robertson D and Irvine DR. Plasticity of frequency organization in auditory cortex of guinea pigs with partial unilateral deafness. *J Comp Neurol* 282: 456–471, 1989.

269. Rose JE, Galambos R, and Hughes JR. Microelectrode studies of the cochlear nuclei in the cat. *Bull Johns Hopkins Hosp* 104: 211–251, 1959.

270. Rosowski JJ. The effects of external- and middle-ear filtering on auditory threshold and noise-induced hearing loss. *J Acoust Soc Am* 90: 124–135, 1991.

271. Rubel EW, Popper AN, and Fay RR. *Development of the auditory system*. New York: Springer, 1998.

272. Ruben RJ, Hudson W, and Chiong A. Anatomical and physiological effects of chronic section of the eighth cranial nerve in the cat. *Acta Otolaryngol (Stockh)* 55: 473–484, 1962.

273. Rubinstein JT, Tyler RS, Johnson A, and Brown CJ. Electrical suppression of tinnitus with high-rate pulse trains. *Otology & Neurotology* 24: 478–485, 2003.

274. Rybak LP and Kelly T. Ototoxicity: bioprotective mechanisms. *Curr Opin Otolaryngol Head Neck Surg* 11: 328–333, 2003.

275. Sachs MB and Kiang NYS. Two tone inhibition in auditory nerve fibers. *J Acoust Soc Am* 43: 1120–1128, 1968.

276. Sachs MB and Young ED. Encoding of steady-state vowels in the auditory nerve: Representation in terms of discharge rate. *J Acoust Soc Am* 66: 470–479, 1979.

277. Saito S and Møller AR. Chronic electrical stimulation of the facial nerve causes signs of facial nucleus hyperactivity. *Neurol Res* 15: 225–231, 1993.

278. Salt AN. Regulation of endolymphatic fluid volume. *Ann N Y Acad Sci* 942: 306–312, 2001.

279. Salvi RJ, Henderson D, Fiorino F, and Colletti V. *Auditory system plasticity and regeneration*. New York: Thieme Medical Publishers, 1996.

280. Salvi RJ, Saunders SS, Gratton MA, Arehole S, and Powers N. Enhanced evoked response amplitudes in the inferior colliculus of the chinchilla following acoustic trauma. *Hear Res* 50: 245–258, 1990.

281. Salvi RJ, Wang J, and Ding D. Auditory plasticity and hyperactivity following cochlear damage. *Hear Res* 147: 261–274, 2000.

282. Sando I. The anatomical interrelationships of the cochlear nerve fibers. *Acta Otolaryng (Stockh)* 59: 417–436, 1965.

283. Sass K. Sensitivity and specificity of transtympanic electrocochleography in Ménière's disease. *Acta Otolaryngol (Stockh)* 118: 150–156, 1998.

284. Sass K, Densert B, and Arlinger S. Recording techniques for transtympanic electrocochleography in clinical practice. *Acta Otolaryngol (Stockh)*: 17–25, 1998.

285. Sataloff RT and Sataloff J. *Hearing loss*. New York: Marcel Dekker, 1993.

286. Scharf B, Magnan J, and Chays A. On the role of the olivocochlear bundle in hearing: 16 case studies. *Hear Res* 103: 101–122, 1997.

287. Schroeder M. Vocoders: Analysis and synthesis of speech. *Proc IEEE* 54: 720–734, 1966.

288. Schulman A, Tonndorf J, and Goldstein B. Electrical tinnitus control. *Acta Otolaryngol (Stockh)* 99: 318–325, 1985.

289. Selters WA and Brackmann DE. Acoustic tumor detection with brainstem electric response audiometry. *Arch Otolaryngol* 103: 181–187, 1977.

290. Sen CN and Møller AR. Signs of hemifacial spasm created by chronic periodic stimulation of the facial nerve in the rat. *Exp Neurol* 98: 336–349, 1987.

291. Sha SH and Schacht J. Stimulation of free radical formation by aminoglycoside antibiotics. *Hear Res* 128: 112–118, 1999.

292. Shambaugh GE. Surgery of the endolymphatic sac. *Arch Otolaryngol Head and Neck Surg* 83: 302, 1966.

293. Shannon RV, Zeng F-G, Kamath V, Wygonski J, and Ekelid M. Speech recognition with primarily temporal cues. *Science* 270: 303–304, 1995.

294. Shapiro SM. Binaural effects in brainstem auditory evoked potentials of jaundiced Gunn rats. *Hear Res* 53: 41–48, 1991.

295. Sharma A, Dorman MF, and Kral A. The influence of a sensitive period on central auditory development in children with unilateral and bilateral cochlear implants. *Hear Res* 203: 134–143, 2005.

296. Shepherd GM. *Neurobiology*. New York: Oxford University Press, 1994.

297. Sie KCY and Rubel EW. Rapid changes in protein synthesis and cell size in the cochlear nucleus following eighth nerve activity blockade and cochlea ablation. *J Comp Neurol* 320: 501–508, 1992.

298. Silverstein H, Thompson J, Rosenberg SI, Brown N, and Light J. Silverstein MicroWick. *Otolaryngol Clin North Am* 37: 1019–1034, 2004.

299. Silverstein H, Wanamaker HH, and Rosenberg SI. Vestibular neurectomy. In: *Neurotology*, edited by Jackler RK and Brackmann DE. St Louis, MO: Mosby, 1994, p. 945–954.

300. Simmons FB. Sudden idiopathic sensory-neural hearing loss: some observations. *Laryngoscope* 83: 1221–1227, 1973.

301. Simmons FB, Mongeon CJ, Lewis WR, and Huntington DA. Electrical stimulation of acoustical nerve and inferior colliculus. *Arch Otolaryngol* 79: 559–567, 1964.

302. Skellett RA, Cullen Jr. JK, Fallon M, and Bobbin RP. Conditioning the auditory system with continuous vs. interrupted noise of equal acoustic energy: Is either exposure more protective? *Hear Res* 116: 21–32, 1998.

303. Smoorenburg GF. Speech reception in quiet and in noisy conditions by individuals with noise-induced hearing loss in relation to their tone audiogram. *J Acoust Soc Am* 91: 421–437, 1992.

304. Snyder RL, Rebscher SJ, Cao K, and Leake PA. Effects of chronic intracochlear stimulation in the neonatally deafened cat: I. Expansion of central spatial representation. *Hear Res* 50: 7–33, 1990.

305. Spandow O, Anniko M, and Møller AR. The round window as access route for agents injurious to the inner ear. *Am J Otolaryngol* 9: 327–335, 1988.

306. Sparano A, Leonetti JP, Marzo S, and Kim H. Effects of stapedectomy on tinnitus in patients with otosclerosis. *Int Tinnitus J* 10, 2004.

307. Sperling N, M., Franco Jr RA, and Milhorat TH. Otologic manifestations of Chiari I malformation. *Otol Neurotol* 22: 678–681, 2001.

308. Spoendlin H. Anatomical changes following various forms of noise exposure. In: *Effects of noise on hearing*, edited by Henderson D, Hamernik RP, Dosanjh DS, and Mills JH. New York: Raven Press, 1976, p. 69–90.

309. Spoendlin H and Schrott A. Analysis of the human auditory nerve. *Hear Res* 43: 25–38, 1989.

310. Spoor A. Presbycusis values in relation to noise induced hearing loss. *Int Audiol* 6: 48–57, 1967.

311. Stahle J, Stahle C, and Arenberg IK. Incidence of Ménière's disease. *Arch Otolaryngol* 104: 99–102, 1978.

312. Starr A, Picton TW, Sininger Y, Hood LJ, and Berlin CI. Auditory neuropathy. *Brain* 119: 741–753, 1996.

313. Strouse A, Ashmead DA, Ohde RN, and Grantham W. Temporal processing in the aging auditory system. *J Acoust Soc Am* 104: 2385–2399, 1998.

314. Sunderland S. Microvascular relations and anomalies at the base of the brain. *J Neurol Neurosurg Psychiatry* 11: 243–257, 1948.

315. Syka J. Plastic changes in the central auditory system after hearing loss, restoration of function, and during learning. *Physiol Rev* 82: 601–636, 2002.

316. Syka J and Popelar J. Noise impairment in the guinea pig. I. Changes in electrical evoked activity along the auditory pathway. *Hear Res* 8: 263–272, 1982.

317. Syka J, Popelar J, and Kvasnak E. Response properties of neurons in the central nucleus and external and dorsal cortices of the inferior colliculus in guinea pig. *Exp Brain Res* 133: 254–266, 2000.

318. Syka J and Rybalko N. Threshold shifts and enhancement of cortical evoked responses after noise exposure in rats. *Hear Res* 139: 59–68, 2000.

319. Syka J, Rybalko N, and Popelar J. Enhancement of the auditory cortex evoked responses in awake guinea pigs after noise exposure. *Hear Res* 78: 158–168, 1994.

320. Szczepaniak WS and Møller AR. Evidence of neuronal plasticity within the inferior colliculus after noise exposure: a study of evoked potentials in the rat. *Electroenceph Clin Neurophysiol* 100: 158–164, 1996.

321. Szczepaniak WS and Møller AR. Interaction between auditory and somatosensory systems: a study of evoked potentials in the inferior colliculus. *Electroencephologr Clin Neurophysiol* 88: 508–515, 1993.

322. Szymanski M, Golabek W, and Mills R. Effect of stapedectomy on subjective tinnitus. *J Laryngol Otol* 117: 261–264, 2003.

323. Terrence C, Sax M, Fromm GH, Chang C–H, and Yoo CS. Effect of baclofen enantiomorphs on the spinal trigeminal nucleus and steric similarities of carbamazepine. *Pharmacology* 27: 85–94, 1983.

324. Tos M and Stangerup SE. Secretory otitis and pneumatization of the mastoid process: sexual differences in the size of mastoid cell system. *Am J Otolaryngol* 6: 199–205, 1985.

325. Tos M, Stangerup SE, and Andreassen UK. Size of the mastoid air cells and otitis media. *Ann Otol Rhinol Laryngol* 94: 386–392, 1985.

326. Tos M, Thomsen J, and Charabi S. Incidence of acoustic neuromas. *Ear, Nose, & Throat Journal* 71: 391–393, 1992.

327. Trune DR. Influence of neonatal cochlear removal on the development of mouse cochlear nucleus: I. Number, size, and density of its neurons. *J Comp Neurol* 209: 409–424, 1982.

328. Turner JG and Willott JF. Exposure to an augmented acoustic environment alters auditory function in hearing-impaired DBA/2J mice. *Hear Res* 118: 101–113, 1998.

329. Vasama JP, Møller MB, and Møller AR. Microvascular decompression of the cochlear nerve in patients with severe tinnitus. Preoperative findings and operative outcome in 22 patients. *Neurol Res* 20: 242–248, 1998.

330. Vernon J. The loudness of tinnitus. *Hear Speech Action* 44: 17–19, 1976.

331. Vernon JA and Fenwick JA. Attempts to suppress tinnitus with transcutaneous electrical stimulation. *Otolaryngol Head Neck Surg* 93: 385–389, 1985.

332. Vernon JA and Meikle MB. Masking devices and alprazolam treatment for tinnitus. *Otolaryngol Clin North Am* 36: 307–320, 2003.

333. Voss SE, Rosowski JJ, Merchant SN, and Peake WT. Middle-ear function with tympanic-membrane perforations. I. Measurements and mechanisms. *J Acoust Soc Am* 110: 1432–1444, 2001.

334. Voss SE, Rosowski JJ, Merchant SN, and Peake WT. Middle-ear function with tympanic-membrane perforations. II. A simple model. *J Acoust Soc Am* 110: 1445–1452, 2001.

335. Voss SE, Rosowski JJ, and Peake WT. Is the pressure difference between the oval and round windows the effective acoustic stimulus for the cochlea? *J Acoust Soc Am* 100: 1602–1616, 1996.

336. Wable J, Collet L, and Croze SC. Age-related changes in peri-lymphatic pressure: preliminary results. In: *Intracranial and intralabyrinthine fluids*, edited by Ernst A, Marchbanks R, and Samii M. Berlin: Springer Verlag, 1996, p. 191–198.

337. Wada JA. *Kindling 2.* New York: Raven Press, 1981.

338. Wall JT, Kaas JH, Sur M, Nelson RJ, Felleman DJ, and Merzenich MM. Functional reorganization in somatosensory cortical areas 3b and 1 of adult monkeys after median nerve repair: possible relationships to sensory recovery in humans. *J Neurosci* 6: 218–233, 1986.

339. Wall PD. The presence of ineffective synapses and circumstances which unmask them. *Phil Trans Royal Soc (Lond)* 278: 361–372, 1977.

340. Wang J, Ding D, and Salvi RJ. Functional reorganization in chinchilla inferior colliculus associated with chronic and acute cochlear damage. *Hear Res* 168: 238–249, 2002.

341. Wang J, Salvi RJ, and Powers N. Plasticity of response properties of inferior colliculus neurons following acute cochlear damage. *J Neurophysiol* 75: 171–183, 1996.

342. Warrick JW. Stellate ganglion block in the treatment of Ménière's disease and in the symptomatic relief of tinnitus. *Br J Otol* 41: 699–702, 1969.

343. Wayman DM, Pham HN, Byl FM, and Adour KK. Audiological manifestations of Ramsay Hunt syndrome. *J Laryngol Otol* 104: 104–108, 1990.

344. Weber H, K. P, Stohr M, and Rosler A. Central hyperacusis with phonophobia in multiple sclerosis. *Mult Scler* 8: 505–509, 2002.

345. Webster DB and Webster M. Neonatal sound deprivation affects brain stem auditory nuclei. *Arch Otolaryngol Head & Neck Surg* 103: 392–396, 1977.

346. Wever EG and Lawrence M. *Physiological acoustics.* Princeton, NJ: Princeton University Press, 1954.

347. White M, Merzenich M, and Gardi J. Multichannel cochlear implants: channel interaction and processor design. *Arch Otolaryngol* 110: 493–501, 1984.

348. Willer JC. Relieving effect of TENS on painful muscle contraction produced by an impairment of reciprocal innervation: an electrophysiological analysis. *Pain* 32: 271–274, 1988.

349. Willott JF, Chisolm TH, and Lister JJ. Modulation of presbycusis: current status and future directions. *Audiol Neurotol* 6: 231–249, 2001.

350. Willott JF and Lu SM. Noise-induced hearing loss can alter neural coding and increase excitability in the central nervous system. *Science* 216: 1331–1334, 1982.

351. Wilson PH, Henry JL, Andersson G, Hallam RS, and Lindberg P. A critical analysis of directive counselling as a component of tinnitus retraining therapy. *Brit J Audiol* 32: 273–286, 1998.

352. Wladislavorsky-Wasserman P, Facer GW, Mokri B, and Kurland LT. Ménière's disease: a 30 year epidemiologic and clinical study in Rochester, MN, 1951–1980. *Laryngoscope* 94: 1098–1102, 1984.

353. Yoshida N and Liberman MC. Sound conditioning reduces noise-induced permanent threshold shift in mice. *Hear Res* 148: 213–219, 2000.

354. Young ED and Sachs MB. Representation of steady-state vowels in the temporal aspects of the discharge patterns of

populations of auditory nerve fibers. *J Acoust Soc Am* 66: 1381–1403, 1979.

355. Zadeh MH, Storper IS, and Spitzer JB. Diagnosis and treatment of sudden-onset sensorineural hearing loss: a study of 51 patients. *Otolaryngol Head Neck Surg* 128: 92–98, 2003.

356. Zakrisson JE, Borg E, Diamant H, and Møller AR. Auditory fatigue in patients with stapedius muscle paralysis. *Acta Otolaryngol (Stockh)* 79: 228–232, 1975.

357. Zwicker E. On a psychoacoustical equivalent of tuning curves. In: *Facts and models in hearing*, edited by Terhardt E. Berlin: Springer–Verlag, 1974, p. 132–141.

358. Zwislocki JJ. What is the cochlear place code for pitch? *Acta Otolaryngol (Stockh)* 111: 256–262, 1991.

APPENDIX

A

Definitions in Anatomy

When describing anatomy of body parts such as the ear, it is important to have unambiguous and clear definitions of directions and planes of the body. Several different methods are in use. In this book I will use the ones illustrated in Fig. A.1. The direction from head to tail is caudal (means pertaining to the tail) and the opposite is rostral (relating to the beak). Ventral and dorsal give the directions from the belly to the back. The direction from the midline and out is called lateral, and the opposite direction is medial. A plane extending in the caudal-rostral direction and oriented ventrally-dorsally is the saggital plane. A saggital plane that divides the body into two identical halves is called the mid-saggital plane. A rostral-caudal plane that is perpendicular to the saggital plane is the coronal plane. A plane that is perpendicular to the saggital and coronal planes is the transverse or horizontal plane. It is common in medicine and surgery to use names like posterior and anterior, superior and inferior, which for humans is equivalent to dorsal and ventral, but that terminology becomes ambiguous when used in animals. The description given above can be used for animals as well as for humans.

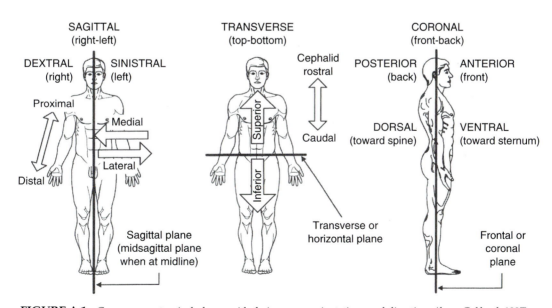

FIGURE A.1 Common anatomical planes with their names, orientations and directions (from Gelfand, 1997).

APPENDIX

B

Hearing Conservation Programs

1. INTRODUCTION

Hearing conservation programs have been an effective means of reducing the risk of noise induced hearing loss (NIHL) in industries where the noise level amounts to a risk of causing hearing loss. Hearing conservation programs include establishment of limits for exposure (noise standards) and monitoring of hearing in individuals who are exposed to noise in their occupation that may involve a risk of causing NIHL. Creation and enforcing of regulations regarding allowable noise exposure (noise standards) through legislation has reduced the incidence and the extent of NIHL. Noise standards state the limits of exposure to occupational noise based on measurement of noise levels in the workplace and on the basis of personal exposure using dosimeters. Complying with noise standards will ensure that no more than a certain (small) percentage of individuals who are exposed to the noise level that is stated in the regulations will get a certain degree of hearing loss. Monitoring of hearing (audiometry) in individuals who are exposed to noise levels that are deemed to involve a risk of causing NIHL is an important component of hearing preservation programs. Knowledge about noise standards and promotion of noise reduction at the source and use of personal protections (ear protectors) are important for reducing the risk of acquiring NIHL. Hearing conservation programs provide such information to individuals who are at risk of acquiring NIHL.

The individual variation in the susceptibility of NIHL is an obstacle in the prevention of hearing loss through regulations of noise exposure. Since it is not possible to identify the individuals who will acquire NIHL from

exposure to noise, the level of which is below the stated limit before they are exposed to noise that can damage hearing, all individuals who are exposed to noise above a certain level must be monitored.

Hearing conservation programs are often referred to as "hearing conservation programs in the work place" because they are aimed at protecting workers in the workplace from NIHL, but many people acquire NIHL in other situations, such as through recreational activities. Visiting rock concerts and exposure to other kinds of loud music can cause hearing loss. Those who perform at such concerts are at a higher risk of acquiring NIHL than the audience. Such activities are not included in hearing conservation programs.

In this appendix, I use the word noise to describe sound that may be damaging to hearing. This word has traditionally had negative connotations and thus will be identified more readily with health hazards. Any sound of sufficient intensity has a potential to cause hearing loss.

Hearing conservation programs are aimed at reducing the risk of NIHL, which is the best-known adverse health effect of noise. Noise induced hearing loss is of two kinds, temporary threshold shift (TTS) and permanent threshold shift (PTS) (see Chapter 2). Immediately after exposure to noise, the NIHL is likely to consist of both PTS and TTS, but after some time without noise exposure the hearing loss is mainly PTS. The amount of NIHL (PTS and TTS) that is acquired is related to the intensity and duration of the exposure to noise and to the character of the noise (spectrum, and time pattern– whether it is continuous or transient). Therefore, different types of noise pose different degrees of risk to hearing, even though the overall intensity of the noise is the same; impulsive sounds such as that from gunshots generally pose a greater risk than continuous noise. These factors are only partly covered in current legislation regarding allowable noise exposure. The NIHL that an individual person acquires also depends on many other factors, most of which are poorly understood and not covered by noise standards.

The other commonly occurring adverse effects of noise exposure, tinnitus and hyperacusis, are not included in hearing conservation programs and have received little attention despite the fact that both tinnitus and hyperacusis (see Chapter 10) reduce the quality of life to an extent that may be greater than that of hearing loss.

2. PURPOSE AND DESIGN OF HEARING CONSERVATION PROGRAMS

The purpose of hearing conservation programs is to reduce the risk of NIHL. This is done by enforcing regulations regarding allowable noise exposure, by promoting the use of personal protection devices, by monitoring hearing in those who are exposed to noise above a certain level, and by education. Hearing conservation programs are directed to occupational noise, produced by machinery and handling of material. Separate programs are directed to the military (in the USA).

Hearing conservation programs promote reduction in occupational noise exposure, which can be done by reducing the emission of noise, reducing the need of having people be close to noise sources and changing the way material is handled. Reducing the time that people are exposed to noise can also reduce the risk of NIHL and changes in work procedures are effective in reducing the risk of NIHL. An example is the change from to riveting to welding in shipbuilding.

Hearing conservation programs have brought awareness of the risks of acquiring NIHL and promoted compliance with noise standards, which have resulted in reduced noise exposure. Promotion of personal protection has also lowered the risk of NIHL considerably. Monitoring hearing has been effective in identifying individuals who have a higher risk than normal for acquiring NIHL.

2.1. Basis for Hearing Conservation Programs

Hearing conservation programs are based on valid risk criteria. Establishing such criteria rests on knowledge about the relationship between acquired NIHL

and the level, duration of, and the character of the noise to which a person is exposed. Assessing other factors such as individual susceptibility to NIHL has been less successful. An important component of hearing conservation programs is monitoring hearing in individuals who are exposed to noise that may cause NIHL.

The risk of NIHL increases when the intensity of the noise is increased and when the duration of the exposure is prolonged, thus increasing the energy of the noise exposure (the product of noise intensity and duration). The risk of PTS is, however, not directly proportional to the total energy of the noise exposure. If the risk of PTS were proportional to the total energy of the noise exposure, doubling of the exposure time would have the same effect as an increase of the noise level by 3 dB (i.e., a doubling of the energy). Some data on NIHL have been interpreted to indicate that a doubling of exposure time instead increases the risk of hearing loss with the same amount as an increase of the noise level of 5 dB.

The character of the noise to which different individuals are exposed also affects the risk of acquiring PTS. The greatest hearing loss from exposure to tones or narrow band of noise occurs at approximately one-half octave above the frequency of the noise (see Chapter 9). The energy around 4 kHz is enhanced by the ear canal, which acts as a resonator that amplifies sounds in the frequency range of 3 kHz (see Chapter 2). The half-octave shift makes the greatest hearing loss occur at approximately 4 kHz (see p. 20) (the exact frequency of the largest hearing loss in an individual person depends on the length of the ear canal, which varies among individuals) [21] (see p. 220). The amount of PTS that a certain individual acquires is therefore not only dependent on the intensity of the noise and the exposure time but it also depends on its spectrum.

The hearing loss shown in Fig. B.1 is typical for individuals who have been exposed to noise in various manufacturing industries where the noise tends to be of a broad spectrum and continuous in nature. Most of the hearing loss that is expected after 40 years of noise exposure is already acquired during the first 10 years of the exposure (Fig. B.1).

The individual variation in the hearing loss from exposure to noise is an obstacle for establishing valid risk criteria. Different people who are exposed to noise of the same intensity and for exactly the same length of time may suffer different degrees of hearing loss. Some people can tolerate high-intensity noise for a lifetime and not suffer any noticeable degree of hearing loss while other people may acquire substantial hearing loss from exposure to much less intense noise (see Fig. 9.14). The average hearing loss from exposure to continuous noise of 85 dB(A) for 20 years in the study in Fig. 9.14

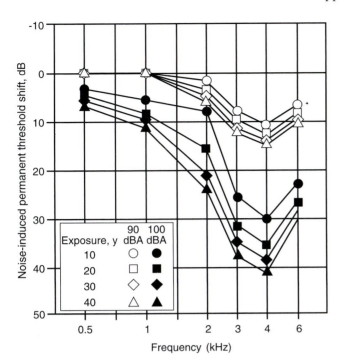

FIGURE B.1 Median estimated noise-induced permanent threshold shift plotted as a function of frequency for two exposure levels (assuming 8-h daily exposure) and four durations of exposure (reproduced from Dobie, 1995; after ISO-1999, Annex E, with permission from *Arch. Otolaryngol. Head Neck Surg.*).

is less than 5 dB at 4 kHz, but many people in this study experienced 30- to 40-dB hearing loss, and some people acquired very little loss.

The noise intensity and the exposure time have been combined in a single value, known as the noise immission value (in decibels, see Fig. 9.14 and p. 220). The effect of the exposure time is probably different for different kinds of noise and it is also most likely different for different intensities of the noise. This makes description of noise by a single number insufficient to characterize its ability to cause NIHL and consequently contribute to the uncertainty in the evaluation of risks of NIHL.

Because of the large individual variation in noise-induced hearing loss (NIHL), only the average probability for acquiring a hearing loss can be predicted on the basis of knowledge about the physical characteristics of noise and the duration of exposure to noise.

Individual variations in noise susceptibility have earlier been ascribed to genetic factors but that cannot explain all the individual variations and there is evidence that epigenetics and other variations in gene expressions play a role (see Chapter 9).

Since attempts to estimate an individual person's susceptibility to PTS have been unsuccessful it is important to monitor hearing in individuals who are exposed to noise that involves a risk of causing NIHL. The only way to determine an individual's susceptibility to noise-induced hearing loss is to test those who are exposed to loud noise at frequent intervals. Audiometric testing is therefore an important component of hearing

---

BOX B.1

CAUSES OF VARIATIONS IN NIHL

The effect of genetics in NIHL is apparent from studies in animals. The variation in NIHL for the same exposure is much less in inbred animals than in normal animals. The NIHL to the same noise exposure in inbred animals that are assumed to be genetically identical, however, has some variations that must be ascribed to epigenetics or random variations in gene expression (see Chapter 9, Fig. 9.15). That genetics is involved is supported by studies that have shown that rats that are genetically predispositioned for high blood pressure also acquired more hearing loss from noise exposure than normal rats when both groups were exposed to noise for their entire lifetimes [8]. Although these findings have not been duplicated in humans, the results of some studies in humans support a relationship between

high blood pressure and hearing loss from noise exposure [6].

Individual differences in cochlear blood flow may also affect the NIHL that individuals acquire [7] (see Chapter 9). While research along these lines has provided important knowledge, it has not resulted in the development of efficient ways to assess an individual's susceptibility to NIHL or to effectively decrease a person's risks of acquiring PTS.

Attempts to determine individuals' likelihood to acquire NIHL have been made by determining the degree of TTS a person acquires from exposure to test sounds that are not loud enough to cause PTS. The results of such tests have, however, been discouraging and there is only a weak correlation between PTS and the degree of TTS in any individual person.

conservation programs. NIHL usually first affects the hearing threshold at frequencies around 4 kHz (see Chapter 9), thus above the frequency range that is essential for comprehension of speech. Common hearing tests (pure tone audiometry) can therefore reveal hearing loss before it is noticed by an individual and before it affects the ability to understand speech. Hearing loss that occurs days or weeks after a person has begun to be exposed to noise of moderate intensity may indicate that the person in question is more susceptible to noise-induced hearing loss than normal. Audiometric tests must therefore be carried out at short intervals during the first period of a person's exposure to noise and continued at longer intervals for the time the person is exposed to noise that can cause PTS.

It is important to consider at what time, in relation to the noise exposure, hearing tests should be performed.

NIHL include both TTS and PTS where the TTS decreases after the end of noise exposure. If hearing tests are done a short time after the end of a workday, the measured hearing loss will include both PTS and TTS. If it is performed after a weekend, less TTS will be included. However, it is probably not possible to obtain accurate estimates of PTS unless the hearing test is performed several weeks after the last noise exposure (Fig. B.2). It is therefore important to consider how hearing tests are administered.

OSHA states that the employer must arrange hearing tests and that such hearing tests must be "performed by a licensed or certified audiologist, otolaryngologist, or other physician, or by a technician certified by the Council of Accreditation in Occupational Hearing Conservation, or who has demonstrated competence in administering audiometric examination." This means that essentially anybody can be assigned by an

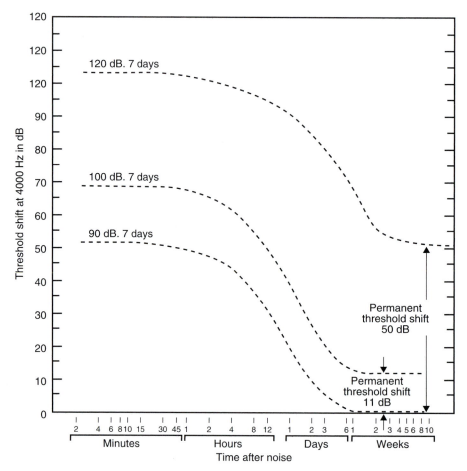

FIGURE B.2 Schematic diagram illustrating how noise can affect hearing. The graph shows the hearing loss (*threshold shift*) at 4 kHz a certain time (*horizontal axis*) after noise exposure. Noise with intensity below a certain value is expected to give rise to a temporary threshold shift (*90 dB, 7 days curve*), while a louder noise (*100 dB, 7 days*) results in a permanent threshold shift. A very intense noise (*120 dB, 7 days*) gives rise to a considerable permanent shift in threshold (modified from Miller, 1974, with permission from the American Institute of Physics).

employer to do the testing. It is surprising that the professional organizations of audiologists have not protested such a statement.

Testing of hearing in hearing conservation programs is often limited to obtaining pure tone audiograms. It is also important to determine a person's ability to understand speech because there is a great individual variation in the relationship between the tone audiogram and the ability to understand speech (see Chapter 9). However, determination of speech discrimination is not standardized and the outcome depends on whether the tests are done in quiet or with a background of noise [26]. Also, other effects of noise exposure on people with noise-induced hearing loss, such as tinnitus and hyperacusis, should be assessed in hearing conservation programs [19].

3. ESTABLISHMENT OF NOISE STANDARDS

Recommendations of acceptable noise levels have been established for the purpose of reducing the risk of noise-induced hearing loss. These recommendations appear in the form of "noise standards." Noise standards are regulations regarding the permitted exposure to noise. Establishment of the correlation between noise exposure levels (sound intensity and duration of exposure) to the risk of NIHL is the basis for such noise standards (noise criteria).

Noise standards provide estimates of the hearing loss a person will acquire from exposure to noise of different intensity and for different lengths of time. Noise standards play important roles in legal matters such as litigation regarding workmen's compensation [11]. Different countries have adopted different standards, and the ways in which the standards are enforced also differ.

The considerable individual variation in susceptibility to PTS makes it unrealistic to protect everybody from any detectable degree of NIHL. The limits of noise exposure that are deemed to be "safe," according to current noise standards, in fact allow a certain percentage of individuals to acquire hearing loss that is greater than what is defined as hearing loss (10dB average at 2, 3, and 4 Hz).

Noise standards that are in use thus have similarities and differences, but all presently accepted standards use a single-value that is a combination of noise level and the duration of the exposure to calculate the risk of noise-induced permanent hearing loss. In the United States, standards have been tightened so

---

BOX B.2

NOISE STANDARDS

In the United States, legislation that covers occupational noise hazards includes the Federal Aviation Act of 1958, the 1969 Amendment of the Walsh-Healy Public Contracts Act, the Occupational Safety and Health Act of 1970, the Noise Control Act of 1972, and the Mine Safety and Health Act of 1978. These acts require certain agencies to regulate exposure to noise. In the USA, also the three branches of the military have established criteria regarding risks of NIHL from noise exposure. In Europe, legislation in various countries regarding the limitations on industrial noise has largely been guided by recommendations made by the International Organization for Standardization (ISO) [2].

The maximal noise level and duration accepted in most industrial countries is either 85 or 90-dB (A)[1] for 8 hours a day, 5 days a week. In the United States 90 dB (A) is the accepted level stated by the Occupational Safety and Health Administration (OSHA). Certain measures must be taken if workers are exposed to noise levels above 85 dB (A). These acts also state that hearing conservation programs must be in place when people are exposed to noise levels of 85 or more dB (A) (8-hour weighted average) [4]. If action taken to reduce the noise exposure to 90 dB (A) or lower is not successful employers must make personal hearing protection devices (ear protectors) available to such workers and perform hearing tests at specified intervals during employment through a hearing conservation program. If hearing loss of 10 dB average over frequencies 2, 3, and 4 kHz is detected, then the person must be referred for further evaluation and action must be taken to avoid further deterioration of hearing. Such action may include moving the person to a less noisy environment to prevent the progress of the hearing loss before it becomes a social handicap.

The National Institute for Occupational Safety and Health (NIOSH) has recently issued a recommendation that has 85 dB (A) as the limit of accepted exposure level [3].

[1]dB (A) refers to A weighting used in sound level meters (see p. 297).

that no worker should be exposed to continuous noise above 115 dB (A) or impulsive noise above 140 dB (A), independent of the duration of exposure. This action sets a ceiling for acceptable combinations of noise intensity and exposure time. Some standards use correction factors regarding the nature of the sound (for instance, impulsive versus continuous sounds). Some standards take normal age-related hearing loss (presbycusis) into account while others do not (see p. 216).

3.1. Noise Level and Exposure Time

Noise standards are based on maximum daily exposure of 8 hours. If the exposure time is shorter than 8 hours a day, a higher level of noise can be tolerated. To estimate how much higher level of noise can be tolerated when the duration of the exposure to noise is less than 8 hours per day, a conversion factor is used. Europe has used a 3-dB "doubling factor" for a long time while the United States has used a 5-dB doubling factor. Research indicates that a doubling factor of 5 dB may be adequate for relatively low noise levels, but that a smaller doubling factor (3 dB, i.e., equal energy) more correctly reflects the hazards presented by noise of a high level. NIOSH also now recommend a 3-dB doubling factor for calculation of the time weighted average exposure to noise [3].

A 3-dB doubling factor implies that a reduction of the exposure time by a factor of 2 (e.g., from 8 to 4 hours), can allow a 3-dB higher sound level to be accepted. Thus 88 dB (A) for 4 hours is assumed to have the same effect on hearing as 85 dB (A) for 8 hours. If the exposure time to noise is 2 hours per day, a 6-dB higher sound level is assumed to be acceptable, and so on. This way of calculating an acceptable noise level reflects "the equal energy principle," and assumes that it is the total energy of the noise that determines the risk of permanent hearing loss.

Because the level of noise exposure usually varies during a workday, noise exposure is often described by its equivalent level (Leq), which is defined as the level of noise that has the same average energy as the noise that is measured during a workday. The equivalent level is determined by adding the noise energy to which a person is exposed and dividing it by the duration of exposure. The calculation of the Leq assumes that the equal energy principle is valid for estimating risks of acquiring NIHL.

The fact that the present noise standards are based on a simplified measure of noise, namely the A-weighted measure dB (A), adds to the uncertainty in predicting the risk of acquiring a hearing loss that may result from exposure to a certain noise.

3.2. Effect of Age-related Hearing Loss

Hearing loss from causes other than noise interacts with NIHL in a complex way. It may seem natural to correct for hearing loss from other causes when NIHL is estimated and it has been suggested that such hearing loss should be subtracted from the total hearing loss acquired by a person who is exposed to noise. At first glance it may seem correct to subtract hearing loss that can be attributed to age (presbycusis) from the total hearing loss to arrive at the value of pure NIHL. However, such a "correction" of the hearing loss in an elderly person with a long history of noise exposure may cause the calculated NIHL to decrease with time of noise exposure. The reason for this paradoxical result is that hearing loss from presbycusis and noise exposure do not add linearly. Above a certain age, presbycusis increases more than the NIHL. That means that 2 + 2 is not 4 but rather less, perhaps 3. When one factor is subtracted from such a "sum" in order to obtain the other factor, a paradoxical result occurs. The 1998 NIOSH criterion [3] no longer recommend age correction to take into account presbycusis.

3.3. What Degree of Hearing Loss is Acceptable?

Because of the great individual variation in NIHL from the same noise exposure, it is impossible to predict what hearing loss an individual will acquire when exposed to a certain noise. Noise standards therefore at best predict the percentage of people who will acquire less than a certain specified hearing loss when exposed to noise no louder than a certain value [15, 18]. This "specific hearing loss" is then regarded to be acceptable. The presently applied noise standards allow that a certain (small) percentage of the population will acquire PTS that is greater than a certain value.

In the beginning of the era of efforts to reduce the occurrence of NIHL, the "acceptable hearing loss" was defined as the hearing loss at which an individual begins to experience difficulty in understanding everyday speech in a quiet environment. This definition was based on the American Academy of Ophthalmology and Otolaryngology (AAOO) guidelines for evaluation of hearing impairment (revised in 1979 by AAO, from 1959 and 1973) [1] which state that the ability to understand normal everyday speech at a distance of about 1.5 m (5 ft) does not noticeably deteriorate as long as the hearing loss does not exceed an average value of 25 dB at frequencies 0.5, 1, and 2 kHz. This amount of hearing loss was regarded as a just-noticeable handicap for which a worker in the United States was entitled to receive workmen's compensation for loss of earning

TABLE B.1 Estimated Risk of Hearing loss after 40 Years Working Lifetime[a]

Reporting organization	Average daily exposure (dBA)	Excess risk[b]
ISO	90	21
	85	10
	80	0
EPA	90	22
	85	12
	80	5
NIOSH	90	29
	85	15
	80	3

[a]Data from NIOSH (3)

[b]Percentage of individuals who acquire greater hearing loss than 25 dB at 0.5, 1, and 2 kHz of noise on other bodily functions.

power. These recommendations have not been updated by the American Academy of Otolaryngology (AAO). The American Medical Association (AMA) [10] has recently provided its own guidelines. These guidelines, however, follow the AAO 1979 guidelines. It has been argued that these guidelines should be modified to include hearing loss at 3 kHz [10. 23].

It is puzzling that this degree of hearing loss given in the AAO (1979) recommendation to describe the hearing level at and above which disability occurs was later designated as an acceptable degree of hearing loss. The estimated percentage of individuals who acquire hearing loss in excess of such hearing loss (Table B.1) naturally depends on the noise exposure.

4. MEASUREMENT OF NOISE

Hearing conservation programs depend on accurate measurements of the noise to which people are exposed. Ideally, the units of measurement should be related to the risk of NIHL but that is not possible because the risk of PTS is not a simple function of the energy of noise as it is measured physically. The spectrum of the noise and its temporal pattern are important factors, which are difficult to account for in practical measurements of noise. To somehow take the spectrum into consideration, the spectrum is weighted before being measured by noise level meters. Since low-frequency sounds are considered to be less damaging than high-frequency sounds of the same physical intensity, low-frequency sounds are attenuated when noise intensity is measured for predicting its effect on hearing. The commonly used weighting (A-weighting) gives energy at low frequencies less weight than energy at high frequencies. The temporal pattern of noise also affects its ability to cause NIHL, but this factor is more difficult to represent in standard measurement of noise level. Noise exposure is often expressed in noise immission level (see Chapter 9), which is a measure that combines duration of exposure and intensity of the noise importance.

4.1. Sound Level Meters

Sound level meters are available in many different forms, starting with simple devices consisting of basic components, namely a microphone, an amplifier with circuits that allow integration of the output of the amplifier and display. Most sound level meters have at least one spectral weighting, namely A-weighting. The most sophisticated sound level meters have many options regarding weighting functions, spectral filtering (1/3 and 1 octave wide), and integration times. Such sound level meters also provide readings of equivalent sound level (Leq, steady A-weighted sound level averaged over a specified time).

Noise level meters are now standardized by the International Electrotechnical Commission (IEC) (IEC 61672:1999), the International Organization for

---

BOX B.3

IMMISSION LEVEL

The noise immission level E is defined as $L + 10 \log(T)$, where L represents the sound level (measured with A-weighting) that is exceeded during 2% of the exposure time, T, which is the time in months. For example, exposure to 85-dB noise during 20 years of work corresponds to $85 + 10 \log(20 \times 12) = 85 + 10 \log 240 = 85 + 24 = 109$. For continuous noise, the measure L deviates only slightly from the A-weighted intensity of the noise. The difference between these two values can be large for noise that is transient or intermittent (i.e., noise that varies considerably in intensity).

BOX B.4

SPECTRAL WEIGHTING IN NOISE LEVEL METERS

The earliest design of noise level meters had built-in filters that weighted the spectrum of the sound to provide measurements that were as close as possible to how noise was perceived. These sound level meters used three different weightings of the spectrum known as A, B and C weighting. The A-weighting was designed to follow the 40 phon curve[2] and was used for measuring sounds of low intensity. The B-weighting mimicked the 70 phon curve and was designed for measurement of sounds of greater intensity. The C-weighting was nearly flat and was used for high intensity sounds.

[2]Phon curves: contours that represent equal loudness. The curves show the sound intensity (in dB SPL) at different frequencies that have the same loudness as a tone at 1000 Hz. The curves are labeled in accordance to that, so that for example, the 40 phon curve shows the intensity of tones in dB SPL that have the same loudness as a tone at 1 kHz and 40 dB SPL.

Standardization (ISO) and the American Standards Institute (ANSI) S1.4-1971 (R1976) or S1.4-1983 (R 2001) with Amd.S1.4A-1985, S1.43-1997 (R2002) (Type 0 is used in laboratories, Type 1 is used for precision measurements in the field, and Type 2 is used for general-purpose measurements). (Complete descriptions of these standards are available from http://www.ansi.org.)

The risk of noise induced hearing loss is related to the spectrum of sounds in a rather different way than perception and the properties of modern sound level meters are therefore modified from early designs to better reflect the risk of hearing loss. Experience has shown that the A-weighting reflected the ability of sounds to cause NIHL for sound intensities much higher than 40 phon. This is why the most commonly used weighting is the so-called A-weighting. The unit of measurement of noise using that weighting is dB (A).

Industrial noise that can cause hearing loss typically varies over time. It can be slowly varying or it can be impulsive in nature, such as gun shot noise. Noise level meters were originally designed to integrate sound over about 200 (or 100) ms in order to provide a reading that was in accordance with the perceived loudness of sounds. (The integration time of the auditory system above threshold is approximately 100 ms.) This (long) integration time is not appropriate for assessing the risk of hearing loss because injury from noise exposure occurs in the cochlea, which has a much shorter integration time than that of perception of sound. Sound level meters should therefore have an integration time that corresponds to that of the cochlea, which is only a few milliseconds. More sophisticated sound-level meters (so-called impulse sound-level meters) have a choice of integration time and it is possible to choose an integration time that is appropriate for measurement of impulsive sounds.

4.2. Noise Dosimeters

Measurements of sound levels are usually made at a location where people work, but the sound level at the entrance of the ear canal will be different because the head and the outer ear amplify sounds within frequencies between 2 and 5 kHz by as much as 6 dB, depending on the direction to the sound source [24, 25] (these data were obtained from studies using a spherical model of the head; see Chapter 2, Fig. 2.5). If the noise to which a person is exposed contains much energy in that frequency range, the sound that actually reaches the ear may be as much as 6 dB more intense than the actual reading on a sound level meter placed in the person's location when the person is not present. At and above 5 kHz the outer ear causes an additional increase in sound pressure of 5 to 10 dB depending on the angle to the sound source in the horizontal and the vertical plane. This problem can be avoided by using noise dosimeters, which measure the noise level near the ear of workers.

The noise level is often different at different locations, and when a person moves around the exposure varies and it becomes difficult to estimate the total exposure during a workday. Noise dosimeters integrate the noise level near the ear of a person over an entire day and provide information about the total noise exposure in a single measure, thus similar to radiation dosimetry. The increase in sound level at the tympanic membrane caused by the resonance of the ear canal (between 2 and 6 kHz) is not included in the measurement of sound using dosimeters (see Chapter 2).

While noise dosimetry has advantages over the conventional way of measuring noise exposure, it relies on the conversion between exposure time and sound level. Noise dosimetry does not eliminate the uncertainty that is related to individual difference in susceptibility to NIHL.

5. PERSONAL PROTECTION

When the possibilities of reducing the noise at the source and moving people away from the noise source have been exhausted, personal protection is justified. Three main kinds of personal protection devises are in common use, namely earplugs, earmuffs, and active noise reduction devices.

5.1. Earplugs and Earmuffs

Use of ear protectors is an option for reducing the risk of acquiring NIHL. The use of such personal protection is justified in situations where the exposure to noise cannot be reduced sufficiently, and in situations where people only spend short times in noisy environments, such as on the tarmac of airports. The most commonly used ear protectors are earplugs that are inserted in the ear canal and earmuffs, which are attached to a helmet or worn on a headband. Earmuffs have the advantage over earplugs that they can be removed easily and are therefore suited for intermittent use such as in situations when people are walking in and out of noisy areas. Earplugs are more suitable for use by people who spend long periods of time in noisy environments.

Laboratory studies show that earplugs attenuate sound more than earmuffs. The achieved sound attenuation of different types of earplugs and earmuffs depends not only on the type of device but also on how well the devices fit the individual person.

BOX B.5

ATTENUATION OF EAR PROTECTORS

Manufactures' laboratory studies of attenuation of different types of ear protectors show that insert ear protectors (earplugs) provide approximately 20 dB attenuation at 0.125 and 0.25 kHz, 20–25 dB for frequencies from 0.5–2 kHz and approximately 40 dB at 4 and 8 kHz. Some types of earplugs provide 4–5 dB more attenuation. Similar studies show that ear muffs provide 10–15 dB attenuation for 0.125 and 0.25 kHz, 20–25 dB at 0.5 kHz and 35–40 dB for 1–8 kHz. Real world data show slightly different values, and field data generally show less attenuation than laboratory data (Fig. B.3) [5]. The reduction in the risk of NIHL that is achieved in practice from wearing ear protection may be less than anticipated from laboratory studies.

Field studies of workers who were exposed to high intensity noise (shipyard) have shown that earplugs provided better protection than earmuffs [13]. Studies of hearing loss from noise exposure in groups of people wearing earplugs have shown that under certain circumstances earplugs provide better protection than earmuffs although earmuffs have greater attenuation [13]. These results were confirmed in other studies [12]. One reason for the difference in protection against NIHL may have been related to the fact that they may not always be worn when indicated [20]. Studies of the efficacy of ear protectors done in shipyards with intense continuous noise and superimposed impulsive noise have shown that those who were exposed to low-intensity noise suffered more hearing loss than did those in the high-intensity noise group. This may be surprising but it may be due to workers' different habits of wearing ear protectors and that many more workers exposed to high-intensity noise wore ear protectors than those who were exposed to low-intensity noise.

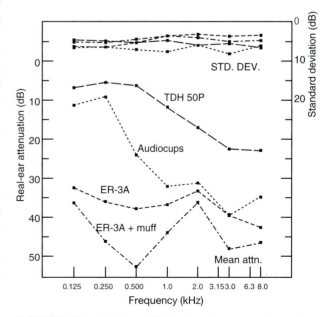

FIGURE B.3 "Real-world" data for four types of earplugs and one kind of earmuff (data after Berger and Royster, 1996).

The difference between laboratory data and experience from the field can have several reasons. One is that the efficacy of ear protectors depends on compliance with use of the devices, which is difficult to control and poorly documented. Another reason for less efficacy of protective devices can be individual variations in anatomy causing the fit of ear protectors and earmuffs to vary among individuals. The beneficial effect is much reduced if the protective devices are only worn part of the time that an individual is exposed to noise [13]. People may avoid wearing ear protectors for long periods because it is inconvenient, especially in hot environments. Ear protectors impair speech communication, making it more difficult for people to hear alarm signals or other acoustic signs of danger. This may be yet another reason for poor compliance. Lack of understanding of the importance of wearing ear protectors can also cause poor compliance.

5.2. Active Noise Cancellation

Active noise cancellation is realized by having a microphone and a small loudspeaker placed inside an earmuff. The microphone picks up sound that reaches the ear; the output of the microphone is amplified and inverted before applied to the loudspeaker. A computer adjusts the amount of sound that is presented through the loudspeaker so that it cancels out the sound that reaches the inside of the headset as indicated by the output from the microphone. Such active noise cancellation is most effective for low frequencies. The noise cancellation earmuffs can also have small loudspeakers that can provide communication or music. Active noise cancellation is often used in helicopters where the noise may be sufficient to cause a risk of NIHL. The excessive noise makes it difficult to communicate and the active noise cancellation in connection with their built-in loudspeakers makes voice communication easier. These devices have found use other than reducing the risk of NIHL. Noise cancellation earphones are popular on commercial airplanes where they effectively suppress noise, making it easier to work or sleep, and allowing music to be played with less disturbing airplane noise.

5.3. Other Means of Reducing the Risk of Noise Induced Hearing Loss

The finding that pre-exposure to noise can reduce the risk of NIHL [9, 17] (see p. 226) is interesting but is yet to have practical use. Attempts to find drugs that can reduce risk of NIHL have so far not yielded any results that have achieved practical use, although a few studies in animals have produced promising results [27].

6. NON-OCCUPATIONAL NOISE EXPOSURE

Until relatively recently the focus of NIHL concerned industry and the military. Many sources of noise to which people are exposed during recreational activities can cause NIHL. Most notable is perhaps the risk from rock concerts but other recreational activities, such as target shooting, can cause NIHL. Explosions such as fireworks and sirens used in fire alarm and emergency vehicles involve risks of acquiring NIHL. Exposure to these sources of noise is also likely to cause tinnitus.

7. EFFECT OF NOISE ON BODILY FUNCTIONS

While the greatest risk to health from noise exposure is hearing impairment (PTS), there are other adverse effects of noise such as tinnitus and stress. These symptoms are usually not included in hearing conservation programs. The effects of noise on bodily functions other than hearing are poorly understood. It has been reported that noise exposure can cause increase in blood pressure and changes in other body functions such as change (usually increase) in the secretion of pituitary hormones. Some retrospective studies [14] found that workers who were exposed to industrial noise had higher systolic and diastolic blood pressures, while others [22] found no relationship between noise-induced hearing loss and blood pressure in shipyard workers. Studies in rats have shown that animals with a hereditary predisposition for hypertension developed greater degrees of hearing loss from exposure to noise than did rats without this hereditary predisposition to high blood pressure [6]. If these results can be applied to humans, results of studies of hypertension in individuals who were exposed to industrial noise [14] may have to be reevaluated because these studies used hearing loss as a measure of the noise exposure. They may thereby have inadvertently selected workers who were predisposed to hearing loss because of their hypertension and not vice versa, as was intended.

7.1. Effect of Ultrasound and Infrasound

Ultrasound and infrasound are sounds that are not audible to humans because their frequencies are above or below our audible frequency range. There is no evidence to indicate that exposure to such sounds can damage the ear, and there is little evidence that such sounds can have other untoward effects. Some experiments indicate that infrasound may cause decrease in blood pressure. It is possible that such an effect will be caused by stimulation of the vestibular part of the inner ear.

Ultrasound is rapidly attenuated when transmitted in air. Exposure to high intensity of ultrasound can kill furred animals such as mice, rats, and guinea pigs because sound absorption in the fur causes heat, but this effect cannot occur in humans because bare skin does not absorb enough energy to cause damage. Studies have indicated that exposure to high intensity infrasound can cause diffuse symptoms such as headache, nausea, and fatigue.

APPENDICES REFERENCES

1. Anonymous. American Academy of Ophthalmology and Otolaryngology (AAOO). Committee on Hearing and Equilibrium and the American Council of Otolaryngology, Committee on Medical Aspects of Noise. Guide for Evaluation of Hearing Handicap. *Am J Med Assoc* 241: 2055–2059, 1979.
2. Anonymous. *Determination of occupational noise exposure and estimation of noise-induced hearing impairment, ISO-1999 International Organization for Standardization: Acoustics.* Geneva, Switzerland, 1990.
3. Anonymous. National Institute for Occupational Safety and Health (NIOSH) Criteria for a recommended standard: occupational exposure to noise. revised criteria 1998 Publication No. 98-126. 1998.
4. Anonymous. Occupational Safety and Health Administration (OSHA). Occupational noise exposure: hearing conservation amendment, final rule. *Federal Register* 48: 9738–9785, 1983.
5. Berger EH and Royster LH. In search of meaningful measures of hearing protector effectiveness. *Spectrum Suppl.* 1, 13: 29, 1996.
6. Borg E. Noise, hearing, and hypertension. *Scand Audiol* 10: 125–126, 1981.
7. Borg E, Canlon B, and Engstrom B. Noise-induced hearing loss. Literature review and experiments in rabbits. *Scand Audiol* 24(Suppl 40): 1–147, 1995.
8. Borg E and Møller AR. Noise and blood pressure: Effects on life-long exposure in the rat. *Acta Physiol Scand* 103: 340-342, 1978.
9. Canlon B, Borg E, and Flock A. Protection against noise trauma by pre-exposure to a low level acoustic stimulus. *Hear Res* 34: 197-200, 1988.
10. Demeter SL and Andersson GBJ. Chapter 11. Ear, nose, throat, and related structures. In: *Guides to the evaluation of permanent impairment, fifth edition.* American Medical Association, 2003.
11. Dobie RA. *Medical-legal evaluation of hearing loss.* New York: van Nostrand Reinhold, 1993.
12. Dobie RA. Prevention of noise-induced hearing loss. *Arch Otolaryngol Head and Neck Surg* 121: 385–391, 1995.
13. Erlandsson B, Hakanson H, Ivarsson A, and Nilsson P. The difference in protection efficiency between earplugs and earmuffs. *Scand Audiol (Stockh)* 9: 215–221, 1980.
14. Jonsson A and Hansson L. Prolonged exposure to a stressful stimulus (noise) as a cause of raised blood-pressure in man. *Lancet* 1: 86–87, 1977.
15. Kryter KD. Impairment to hearing from exposure to noise. *Acoust Soc Am* 53: 1211–1234, 1973.
16. Miller JD. Effects of noise on people. *J Acoust Soc Am* 56: 729–764, 1974.
17. Miller JM, Watson CS, and Covell WP. Deafening effects of noise on the cat. *Acta Otolaryng Suppl* 176: 1–91, 1963.
18. Møller AR. Noise as a health hazard. *Ambio* 4: 6–13, 1975.
19. Møller AR. Pathophysiology of tinnitus. In: *Otolaryngologic clinics of north america,* edited by Sismanis A. Amsterdam: W.B. Saunders, 2003, p. 249–266.
20. Nilsson R and Lindgren F. The effect of long term use of hearing protectors in industrial noise. *Scand Audiol (Stockh)* Suppl 12: 204–211, 1980.
21. Pierson LL, Gerhardt KJ, Rodriguez GP, and Yanke RB. Relationship between outer ear resonance and permanent noise-induced hearing loss. *Am J Otolaryngol* 15: 37–40, 1994.
22. Sanden A and Axelsson A. Comparison of cardiovascular responses in noise-resistant and noise-sensitive workers. *Acta Otolaryngol (Stockh)* Suppl 377: 75–100, 1981.
23. Sataloff RT and Sataloff J. *Hearing loss.* New York: Marcel Dekker, 1993.
24. Shaw EAC. The external ear. In: *Handbook of sensory physiology,* edited by Keidel WD and Neff WD. New York: Springer-Verlag, 1974, p. 455–490.
25. Shaw EAC. Transformation of sound pressure level from the free field to the eardrum in the horizontal plane. *J Acoust Soc Am* 56: 1848–1861, 1974.
26. Smoorenburg GF. Speech reception in quiet and in noisy conditions by individuals with noise-induced hearing loss in relation to their tone audiogram. *J Acoust Soc Am* 91: 421–437, 1992.
27. Yamasoba T, Schacht J, Shoji F, and Miller JM. Attenuation of cochlear damage from noise trauma by an iron chelator, a free radical scavenger and glial cell line-derived neurotrophic factor in vivo. *Brain Research* 815: 317–325, 1999.

List of Abbreviations

3 CLT: 3 channel Lissajous' trajectory

5-HT: 5-Hydroxytryptamine (serotonin)

AAE: Augmented acoustic environment

AAF: Anterior auditory field

AAO: American Academy of Otolaryngology

AAOO: American Academy of Ophthalmology and Otolaryngology

ABIs: Auditory brainstem implants

ABL: Basolateral nucleus of the amygdala

ABR: Auditory brainstem responses

ACE: Central nucleus of the amygdala

AGC: Automatic gain control

AI: Primary auditory cortex

AICA: Anterior inferior cerebellar artery

AII: Secondary auditory cortex

AL: Lateral nucleus of the amygdala

AM: Amplitude-modulated

AMA: American Medical Association

AN: Auditory nerve

AP: Action potential

AVCN: Anteroventral cochlear nucleus

BIC: Brachium of the inferior colliculus

CA: Compressed-analog

CAP: Compound action potential

CATS: Caffeine, alcohol, tobacco and stress

CF: Characteristic frequency

CI: Cochlear implants

CIS: Continuous interleaved sampling

CM: Cochlear microphonic

CMR: Common mode rejection

CMV: Cytomegalovirus

CN IX: Glossopharyngeal nerve

CN V: Trigeminal nerve

CN VII: Facial nerve

CN VIII: Eighth cranial nerve

CN X: Vagus nerve

CN: Cochlear nucleus

CNS: Central nervous system

COCB: Crossed olivocochlear bundle

CPA: Cerebellopontine angle

dB: Decibel

d.c.: Direct current

DC: Dorsal cortex of the inferior colliculus

DCN: Dorsal cochlear nucleus

DNLL: Dorsal nuclei of the lateral lemniscus

DPOAE: Distortion product otoacoustic emission

DPV: Disabling positional vertigo

ECoG: Electrocochleogram

EECIS: Enhanced CIS

EMG: Electromyographic

EP: Endocochlear potential

Ep: Posterior ectosylvian area

EPSP: Excitatory postsynaptic potentials

ERP: Event related potentials

F: Force

f: Frequency

FFR: Frequency following response

GABA: Gamma amino butyric acid

GABA$_A$: Gamma amino butyric acid receptor type A

GABA$_B$: Gamma amino butyric acid receptor type B

HFS: Hemifacial spasm

HL: Hearing level

I: Current

IAC: Internal auditory canal

IC: Inferior colliculus

ICC: Central nucleus of the inferior colliculus

ICP: Intracranial pressure

ICX: External nucleus of the inferior colliculus

IEC: International Electrotechnical Commission

ILD: Interaural level differences

ISO: International Organization for Standardization

ITD: Interaural time difference

LDV: Laser Doppler vibrometer

Leq: Equivalent level

LL: Lateral lemniscus

LSO: Lateral superior olivary nucleus

LSO: Superior olivary complex

LTP: Long-term potentiation

LV: Pars lateralis

MGB: Medial geniculate body

MHz: Megaherz

MLR: Middle latency response

MRI: Magnetic resonance imaging

MSO: Medial superior olivary (nucleus)

mV: Millivolts

MVD: Microvascular decompression

NF2: Neurofibromatosis type 2

NIHL: Noise induced hearing loss

NIOSH: National Institute for Occupational Safety and Health

NLL: Nucleus of the lateral lemniscus

NTB: Nucleus of the trapezoidal body

OAE: Otoacoustic emission

OME: Otitis media with effusion

OR: Obersteiner-Redlich

OV: Pars ovoidea

PAF: Posterior auditory field

PE: Polyethylene

PeSPL: Peak-equivalent sound pressure level

PO: Posterior division of the medial geniculate body

pps: Pulses per second

PST: Post stimulus time

PTS: Permanent threshold shift

PVCN: Posterior ventral cochlear nucleus

R: Resistance

RE: Reticular nucleus of the thalamus

RNA: Ribonucleic acid

SC: Superior colliculus

SEM: Scanning electron microscopy

SFOAEs: Stimulus-frequency otoacoustic emission

SH: Stria of Held

SL: Sensation level

SM: Stria of Monaco

SMSP: Spectral Maxima Sound Processor

SN_{10}: Slow negative

SOAE: Spontaneous otoacoustic emission

SP: Summating potential

SPEAK: Spectral peak strategy

SPL: Sound pressure level

SSNHL: Sudden sensorineural hearing loss

TB: Trapezoid body

TEOAE: Transient evoked otoacoustic emission

TGN: Trigeminal neuralgia

TMJ: Temporomandibular joint

TRT: Tinnitus retraining therapy

TTS: Temporary threshold shift

UCOCB: Uncrossed olivocochlear bundle

V: Voltage

VAS: Visual analog scale

VCN: Ventral cochlear nucleus

VNLL: Ventral nuclei of the lateral lemniscus

WBS: Williams-Beuren syndrome

Y: Admittance

Z: Impedance

μs: Microseconds

μV: Microvolts

Index